Devolving English Literature

Books are ... on or before
the '

F
F

DEVOLVING ENGLISH LITERATURE

ROBERT CRAWFORD

CLARENDON PRESS · OXFORD

1992

Oxford University Press, Walton Street, Oxford OX2 6DP
Oxford New York Toronto
Delhi Bombay Calcutta Madras Karachi
Petaling Jaya Singapore Hong Kong Tokyo
Nairobi Dar es Salaam Cape Town
Melbourne Auckland
and associated companies in
Berlin Ibadan

Oxford is a trade mark of Oxford University Press

Published in the United States
by Oxford University Press, New York

British Library Cataloguing in Publication Data
Data available

Library of Congress Cataloging in Publication Data
Crawford, Robert, 1959–
Devolving English literature / Robert Crawford.
Includes index.
1. English literature—History and criticism—Theory, etc.
2. Scottish literature—History and criticism—Theory, etc.
3. American literature—Scottish influences. 4. English literature—
Scottish influences. 5. Literature and anthropology. 6. Scotland—
Intellectual life. 7. Modernism (Literature) 8. Canon
(Literature) I. Title
PR21.C7 1992   820.9—dc20   91-43522
ISBN 0-19-811955-0 (pbk)
ISBN 0-19-811298-X

Typeset by Downdell Ltd, Oxford
Printed and bound in
Great Britain by Bookcraft Ltd
Midsomer Norton, Bath

To Scotland

Acknowledgements

THE research for this book was started during my tenure of the Elizabeth Wordsworth Junior Research Fellowship at St Hugh's College, Oxford, and was continued at the University of Glasgow where for two years I held a British Academy Post-Doctoral Fellowship in the Department of English Literature. The book was largely written at St Andrews University, following my 1989 appointment as Lecturer in Modern Scottish Literature in the Department of English. I would like to thank all these institutions for their support, without which *Devolving English Literature* could not have been written. I owe a great debt to colleagues and students in Oxford, Glasgow, and St Andrews, as well as to librarians at those universities and at the universities of Dundee, Leeds, Princeton, and Yale. The staff of the Bodleian Library in Oxford, the Mitchell Library in Glasgow, the National Library of Scotland, and the British Library were equally helpful.

Part of the material about J. G. Frazer in Chapter 3 of this book appeared in slightly different form in an essay in *Sir James Frazer and the Literary Imagination*, edited by Robert Fraser (Macmillan, 1990). I am grateful to the editor and publisher for their permission to recast the material. Acknowledgement is due also to Faber and Faber and to Farrar Straus & Giroux for permission to quote from Seamus Heaney's collections *Station Island* and *The Haw Lantern*; to Faber and Faber and to Harcourt Brace Jovanovich for permission to quote from *Collected Poems 1909–1962* by T. S. Eliot; to Faber and Faber to quote from Douglas Dunn's *Selected Poems, 1964–1983*; to Faber and Faber and Harcourt Brace Jovanovich for permission to quote from the uncollected prose of T. S. Eliot; to Tony Harrison and Peters, Fraser & Dunlop for permission to quote from 'Them and [uz]'; to Tom Leonard and Galloping Dog Press for permission to quote from *Intimate Voices* and from *Situations Theoretical and Contemporary*; to Oxford University Press for permission to quote from the prefatory matter to the Oxford English Dictionary; and to David Mach and the National Galleries of Scotland for permission to reproduce the jacket illustration. All other materials quoted in the text are deemed to fall within the category of fair use in a critical work.

A book of this nature involves its author in many debts, some of which are specifically recorded in the footnotes. For help, encouragement, and advice at various stages I would like to express particular gratitude to the following individuals: Professor Michael Alexander, Mr A. H. Ashe, Professor Marilyn Butler, Dr Ian Campbell, Professor John Carey, Mr and Mrs R. A. N. Crawford, Professor Harriet Davidson, Professor Douglas Dunn, the late Professor Richard Ellmann and the late Mrs Mary Ellmann, Mr Anthony Esposito, Professor Peter France, Dr Robert Fraser, Dr Douglas Gifford, Professor Steven Helmling, Dr W. N. Herbert, Professor Andrew Hook, Dr Cleo Kearns, Dr Peter McCarey, Dr C. J. M. MacLachlan, Dr Colin Matthew, Professor Karl Miller, Professor Edwin Morgan, Mr Les A. Murray, Professor Norman Page, Dr Neil Rhodes, Dr John Robertson, Ms Mary Sillitoe, the late Mr Martin Spencer, Dr Anne Varty, Mr Hamish Whyte, Mr Patrick Williams.

Dr Margaret Connolly of St Andrews University typed the book with easy efficiency; my co-editors of *Verse*, Professor Henry Hart, Dr David Kinloch, Mr Richard Price, and Dr Nicholas Roe, made sure that the magazine continued to grow as I wrote; at Oxford University Press my editor, Andrew Lockett, was extremely helpful, as were Nicola Pike and Katie Ryde; most importantly, my wife, Dr Alice Crawford, knows what I owe her. I hope I know that too.

R.C.

St Andrews
1991

Contents

Abbreviations

Corsica   James Boswell, *An Account of Corsica* (Glasgow: Robert and Andrew Foulis, 1768).

CP   Walt Whitman, *Complete Poetry and Collected Prose*, ed. Justin Kaplan (Cambridge: Cambridge University Press, Library of America Series, 1982).

EL   Ralph Waldo Emerson, *Essays and Lectures*, ed. Joel Porte (Cambridge: Cambridge University Press, Library of America Series, 1983).

GS   G. Gregory Smith, *Scottish Literature: Character and Influence* (London: Macmillan, 1919).

Heroes   Thomas Carlyle, *On Heroes and Hero-Worship*, in *Works* (30 vols.; London: Chapman and Hall, 1897–9), v.

Journal   James Boswell, *The Journal of a Tour to the Hebrides*, ed. R. W. Chapman (1924; repr. Oxford: Oxford University Press, 1974).

Journey   Samuel Johnson, *A Journey to the Western Islands of Scotland*, ed. Chapman (see previous entry).

L, i   T. S. Eliot, *Letters*, i, ed. Valerie Eliot (London: Faber and Faber, 1988).

Letters   Robert Burns, *Letters*, 2nd edn., ed. G. Ross Roy (2 vols.; Oxford: Oxford University Press, 1985).

Life   *Boswell's Life of Johnson*, ed. George Birkbeck Hill, revised and enlarged by L. F. Powell (6 vols.; Oxford: Clarendon Press, 1934–64).

MCP   Hugh MacDiarmid, *Complete Poems*, ed. Michael Grieve and W. R. Aitken (2 vols.; London: Martin Brian and O'Keeffe, 1978).

PS   Robert Burns, *Poems and Songs*, ed. James Kinsley (Oxford: Oxford University Press, 1969).

SR   Thomas Carlyle, *Sartor Resartus*, in *Works* (30 vols.; London: Chapman and Hall, 1897–9), i.

W   Sir Walter Scott, *Waverley*, ed. Andrew Hook (Harmondsworth: Penguin Books, 1972).

Introduction

DIFFERENCE: no term of literary discourse was more frequently used, appreciated, or paraded in discussions of literature in the 1980s. Since Jacques Derrida's *Writing and Difference* appeared in English in 1978, difference as post-structuralist *différance*, seen both as constituting and as problematizing meaning, has opened up and furthered enormous geographies of debate which have intersected sometimes awkwardly with considerations of how writing occludes, constructs, or distorts racial and sexual difference.[1] During the 1980s many of the most theoretically orientated investigations, and the criticism which followed them, seemed to be geared up to the study or implementation of what Gayle Greene and Coppélia Kahn call, in the title of their book, *Making a Difference*.[2] Nothing seemed more thoroughly investigated, more fruitful for English studies, or more fashionable.

Yet there were also areas of difference which almost all the consciously theorized writing of that period, as well as the more traditionally orientated criticism, obscured or ignored in a gesture which, deliberate or not, curiously reproduced distortions perpetuated by traditional literary criticism or historiography. If in 1879 J. C. Shairp could write a volume on Robert Burns for the series English Men of Letters, edited by John Morley, without demur, then over a century later Hugh Kenner, a widely respected commentator, is capable of producing a book called *A Sinking Island*, where the 'island' in question is England.[3] Now, no doubt, if questioned, Shairp (a Scotsman) would have admitted that there was a certain comedy in calling Burns an English man of letters. Similarly, Kenner, as an American, might concede that, while American usage often uses the word 'England' loosely, it is not true that England is an island.

---

[1] Jacques Derrida, *Writing and Difference*, trans. Alan Bass (London: Routledge and Kegan Paul, 1978).

[2] Gayle Greene and Coppélia Kahn, *Making a Difference: Feminist Literary Criticism* (London: Methuen, 1985).

[3] J. C. Shairp, *Robert Burns* (English Men of Letters Series; London: Macmillan, 1879); Hugh Kenner, *A Sinking Island: The Modern English Writers* (London: Barrie and Jenkins, 1988).

Shairp, after all, was working within the constraints of a particular and prestigious series; surely it was not his place to question the appropriateness of the series title to the life and work of Burns. Kenner was writing a book which ignores Scottish and Welsh writing, so surely he should be free to simplify cultural geography by assuming that England's boundaries extend to Aberystwyth and John o' Groats.

These points may seem pedantic, or the products of the type of unusually thin-skinned sensitivity which Hugh MacDiarmid mocked in *A Drunk Man Looks at the Thistle* when he caricatured the sort of prickly Scots who wrote to the newspapers about 'regimental buttons or buckled shoon, | Or use o' England whaur the U.K.'s meent'.[4] Yet they are points which indicate a noticeable slipperiness in the use of the term 'English', the term which, among other things, labels, or fails to label, the academic discipline of English Literature. The implications of this slipperiness are wider than they might appear initially.

Much attention has been devoted to the question of how we might define, select, or construct the entity known as 'literature'. Until very recently it seemed the word 'English' was left unexamined. In the teaching of English Literature, however, practical problems arose which questioned assumptions about its unitary nature. Most university English departments ran special courses on American Literature. Some also ran courses on Irish Literature, Caribbean Literature, Canadian Literature, Scottish Literature, or Australian Literature. Yet the teaching of these last tended to be confined to geographical areas. Few universities outside Australia run courses in Australian Literature; few universities outside Scotland run particular courses in Scottish Literature. For some, this may be a 'natural' position, reflecting 'obvious' matters of literary value. Scottish Literature might be taught as such in its country of origin, perhaps, but elsewhere it does not merit an independent status, and may be subsumed within 'English' Literature. The same could hold good for Australian Literature, but American Literature escapes such limitations. Is this a question of literary merit, or of American economic and political power? Questions of cultural authority constantly arise in discussions of 'minor' literatures, such as Caribbean or Irish writing, where there is a repeated and troubled interaction with Anglocentric values. Yet in supposedly 'main-line' English courses these issues are easily ignored. Figures as diverse as Smollett, Carlyle, Eliot, and Joyce are

4 *MCP*, 95.

seen too often as unproblematic parts of 'English' writing, with only token attention being paid to their subtly un-English difference.

While some of the leading literary theorists have written on issues of cultural difference—one of the most prominent examples being Edward Said's *Orientalism*—they tend to concentrate on cases where the differences are most striking. The differences chosen may be those between Oriental and Western culture, or between Third World and First World culture—as treated, for instance, in G. C. Spivak's *In Other Worlds*.[5] Even when teasing apart the strands of that 'English Literature' whose unity is an illusion, the tendency is to concentrate on groups most obviously typified as 'other' than the white English male. David Dabydeen's collection, *The Black Presence in English Literature*, is a fine example of a study of this type.[6] Far less attention has been paid to less immediately visible cultural differences within 'English Literature', or, if that attention has been paid, all too often it has been confined to academic ghettos—Scottish Literature specialists, or those especially interested in Anglo-Welsh writing. This is peculiarly ironic, because, as the present book argues, the Scots in particular were crucially instrumental in constructing the university subject of English Literature itself.

There was a time in the twentieth century when it seemed that issues of difference with relation to smaller, 'provincial' cultures might find a major theorist to be their spokesperson. We now know that, as Lindsay Waters points out, 'The young Paul de Man was a cultural nationalist', concerned with preserving the complex, multiple identities of Belgian culture in time of war, and deeply interested in then-fashionable ideas about the relationship between literature and national identity.[7] The wartime journalism which de Man produced shows, among other things, a repeated concern with minority-language literature in Flemish and with the expression of a Belgian identity through the use of French. De Man explored these issues through the mediums of both French and Flemish. A typical 1942 piece, 'Sur les caractéristiques du roman belge d'expression française', singles out as important the achievement of a dis-

[5] Edward Said, *Orientalism* (London: Routledge and Kegan Paul, 1978); Gayatri Chakravorty Spivak, *In Other Worlds: Essays in Cultural Politics* (London: Methuen, 1987).

[6] David Dabydeen (ed.), *The Black Presence in English Literature* (Manchester: Manchester University Press, 1985).

[7] Paul de Man, *Critical Writings, 1953–1978*, ed. Lindsay Waters (Minneapolis: University of Minnesota Press, 1989), p. xv.

tinctly Belgian tone in works of French-language Literature.[8] De Man was interested in how a literature which could be seen as provincial might preserve a sense of independence while being written in the language of another, dominant culture. This present book argues not simply that such an issue is crucial to modern English-language writing world-wide, but that it emerges first, and is seen most subtly and constantly, in the literature of Scotland.

De Man was concerned with cultural imperialism, national identity, and the construction of a pluralistic bilingual native tradition. But, as we now know, he was also prepared to align himself with Fascist killers. His concern with cultural nationalism led him to think that the best hope for Belgium lay on the side of the Third Reich. His wartime writings are rightly notorious because he sided with the exterminators, and his later work constitutes a decisive swerve away from all his earlier preoccupations.[9] Moving to America, it is as if he wished to put the maximum possible distance, physical and intellectual, between himself and his fascination with questions of the literary significance of cultural nationalism. Those questions tended to be avoided by Anglo-American critics and theorists in the post-war era, when it was perceived that the Romantic nationalism of the nineteenth century had ended in the abyss of the Third Reich. The fates of small nations such as the Baltic States were embarrassing to contemplate, while nationalist writers within Britain, such as Hugh MacDiarmid, were easily dismissed as local cranks. Ernest Gellner, writing in a widely praised 1983 study of *Nations and Nationalism*, sees nationalism only in terms of a 'problem', and even a recent series of conferences held in Aberdeen, Nottingham, and Luxembourg in the late 1980s under the title 'The Literature of Region and Nation' largely avoided discussion of the difference between a region and a nation. Some speakers, for instance, were happy to regard Scotland and Wales as regions; for others, they were obviously nations. On the whole, contemporary historians have shown a much more robust interest in nationalism than have commentators on literature. It would be exciting to see more of an attempt to relate to recent writing such a provocative comment as Eric Hobsbawm's that 'The declining

    [8] Paul de Man, *Wartime Journalism, 1939–1943*, ed. Werner Hamacher, Neil Hertz, and Thomas Keenan (Lincoln, Neb.: University of Nebraska Press, 1988), 198.
    [9] One of the best examinations of de Man's wartime activities is Frank Kermode, 'Paul de Man's Abyss', *London Review of Books*, 16 Mar. 1989, 3–7.

historical significance of nationalism is today concealed . . . by the visible spread of ethnic/linguistic agitations.'[10]

Yet major literary theorists have tended to avoid issues of cultural identity. Most strikingly, Derrida, as a French-speaking Algerian of Jewish background growing up in the 1940s and stimulated to pursue a literary career by fellow-Algerian Albert Camus, must have been well aware of questions of cultural assimilation and difference.[11] His own work, though, seems as much an avoidance of the adoption of any stance on these matters as it is a subversion of the discourses of authority. While he is reported to have a strong wish to write about what it means to him to be Algerian, he has not done this to any significant extent.[12] If post-structuralist thought deconstructs ideologies of authority, then, since it does this for *all* ideologies, its practical effect is often to maintain the status quo rather than to support any counter-movement. Often what small or vulnerable cultural groups need is not simply a deconstruction of rhetorics of authority, but a construction or reconstruction of a 'usable past', an awareness of a cultural tradition which will allow them to preserve or develop a sense of their own distinctive identity, their constituting difference.

Tensions between historically minded commentators who wish to reconstruct the past, rewriting literary history with a non-standard accent, and post-structuralists deeply distrustful of any such empirically based historical work are clearly evident in feminist literary studies, where the work of a literary historian such as Elaine Showalter, who seeks to make available to women a sense that they have what her book calls *A Literature of Their Own*, may be set against the largely anti-historical work of post-structuralists such as Julia Kristeva and Hélène Cixous, whose writing often seems far removed from the circumstances and possibilities of actual cultural or social change. These tensions have been outlined incisively by Janet Todd in her study of *Feminist Literary History*.[13] Readers, writers, and critics interested in the

10 Ernest Gellner, *Nations and Nationalism* (Oxford: Basil Blackwell, 1983), 5; E. J. Hobsbawm, *Nations and Nationalism since 1780* (Cambridge: Cambridge University Press, 1990), 170.

11 See Christopher Norris, *Derrida* (London: Fontana, 1987), 11–12, 239–41.

12 I am grateful to Dr R. M. Cummings and Dr Sandra Kemp of Glasgow University for informing me of Derrida's remarks on this topic at the March 1988 conference on 'The Linguistics of Writing', held at the University of Strathclyde.

13 Elaine Showalter, *A Literature of Their Own: British Women Novelists from Brontë to Lessing* (Princeton, NJ: Princeton University Press, 1977); Janet Todd, *Feminist Literary History: A Defence* (Cambridge: Polity Press, 1988).

literatures of small cultures might learn a lot from the example of feminist studies, another area which, in the not so distant past, was often dismissed as an example of 'special pleading'. Certainly, this present book, though it deals mainly with male authors, has been encouraged by Todd's arguments in favour of the need for close, empirical re-examinations of writing produced by a marginalized group and tied to the circumstances of particular cultural struggles. Encouragement has come also from a very different approach. Though the idiom of *Devolving English Literature* is not at all that of Gilles Deleuze and Félix Guattari, their *Kafka: Toward a Minor Literature* contains material likely to be stimulating for anyone interested in questions of how an un-English identity may be preserved or developed within 'English Literature'. Their book addresses Kafka specifically, but its third chapter, entitled 'What Is a Minor Literature?', involves contentions which are both provocative and pertinent. Deleuze and Guattari state that 'A minor literature doesn't come from a minor language; it is rather that which a minority constructs within a major language.' They go on to argue that, partly because of the pressure exerted on the minority by the major language, it is characteristic of minor literatures 'that everything in them is political', and that they have a particularly strong sense of collective cultural identity.[14] This present book has attempted to remain alert to nuances of cultural politics embedded in most of the works discussed, nuances which set them apart from Anglocentric assumptions but which are ignored when these texts are read in an unexamined context of 'English Literature'. For some, my readings may appear eccentric, but this 'ec-centricity' is deliberate, a gesture designed to make us aware of issues too easily suppressed. *Devolving English Literature* has also attempted to be alert to manifestations of a collective identity, to cultural traditions that can only be seen if we are willing to have a devolved rather than a totalitarian or centralist approach to English Literature.

Devolving English Literature aims to foster a receptive approach to cultural differences, to the way in which authors can question or negotiate with Anglocentricity in their writings. It aims to note, particularly, the differences between Scottish Literature, British Literature, and English Literature—all of which are entities that writers within 'English' have attempted to construct for particular

[14] Gilles Deleuze and Félix Guattari, *Kafka: Toward a Minor Literature*, trans. Dana Polan (Minneapolis: University of Minnesota Press, 1986), 16–17.

cultural ends. If, deliberately or not, we totalize all the constituents of English Literature, and if, as both traditional literary history and post-structuralist criticism have done all too often, we ignore matters of local origin, then we perform an act of naïve cultural imperialism, acting as if books grew not out of particular conditions in Nottingham, Dublin, St Lucia, or Salem, Massachusetts, but out of the bland uniformity of airport departure lounges. The act of inscription is not a simple entry into the delocalized, pure medium of language; it is constantly, often deliberately, an act which speaks of its local origins, of points of departure never fully left behind.

The present book's title is not intended simply to comply with the view that writing in English has been a narrowly centralized activity, and that power must now be devolved from the centre to the margins. Rather, it aims to suggest that, while for centuries the margins have been challenging, interrogating, and even structuring the supposed 'centre', the development of the subject 'English Literature' has constantly involved and reinforced an oppressive homage to central-ism. As such, English Literature is a force which must be countered continually by a devolutionary momentum. Creative writers have been more alert to this need than have most critics.

Devolving English Literature devotes considerable space to American writing, and its last chapter looks at connections between a number of poets in Australia, Ireland, the Caribbean, northern England, and Scotland, all of whom wish to use the English language and its varieties to enunciate a cultural identity that is not that of the traditionally dominant London–Oxbridge English cultural centre. I hope that many of the points which emerge in discussing a text or literary tradition may spark off reflections about other kinds of 'provincial', 'colonial', or 'post-colonial' writing elsewhere. Yet I have chosen to concentrate particularly on Scottish culture and the strategies adopted by Scottish authors. These authors can be seen as contributing to an identifiably Scottish cultural heritage, but they are often either too smoothly assimilated into English Literature, or else (like Hugh MacDiarmid) awkwardly marginalized from considera-tions of work whose focus is not purely Scottish. There may be many parallels between the Scottish writing considered in the present book and the Irish writers so ably discussed in Seamus Deane's *A Short History of Irish Literature* (1986), and I would invite readers to develop these; but, though some Irish works are included in the last two chapters of this book, I have deliberately concentrated on the less

academically fashionable, yet undeniably outstanding, work produced
in Scotland since the Enlightenment.

The first reason for choosing Scottish Literature for special attention is that it offers the longest continuing example of a substantial
body of literature produced by a culture pressurized by the threat of
English cultural domination. The second reason, which follows from
this, is that, because of its historical longevity and cultural circumstance, Scottish writing has often formed a model for writers in other
countries concerned to escape from being England's cultural provinces. This book examines the impact of Scottish writing on a number
of American writers, but similar tasks could be undertaken for
Canadian and Australian writing, to cite only two obvious examples.
Because of Scotland's geographical proximity to England, and because
Scotland, while it maintains separate legal, educational, and religious
institutions, lacks political independence, Scottish writing in English
(like Welsh writing in English) is particularly vulnerable to being
subsumed within the English literary tradition with which it was
frequently, but not exclusively, engaged. Scotland, therefore,
becomes a, if not *the*, test case when considering whether or not we
have devolved our view of 'English Literature' in order to take full
account of the various different cultural traditions which are so easily
lumped together under that label. Scotland itself, as was mentioned
above and as is argued in detail in Chapter 1, was crucially instrumental in the development of the university teaching of English
Literature, which has conditioned our view of the subject; so, again,
Scotland appears a fundamental area to examine.

Though I am concerned with delineating a literary genealogy
which seems particularly important in Scottish writing, an anthropologically orientated line, this book is not a history of Scottish
Literature, of which there have been several recent examples.[15] It
does not argue that what it offers is *the* post-Enlightenment Scottish
tradition. Nor is it a study of a particular period of Scottish writing.
Attention has been drawn elsewhere to the problems of cultural
identity which faced eighteenth-century Scottish writers.[16] This book

---

[15] Roderick Watson, *The Literature of Scotland* (Basingstoke: Macmillan, 1984), is an
outstanding one-volume treatment; more diffuse is Cairns Craig (gen. ed.), *The History
of Scottish Literature* (4 vols.; Aberdeen: Aberdeen University Press, 1987–8).

[16] See especially David Daiches, *The Paradox of Scottish Culture* (Oxford: Oxford
University Press, 1964); David Craig, *Scottish Literature and the Scottish People,
1680–1830* (London: Chatto and Windus, 1961); Kenneth Simpson, *The Protean Scot:
The Crisis of Identity in Eighteenth-Century Scottish Literature* (Aberdeen: Aberdeen
University Press, 1988).

differs not only by emphasizing the importance of seeing how these problems moulded the development of the university teaching of English Literature, but also by arguing that the Scots' solution to them was to develop a 'British Literature' throughout both the eighteenth and nineteenth centuries, before a more explicitly nationalist, post-British literary consciousness came to the fore in the twentieth century. *Devolving English Literature* is intended to stimulate further debate by its emphasis on the way in which the 'provincial' energies so important to Scottish writing, and the anthropological viewpoint developed by Scottish writers, fed into American writing and into the essentially 'provincial' movement we know as Modernism. Lastly, through examining the 'provincial' and demotic energies of Modernism, it becomes possible to see how these continue to energize more recent 'provincial' and 'colonial' writers as diverse as Philip Larkin, Seamus Heaney, Douglas Dunn, Les Murray, and Derek Walcott, several of whom are usually accepted as being opposed to Modernism rather than indebted to it. Throughout this book the word 'provincial' often appears in inverted commas; this is to alert the reader to the fact that the term has come to acquire a derogatory meaning, and is often used as a term of cultural imperialism.[17] The chapters which follow make some effort to counter such attitudes, and even, on occasion, to valorize the term 'provincial'.

Devolving English Literature, then, has a loop structure, beginning with a consideration of eighteenth-century Scottish anxieties and annoyances about the country's 'provincial' status, and concluding by showing the extent to which similar anxieties are highlighted by a variety of contemporary poets all too aware of the question of the 'provincial'. These anxieties and issues are ignored or papered over if we maintain a crude, unitary view of English Literature, but become apparent and exciting if we are prepared to achieve a devolved view which attends closely to local accents. Among other things, the achievement of this devolved view allows a provocative rereading of a wide variety of texts, from the works of Smollett and Boswell, through those of Scott, Carlyle, and Frazer, to the writing of the Modernists and of our own contemporaries. Readings of some eighteenth-century texts in this book are particularly detailed because, while today's readers may pay lip-service to the achievements of Smollett and

---

[17] For one view of the development of the use of this term, see John Lucas, 'The Idea of the Provincial', in *Romantic to Modern Literature: Essays and Ideas of Culture 1750–1900* (Brighton: Harvester Press, 1982), 7–29.

Boswell, I suspect that those authors' works are increasingly unfamiliar, and that generous quotation is necessary. In dealing with
more recent writers, such as Eliot or Heaney, I have assumed that the
reader is more likely to be familiar with the material, and have quoted
from it less extensively.

If literary theorists have largely ignored Scottish writing, this may
be another instance of post-structuralism's tendency in practice to do
little to upset the status quo: Scottish Literature has been hardly the
most popular area for critical investigation; literary theorists have
done little to change this. Indeed, though there are a few signs that
questions of Scottish cultural identity may be received into more
theoretically lively areas of criticism, there are also indications that
this may be done in a crude way which demonstrates little or no
attention to Scottish cultural difference, and that there remains a
great need for empirically grounded work to help free Scottish writing
from the Anglocentric tones of conventional literary history and of
newer approaches.

Evidence to support such a contention is provided by a recent book
of essays on *Nation and Narration* which examines—critically and often
provocatively—such nationally tinged conceptions as 'English reading'.[18] Among the international array of contributors to this book
only Gillian Beer seems to have a clear and explicit awareness that the
words 'England' and 'Britain' are not synonymous. When Scottish
Literature appears fleetingly in the person of Walter Scott in an essay
called 'Literature: Nationalism's Other? The Case for Revision', the
treatment of Scott's work in the context of 'nation and narration' is
perversely careless and wrong-headed: 'In 1812 Scott published
Waverley . . . Scott's texts are not motivated. Unlike *Tom Jones* to
which it is much indebted, *Waverley* does not explain the occasion of
its own writing.'[19] The fact that Scott published *Waverley* in 1814, not
1812, is relevant because, only by siting this text accurately in a
Scottish historical and cultural climate, can we understand its motivation, which is bound up with the construction of a Scottish and British
national narrative. This motivation is explicitly and implicitly acknowledged in the novel, as is argued in detail in Chapter 3 of the
present book.

Throughout *Devolving English Literature* I have aimed to be stimulating and readily comprehensible, hoping to interest writers and

[18] Homi K. Bhabha (ed.), *Nation and Narration* (London: Routledge, 1990), 6.
[19] Ibid. 147.

readers, not just teachers and students. This book seeks to use familiar procedures, idioms, and devices from conventional literary history, yet also to undermine the Anglocentric voice of that history. This is not a totalizing story of English Literature which co-opts non-English authors to incorporate them into a unitary English tradition, but a devolutionary history written on the margin. It tries to erode and complicate any ideas of such a unitary tradition, and force 'English' to take account of the other cultures which are in part responsible for the initial construction of 'English Literature' as a subject.

This last-named aspect of the project brings it into contact with a good deal of recent scholarly work, such as Olivia Smith's consideration of *The Politics of Language 1791–1819*. Though it is not specifically about the history of the university subject of English Literature, Smith's study is extremely useful in its demonstration of how language in this period was the site of a conflict between a polished metropolitan ascendancy, whose values were those of the educational system, and provincial lower-class barbarians, whose views could be discounted because they were expressed in terms considered stupid or corrupt. 'The basic vocabulary of language study', writes Smith, '—such terms as "elegant", "refined", "pure", "proper", and "vulgar"— conveyed the assumption that correct usage belonged to the upper classes and that a developed sensibility and an understanding of moral virtue accompanied it.'[20] Drawing its inspiration from developments in black studies and women's studies, Olivia Smith's book nicely complements the generally Marxist essays of Brian Doyle, Peter Brooker, and Peter Widdowson on the development of English studies, which these writers see as contributing to an upper-class, imperialist, metropolitan discourse aimed at taming the lower classes and Celts by educating them in accord with a more polished and humane English tradition.[21] This view accords with the subtler, more extended treatment of the subject by Chris Baldick in his widely praised book, *The Social Mission of English Criticism*, which devotes much attention to the institutional development of English studies in London, Oxford, and Cambridge, from Matthew Arnold to the

---

[20] Olivia Smith, *The Politics of Language 1791–1819* (Oxford: Clarendon Press, 1984), 9.

[21] See Brian Doyle, 'The Invention of English', and Peter Brooker and Peter Widdowson, 'A Literature for England', in Robert Colls and Philip Dodd (eds.), *Englishness, Politics and Culture 1880–1920* (London: Croom Helm, 1986), 89–115, 116–63; see also Brian Doyle, 'The Hidden History of English Studies', in Peter Widdowson (ed.), *Re-Reading English* (London: Methuen, 1982), 17–31.

Leavises, through critics who fostered ideas about the centrality of a certain English tradition. Terry Eagleton follows Baldick's model in his *Literary Theory* (1983). Jo McMurtry's non-Marxist look at some pioneers of literary teaching on both sides of the Atlantic fleshes out these other studies, while books by Gerald Graff and Kermit Vanderbilt examine the growth of the teaching of English and American Literature in the United States, where American Literature courses often reacted against certain norms assumed in the study of English Literature. There have been earlier attempts to chronicle the rise of English Literature, such as D. J. Palmer's very useful *The Rise of English Studies* (1965) and Stephen Potter's sometimes stimulating anti-academic report on *The Muse in Chains* (1937).[22] But the nature and proliferation of recent work in this field (which is often bound up with an examination of changing concepts of Englishness) suggests that increasing self-questioning is causing the discipline to examine and interrogate its own origins in much the same way that anthropology has been doing for some time. As the following chapter suggests, there may be a certain poetic justice in the way in which the roots of the teaching of English and the development of eighteenth-century anthropology are intertwined.

Recent studies of the growth of English Literature as a university subject are insufficient because of the assumption they make that the actual origins of the subject are in nineteenth-century England. It is to be hoped that *Devolving English Literature* will supplement and, ironically, complicate the books mentioned above, just as it may intersect fruitfully with contemporary examinations of Englishness such as those of Robert Colls and Philip Dodd in *Englishness*, Brian Doyle in *English and Englishness*, and John Lucas in *England and English-ness*.[23] Though frequently helpful, these studies tend to ignore the subtle interrelationships between Englishness, Britishness, and Scottishness which have done so much to mould 'English Literature'.

[22] See Chris Baldick, *The Social Mission of English Criticism 1848–1932* (Oxford: Clarendon Press, 1983); Terry Eagleton, *Literary Theory* (Oxford: Blackwell, 1983); Jo McMurtry, *English Language, English Literature: The Creation of an Academic Discipline* (London: Mansell, 1985); Gerald Graff, *Professing Literature: An Institutional History* (Chicago: University of Chicago Press, 1987); Kermit Vanderbilt, *American Literature and the Academy: The Roots, Growth, and Maturity of a Profession* (Philadelphia: University of Pennsylvania Press, 1986); D. J. Palmer, *The Rise of English Studies* (London: Oxford University Press, 1965); Stephen Potter, *The Muse in Chains* (London: Methuen, 1937).

[23] Colls and Dodd, *Englishness*; Brian Doyle, *English and Englishness* (London: Routledge, 1990); John Lucas, *England and Englishness: Ideas of Nationhood in English Poetry, 1688–1900* (London: Hogarth Press, 1990).

Replying to an outline of part of this present book which appeared in the *London Review of Books*, John Lucas made the objection that my argument 'takes for granted a monolithic culture for England . . . a culture of conformity', ignoring 'some versions of non-conformity'.[24] At times there may be some force to Lucas's point, but I have tried to make it clear that the Anglocentricity with which I am dealing is a phenomenon particularly located in the London–Oxbridge nexus of our own time, and, in earlier ages, in London as the seat of the Court, the place to which the eighteenth-century teachers of Rhetoric looked for their standards of language. I hope it is clear when, for instance, I consider Tony Harrison in the final chapter that I do not see English culture as a clear, unified block. England may not have the complex linguistic and cultural divisions of a Scotland split for centuries between Gaelic, English, and Scots, but England is clearly a cultural amalgam to which a fully devolved reading of English Literature must address itself. Again, in speaking of Scottish culture, I concentrate on Lowland as opposed to Highland culture; I would be sad if this were taken as indicating any denial of Highland or of Gaelic cultural difference.

When writing about matters of cultural nationalism and identity, it is always difficult to avoid treading on toes. A nice example of this occurs in the admirable recent series of Field Day pamphlets on Nationalism, Colonialism, and Literature, where Terry Eagleton, addressing an Irish audience on the theme of *Nationalism: Irony and Commitment*, appears guilty of an ironic lapse when he writes that 'Those of us who happen to be British, yet who object to what has been done historically to other peoples in our name, would far prefer a situation in which we could take being British for granted and think about something more interesting for a change.'[25] The first irony here is that the whole of Ireland was for a time (and part of it remains) constitutionally British. For this reason, Ireland, like Scotland, has been—in part at least—a perpetrator as well as a victim of global British imperialism. Eagleton's paper seems to avoid the embarrassment of this situation. A second irony is that the sentence quoted above is likely to be extremely irritating to present-day Scottish or Welsh nationalists who still constitutionally 'happen to be British', yet will have no wish at all to 'take being British for granted' and will

[24] Lucas, *England and Englishness*, 39.
[25] Terry Eagleton, *Nationalism: Irony and Commitment* (Field Day Pamphlet 13; Derry: Field Day Theatre Company, 1988), 8.

be likely to regard Eagleton's statement as one of ideological imperialism.

Yet Eagleton's main argument in this pamphlet remains valid, for he contends that all oppositional nationalist arguments, political or cultural, 'move under the sign of irony, knowing themselves ineluctably parasitic on their antagonists'.[26] There is too often a heart-hardening, parasitical hatred which develops among cultural nationalists, and which Yeats recognized as leading only to 'withered old and skeleton-gaunt' frustrated ideals.[27] It is good that we should be suspicious of nationalisms and their traditions, subjecting them to the sort of sceptical scrutiny which is applied to the Highland and Jacobite traditions of Scotland by historians and critics as different as Hugh Trevor-Roper, Malcolm Chapman, Peter Womack, and Murray Pittock.[28] All cultural constructions may profit from having their shams exposed, from being demystified or deconstructed. *Devolving English Literature* also implies a taking-apart, but I have tried to emphasize as far as possible the positive side of the endeavour: that it is part of a reconstructing of a literary inheritance. Ultimately, the aim of this book is not deconstructive but creative. For, rather than something to be eschewed, the invention or construction of traditions is a key activity in a healthy culture, one whose view of itself and of its own development is constantly altering and under review. There is no fixed, unchanging entity called Englishness, or Britishness, or Jamaicanness. Each of these cultural identities, like that of Scotland, is in constant evolution, continually re-manufacturing itself. In *Imagined Communities* Benedict Anderson, though he grossly over-simplifies complex cultural issues about the languages of Scotland, is right to draw attention to Tom Nairn's 'good nationalist tendency to treat his "Scotland" as an unproblematic, primordial given'. Scotland

---

[26] Ibid. 7.

[27] W. B. Yeats, *The Poems*, ed. Richard J. Finneran (London: Macmillan, 1984), 233 ('In Memory of Eva Gore-Booth and Con Markievicz').

[28] Hugh Trevor-Roper, 'The Invention of Tradition: The Highland Tradition of Scotland', in Eric Hobsbawm and Terence Ranger (eds.), *The Invention of Tradition* (Cambridge: Cambridge University Press, 1983), 15–42; Malcolm Chapman, *The Gaelic Vision in Scottish Culture* (London: Croom Helm, 1978); Peter Womack, *Improvement and Romance: Constructing the Myth of the Highlands* (Basingstoke: Macmillan, 1988); Murray G. H. Pittock, *The Invention of Scotland: The Stuart Myth and the Scottish Identity, 1638 to the Present* (London: Routledge, 1991).

and Scottish culture, like all nations and cultures, require continual acts of re-imagining which alter and develop their natures.[29]

The culture which ceases to do this stagnates, or comes to be dominated by another culture. Writers are constantly aware of this. Scott, attempting to ensure and articulate Scotland's distinctive place in Britain in *Waverley*, is conscious of the need both to construct and to reconstruct images of cultural identity that are other than Anglocentric. The modern poet Derek Walcott has related aims when he blends Caribbean nation and narration in *Omeros*. It is impossible to appreciate the project of either work if one considers the text as simply belonging to English Literature. One should bear in mind that most English-language writing is now produced outside England. To read many anglophone books of the past, and most anglophone books of the present, requires a devolving of the still prevalent idea of 'English Literature', and a specific attention to the slippages which that label encourages and which are all too often brushed aside. The best way to begin to understand how the Anglocentric notion of English Literature achieved such formidable cultural power, being disseminated through our institutions of higher education whose influence indirectly pervades the whole community, is to attend to the sense in which the subject of English Literature was, ironically, a Scottish invention.

[29] Benedict Anderson, *Imagined Communities: Reflections on the Origin and Spread of Nationalism* (London: Verso, 1983), 85.

1

The Scottish Invention of English Literature

WHETHER in Adam Smith's *Wealth of Nations*, Adam Ferguson's *Essay on the History of Civil Society*, John Millar's *Observations Concerning the Distinction of Ranks in Society*, or a plethora of other works, eighteenth-century Scottish writing is filled with comparisons of cultures at various stages of development, and with speculations about how societies 'improve'. The concept and vocabulary of 'improvement', of turning the primitive into the civilized, govern manifold areas of Scottish life at this time, from agriculture, economics, and chemistry to poetry and the fine arts. Some aspects of the primitive, such as nobility and simplicity, were admired even as the general march continued towards mercantile and literary refinement. Discussions of the primitive and the refined were a major element of the Enlightenment, not only in a Montesquieu-fuelled Scotland but also across Europe and in North America.[1] Yet the debate was particularly intense in, and pertinent to, Scottish culture, because, rightly or wrongly, the small country of Scotland could be seen in various ways as strikingly divided between the barbarically primitive and the confidently sophisticated.

Such a perception of Scotland pre-dates the eighteenth century. The *Britannica*, first compiled by the Englishman William Camden in 1586 and reprinted throughout the seventeenth century, declared that, 'With respect to the manners and ways of living, it is divided into the *High-land-men* and the *Low-land-men*. These are more civilized, and use the language and habit of the English; the other more rude and barbarous, and use that of the Irish.'[2] In the eighteenth

---

[1] See Peter France, 'Primitivism and Enlightenment: Rousseau and the Scots', *Yearbook of English Studies 1985*, 69; for Montesquieu's impact on eighteenth-century Scottish thought, see Ronald L. Meek, *Social Science and the Ignoble Savage* (Cambridge: Cambridge University Press, 1976), 32, 109; John MacQueen, *The Enlightenment and Scottish Literature*, i. *Progress and Poetry* (Edinburgh: Scottish Academic Press, 1982), 52–3.

[2] Camden, cited in R. A. Houston, *Scottish Literacy and the Scottish Identity: Illiteracy and Society in Scotland and Northern England 1600–1800* (Cambridge: Cambridge University Press, 1985), 89–90.

century Lowland Scots themselves could regard the Highlanders as barbarously savage, which may sometimes have lent them a primitive nobility but which also set them apart in terms of the progress of society. In 1755 the Edinburgh clergyman and historian William Robertson preached a sermon in which he suggested that society in the Highlands and Islands was society in its most primitive form. Also in 1755 he made the same claim for the society of the North American Indians.[3] Scottish Enlightenment writers, often anxious to play up their own country's increasingly 'civilized' values, rarely spoke of the inhabitants of Scotland in terms which so clearly linked them to North American 'savages'. But when Adam Ferguson discussed the systems of '*tribes* or of *clans*' among primitive peoples, the word 'clan' must have carried with it some local resonance for a writer brought up on the Highland Line, the frontier between Highlands and Lowlands; the twentieth-century historian Duncan Forbes points out that Ferguson, like other Scottish Enlightenment writers, was encouraged by Scotland's divided cultural situation to take 'a long, cool look at both sides of the medal of modern civilization'.[4] William Robertson's Edinburgh friend Professor Hugh Blair linked ancient Ossianic Highlanders with American Indians in his 1763 *Critical Dissertation on the Poems of Ossian*. James Boswell and Samuel Johnson both wrote of modern Highlanders in terms of native Americans. Unsure if Ossian's poems were ancient or modern, the Glaswegian Professor William Richardson moved easily from his long poem on North American Indians to his Ossianic drama *The Maid of Lochlin* and his 1778 dramatic poem *Agandecca*. Sir Walter Scott had an English judge liken Highlanders to North American Indians in his 1827 story *The Two Drovers*.[5] A constant interest in cultural parallels and comparisons which developed during the Scottish Enlightenment, and which was encouraged by the divided nature of Scotland, would have important repercussions for the development of Scottish literary culture.

[3] Robertson, cited in Meek, *Social Science and the Ignoble Savage*, 137.

[4] Adam Ferguson, *An Essay on the History of Civil Society* (1767), ed. Duncan Forbes (Edinburgh: Edinburgh University Press, 1966), 220, xiii.

[5] Hugh Blair, *A Critical Dissertation on the Poems of Ossian, the Son of Fingal* (London: T. Becket and P. A. de Hondt, 1763), 23; on Boswell and Johnson, see Chapter 2 of this present book; William Richardson, *Poems and Plays*, A New Edition (2 vols.; Edinburgh: Mundell and Son, 1805), ii; on *Agandecca*, see also Davis D. McElroy, *Scotland's Age of Improvement: A Survey of Eighteenth-Century Literary Clubs and Societies* (Washington, Wash.: Washington State University Press, 1969), 43; Scott, cited in Peter France, 'Western European Civilisation and its Mountain Frontiers (1750–1850)', *History of European Ideas*, 6/3 (1985), 302.

This chapter, though, centres on another aspect of 'improvement', and on the academic and cultural consequences of another internal Scottish division, one that depended on the English language. Olivia Smith has pointed out the extent to which, in the second half of the eighteenth century, 'Civilization was largely a linguistic concept, establishing a terrain in which vocabulary and syntax distinguished the refined and civilized from the vulgar and the savage.'[6] In Scotland the general eighteenth-century concern with linguistic propriety was particularly intense because it was bound up with a conflict between the urge to treasure the language of Lowland Scotland—Scots—in which most of the great popular and learned poetry of Scotland, from Gavin Douglas to the ballads, had been created, and a contrary impulse to develop a Scotland which would take complete advantage of the 1707 Act of Union by playing its part in the newly united political entity of Britain. To play a full part, Scottish people would have to move from using Scots to using English, an English which was fully acceptable to the dominant partner in the political union. This English, it was argued, both had to replace Scots and had to be purged of what we would now call 'markers of Scottish cultural difference', purged of Scotticisms. The growing wish for a 'pure' English in eighteenth-century Scotland was not an anti-Scottish gesture, but a pro-British one. If Britain were to work as a political unit, then Scots should rid themselves of any elements likely to impede their progress within it. Language, the most important of bonds, must not be allowed to hinder Scotland's intercourse with expanding economic and intellectual markets in the freshly defined British state. Scots who, like Alexander Wedderburn in the 1755 *Edinburgh Review*, wrote about 'North Britain' rather than 'Scotland' in the context of 'improvement' were emphasizing the new opportunities open to a post-Union and post-Culloden Scotland loyal to the British constitution. Such figures were pro-British because they were pro-Scottish: it was in the promise of 'Britain' that they saw the richest future for a Scotland which would soon 'improve'. Linguistic 'improvement' would be a major step towards that. Throughout the eighteenth century an obsession with cultural comparison between improved and unimproved societies often feeds into, and is spurred by, that pursuit of improving linguistic studies which Nicholas

---

[6] Olivia Smith, *The Politics of Language 1791–1819* (Oxford: Clarendon Press, 1984), p. vii.

Phillipson has called the 'most characteristic preoccupation of the Scottish Enlightenment'.[7]

So, for instance, Thomas Blackwell, Professor of Greek at Marischal College, Aberdeen, in 1735 devoted himself to presenting the poet Homer as poised between the societies which moulded him: one was a basic society of '*Warriors*, and *Shepherds*, and *Peasants* such as he drew'; the other was of 'Cities blessed with Peace, spirited with Liberty, flourishing in Trade, and increasing in Wealth'. In this, Blackwell's Janus-faced Homer, born between a time of 'Nakedness and Barbarity' and one of 'Order and Established Discipline', was an eighteenth-century Aberdonian and, more generally, an eighteenth-century Scot.[8] In Aberdeen Blackwell's work appears to have influenced the student James Macpherson, whose Ossianic epics were to valorize Highland warriors (made safely antique) and balance the primitive with a refinement suited to the 'improving' tastes of Hugh Blair, Alexander Carlyle, and Adam Ferguson, who compared Ossian with Homer. James Beattie, George Campbell, and Alexander Gerard, also students of Blackwell's, were all improvers who furthered the discipline encouraged by Blackwell's widow when, in 1796, she left money for the award of a prize to 'the person who should compose and deliver, in the English language, the best discourse upon a given literary subject'.[9] By 1796 the term 'literary subject' in a Scottish university referred not only to classical, but also to modern literature, which had been taught for almost half a century; the eighteenth-century sense of the word 'literature' encompassed most fields of educated discourse. This Scottish pedagogical development was quite different from the scope and design of the Oxford Professorship of Poetry founded in 1708, whose early incumbents, such as Joseph Trapp, Robert Lowth, and Thomas Warton the Younger, lectured in Latin on Classical and Hebrew subjects, though some of their lectures were later published in translation for a wider audience. The new

---

[7] Nicholas Phillipson, 'Culture and Society in the 18th Century Province: The Case of Edinburgh and the Scottish Enlightenment', in Lawrence Stone (ed.), *The University in Society* (2 vols.; Princeton, NJ: Princeton University Press, 1975), ii. 443 (Wedderburn), 438.

[8] Thomas Blackwell, *Enquiry into the Life and Writings of Homer* (1735), cited in Lois Whitney, 'Thomas Blackwell, a Disciple of Shaftesbury', *Philological Quarterly*, 5/2 (Apr. 1926), 197–9. See also Josef Bysveen, *Epic Tradition and Innovation in James Macpherson's 'Fingal'* (Uppsala: University of Uppsala Press, 1982), 37.

[9] Sir John Sinclair, *The Statistical Account of Scotland 1791–1799 . . .* ed. Donald J. Withrington and Ian R. Grant (Wakefield: EP Publishing, 1983), i. 306.

Scottish university subject, like Blackwell's Homer, arose from the
growing awareness of, and interest in, the gap between primitive or
barbarous societies and 'civil', highly developed ones. The collection
of data about both types, and the speculation about how one might
relate to the other, were activities common to other Enlightenment
countries, but Peter France has found particular interest in questions
of the 'civil' and 'savage' in provincial areas on Europe's mountain-
ous frontiers, as in Switzerland or Scotland.[10] In his epoch-making
study of *Victorian Anthropology* George Stocking mentions the Scottish
roots of anthropology when he points out that, in the winter of
1750–1, Adam Smith in Edinburgh and Baron Turgot at the Sor-
bonne each gave lectures attempting a more general or scientific
formulation of the idea of progress in civilization.[11] Though Smith's
lectures from this period do not survive, the painstaking research of
Ronald Meek has argued convincingly that they closely resembled
Smith's 1762–3 Glasgow lectures on jurisprudence, and that they
contained Smith's first exposition of the celebrated 'four stages
theory' of social evolution.[12] What is particularly interesting is that
we know for certain that Smith was also lecturing in Edinburgh from
1748 until 1751 on 'Rhetoric and Belles Lettres'.[13] Because both were
very much concerned with the process of 'improvement', it comes as
no surprise to find that the roots of the modern university disciplines
of anthropology and of English Literature are mutually entwined,
and that both are linked to the development of modern economics.

To appreciate this further, and to understand the social and
economic motivations which governed the emergence of university-
level English Literature studies, we need a sketch of the Scottish
cultural context. The growing network of convivial and intellectual
clubs in eighteenth-century Scotland—starting with those like Allan
Ramsay's Easy Club of 1712 which, though uneasy about the politics
of the 1707 Union, nevertheless promoted 'Mutual Improvement in
Conversation'—shows a strong accent on 'correct' (southern) taste,
as demonstrated by the widespread reading of the *Spectator*. The clubs
also indicate a concern with Scottish heritage, as is evidenced by their
interest in antiquities, including literary antiquities. Though they
tried to learn 'Improvement in Conversation' from the *Spectator*, the

[10] France, 'Western European Civilisation', 297–310.

[11] George W. Stocking, jun., *Victorian Anthropology* (London: Collier Macmillan,
1987), 14.   [12] Meek, *Social Science and the Ignoble Savage*, 90–117.

[13] On this, see especially Adam Smith, *Lectures on Rhetoric and Belles Lettres*, ed. J. C.
Bryce (Oxford: Clarendon Press, 1983), 7.

Easy Club's members soon changed their pseudonyms from those of English to Scottish literary figures—Allan Ramsay became 'Gavin Douglas'. There was a general difficulty about how one might preserve a Scottish identity while (as was increasingly encouraged by the upper classes) adopting English linguistic mores. 'Fergus Bruce' in 1724 prompted the retort that the efforts of Edinburgh people to 'refine their Conversation . . . will contribute . . . to the Pleasure of the *South* part of this Kingdom'. Scots was being replaced by English in schools, and students at Edinburgh University were welcoming help that would improve their taste in literature through better English composition.[14] John Stevenson, Professor of Logic at Edinburgh from 1730 until 1775, delivered in English what one student remembered as a 'Belles Lettres' hour. Stevenson used passages from Homer and Classical critics as well as modern English and French writers, including Addison and Pope. Thomas Somerville, one of the Edinburgh literati, recalled that Stevenson

also occasionally read lectures on the cardinal points of criticism suggested by the text-books . . . his lectures included some judicious philological discussions, as well as many excellent examples and useful practical rules of composition. I derived more substantial benefit from these exercises and lectures than from all the public classes which I attended at the University.[15]

The historian William Robertson also found these classes uniquely beneficial. Among Stevenson's other students were John Witherspoon, later President of the College of New Jersey (now Princeton University), and Hugh Blair, later Professor of Rhetoric and Belles Lettres at Edinburgh, who wrote an essay on 'The Beautiful' for Stevenson, and spoke of Stevenson's praise of this as making him decide to study 'polite literature'.[16]

The phrase 'Rhetoric and Belles Lettres' probably came from Rollin, Professor of Rhetoric in Paris, whose lectures were published in France in 1726–8, and in a 1734 English translation as *The Method*

14 McElroy, *Scotland's Age of Improvement*, 14–16, 21; David Daiches, *Literature and Gentility in Scotland* (Edinburgh: Edinburgh University Press, 1982), 44; Alexander Law, *Education in Edinburgh in the Eighteenth Century* (London: University of London Press, 1965), 33, 222.

15 Thomas Somerville, *My Own Life and Times 1741–1814* (Edinburgh: Edmonston and Douglas, 1861), 13.

16 On Robertson, see Somerville, *Life and Times*, 14; on Witherspoon, Blair, and Stevenson, see Varnum Lansing Collins, *President Witherspoon: A Biography* (2 vols.; Princeton, NJ: Princeton University Press, 1925), i. 14; on Blair and Stevenson, see also Henry Duncan, 'Blair', in *Edinburgh Encyclopedia*, ed. David Brewster (Edinburgh: Blackwood, 1830), iii. 567.

of Teaching and Studying the Belles Lettres, a book frequently quoted by British schoolteachers.[17] In Edinburgh in 1737 a collection of extracts from the *Spectator*, *Tatler*, and *Guardian* used 'Mr Rollin's Method of teaching and studying the belles lettres'.[18] Such developments were not uniquely Scottish. In some of the English Dissenting Academies (set up after 1689 to educate dissenters, who were excluded by law from the two English universities) there was, increasingly, some teaching of English studies in the eighteenth century. As their name suggests, the Dissenting Academies catered for (male) students, largely of non-aristocratic background, who were not part of the ruling English establishment. The location of the Academies tended to be provincial, northern, non-metropolitan; their links were less with Oxbridge than with the Scottish universities which conferred honorary degrees on Andrew Kippis, John Aikin, Joseph Priestley, and William Enfield, pioneers of the teaching of English studies.[19] But the Scottish universities matter most because they were *universities*, the dominant, established, mainstream (not dissenting) channels of higher education. Their development of English studies in a particular cultural climate shows most clearly how the subject involved an attempted suppression of native tradition in a process of cultural conversion that was thought of as a move from the barbarous Scottish to the polite British—thought of, in short, as 'improvement'.

Particularly after the Jacobite Rebellions, which had brought civil war to Scotland, many influential Scots believed that common British norms—which meant, generally, Anglocentric norms—were needed if Scotland were to prosper and win English approval. A leader of this group was Henry Home, later Lord Kames, whose *Essays upon Several Subjects Concerning British Antiquities* (1747) attempted to purge Scotland of cultural distinctions which might invite English disfavour: 'When one dives into the Antiquities of *Scotland* and *England*, it will appear that we borrowed all our Laws and Customs from the *English*.'[20] Kames spoke Scots, but wished to encourage the use of English in Scotland. In 1762 his *Elements of Criticism* advocated the fine arts, which, 'by cherishing love of order . . . inforce submission to government', and he subscribed that year (along with Hugh Blair, James

[17] Ian Michael, *The Teaching of English from the Sixteenth Century to 1870* (Cambridge: Cambridge University Press, 1987), 163; D. J. Palmer, *The Rise of English Studies*, 12.

[18] Law, *Education in Edinburgh*, 150.

[19] Palmer, *The Rise of English Studies*, 9.

[20] [Henry Home], *Essays upon Several Subjects Concerning British Antiquities* (Edinburgh: A. Kincaid, 1748), 4.

Boswell, and others) towards the publication of Thomas Sheridan's *Course of Lectures on Elocution*, delivered the previous year to an Edinburgh audience anxious 'to cure themselves of a provincial or vicious pronunciation'.[21] Sheridan took it as axiomatic that the only respectable form of speech was that of the London Court. 'All other dialects, are sure marks, either of a provincial, rustic, pedantic, or mechanic education; and therefore have some degree of disgrace annexed to them.'[22] When delivered in Edinburgh, each of Sheridan's lectures was followed by 'select readings from the English classics'. An Edinburgh student recalled that 'Among other results of Mr. Sheridan's visit, a society, consisting of literary men, was formed, for the purpose of concerting measures for the instruction of the young in this hitherto neglected, but, as was now supposed, primary branch of education.' Hugh Blair's *Lectures on Rhetoric and Belles Lettres* would later cite Sheridan with approval.[23] Edinburgh's students, aspiring lawyers, preachers, and public men wished to shed tell-tale contraventions of standard English norms so as to succeed in their careers. A later authority on Scotticisms recalled being told that the 'Duke of Argyle' (a Scotsman of great political influence in London) 'thought a resemblance or identity of language of such real national importance, that he is said to have furnished Mr. Hume with the materials of his printed collection'.[24] This collection was the list of 'Scotticisms' appended to David Hume's 1752 *Political Discourses*. The list of proscribed expressions—such as the Scottish 'mind it' instead of English 'remember it'—may seem trivial, but it is an indicator of the way in which discourse itself was politically important, a key to gaining power in Britain. Like Kames's, Hume's speech was filled with native forms, but he attempted to excise these when writing, complaining of his 'corrupt Dialect'.[25] Eighteenth-century terms such as 'Scotchism' and 'Scotchery' show clear signs of English prejudice

[21] [Henry Home], *Elements of Criticism* (2 vols.; Edinburgh: A. Kincaid and J. Bell, 1762), i, p. iii; Thomas Sheridan, *A Course of Lectures on Elocution* (London: Strahan, 1762), pp. xv, xvii ('Subscribers'), 30.

[22] Sheridan, *Lectures on Elocution*, 30.

[23] Somerville, *Life and Times*, 56; Hugh Blair, *Lectures on Rhetoric and Belles Lettres* (2 vols.; London: W. Strahan and T. Cadell, 1783), i. 177, and ii. 218.

[24] Sir John Sinclair, *Observations on the Scottish Dialect* (London: Strahan and Cadell, 1782), 3.

[25] 'Scotticisms', appended to some copies of David Hume, *Political Discourses* (Edinburgh: A. Kincaid and A. Donaldson, 1752), 3; *The Letters of David Hume*, ed. J. Y. T. Greg (Oxford: Oxford University Press, 1961), i. 255 (letter to Gilbert Elliott of Minto, 2 July 1757).

against power-hungry Scots and their linguistic peculiarities.[26] Aspiring Scots were encouraged to kill off what William Robertson called 'those vicious forms of speech which are denominated Scotticisms'.[27] Scotsmen like Robertson and Hume won respect for Scottish achievement, not least in writing. To many it appeared that the way to advance as a Scot was to appear as English as possible, while at the same time upholding an ideal of Britishness in which Scotland would be able to play her full part.

Sometimes the efforts of Scots to Anglicize their tongues gave rise to anxiety about loss of identity. Actors, used to elocution, were among those who gave lectures and demonstrations of 'proper' English. But occasionally we hear another note:

> At a period when the attention of the public is so laudably engaged in the study of the language of our sister kingdom; it is hoped it will not be deemed improper to pay some regard to that of our own; and that an effort to keep alive some of the first pieces of poetry that can adorn any language will meet with the approbation of those possessed in any degree of the *Amor Patriae*, or who do not wish the Scotch name to sink into utter oblivion. Therefore on Friday next, 15th March, in St Mary's Chapel, Mr Young will deliver a lecture on the Scottish language.[28]

Such lectures in the mid-eighteenth century were building on the earlier efforts of anthologists like Allan Ramsay to raise public awareness of the literary importance of Scots. Yet Scots, as well as Scotticisms, was being pushed firmly out of the discourse of institutional power. Through the networks of gentlemen's 'improving' clubs in Edinburgh and other Scottish towns and cities, ideas were advanced about the improvement of language and speech, the removal of distinctive Scottish linguistic indicators.[29]

There is no shortage of examples of how such attitudes pervaded the ruling class. One typical but interesting case must suffice. John (later Sir John) Sinclair, a Scottish MP in London, published his *Observations on the Scottish Dialect* in London in 1782 to be 'of use to my countrymen . . . particularly those whose object it is to have some share in the administration of national affairs'. For those who wished

---

[26] See the entries for these words in the *OED*.

[27] See entry for 'Scotticism' in *OED*. For further examples of Scottish linguistic sensitivities, see Pat Rogers, 'Boswell and the Scotticism', in Greg Clingham (ed.), *New Light on Boswell* (Cambridge: Cambridge University Press, 1991), 56–71.

[28] Law, *Education in Edinburgh*, 159.

[29] See McElroy, *Scotland's Age of Improvement, passim*.

to get on, 'new manners must be assumed, and a new language adopted. Nor does this observation apply to Scotchmen only: the same remark may be extended to the Irish, to the Welsh, and to the inhabitants of several districts in England.' The London model had to dominate all. Like Kames, Sinclair stressed the need for Scots to assimilate themselves to the English model:

the time, it is hoped, will soon arrive, when a difference, so obvious to the meanest capacity, shall no longer exist between two countries by nature so intimately connected. In garb, in manners, in government, we are the same; and if the same language were spoken on both sides of the Tweed, some small diversity in our laws and ecclesiastical establishments excepted, no striking mark of distinction would remain between the sons of England and Caledonia.

The Author of this little performance, with pleasure contributes his mite to a purpose so truly desirable.[30]

Sinclair sees the repression of linguistic difference as essential to his political and social scheme of the cultural assimilation of successful Scots into an Anglocentric British mould.

Like those of several Scottish 'improvers', Sinclair's linguistic efforts were only part of wider cultural and agricultural concerns. Tutored by Hugh Blair's literary friend John Logan, Sinclair published his first works in 1769.[31] These were letters to the *Caledonian Mercury*, written in response to complaints about the unjust treatment of Highlanders driven to emigrate to America during the Clearances. Sinclair protests himself a patriot, yet one who finds it hard to believe 'that the departure *of a few factious and idle Highlanders* would prove detrimental to these united kingdoms . . . let them be transported to a country, where they may find a nation perhaps as savage as themselves, and, if possible, equally destitute of the least appearance of religion and virtue'.[32] If Sinclair saw the worst of the Highlanders as savages, he saw himself as an improver who wished, as he put it, to 'Kentify Caithness'.[33] As well as 'improving' the Highlands, Sinclair was the moving force behind the vast *Statistical Account of Scotland*, which, through the offices of the General Assembly of the Church of Scotland, anthologized a great data bank, a general assembly of geographical, commercial, demographic, and ethnological information

---

[30] Sinclair, *Scottish Dialect*, pp. iii, 2, 9–10.

[31] John Dwyer, *Virtuous Discourse: Sensibility and Community in Late Eighteenth-Century Scotland* (Edinburgh: John Donald, 1987), 22.

[32] Sinclair, *Statistical Account*, i. 37–8.

[33] Rosalind Mitchison, *Agricultural Sir John* (London: Geoffrey Bles, 1962), 27.

about Scottish parishes. This careful setting-in-order of Scottish material was conducted with a view to the better ordering of an improved Scotland. As an informational anthology, Sinclair's work is a major cultural achievement, fit to stand beside the *Encyclopaedia Britannica* as a highlight of eighteenth-century Scottish culture. His endeavours in furthering the Protestant, English-speaking Society for the Promotion of Christian Knowledge and his proscribing of Scotticisms were other aspects of the virtuous desire to improve; all helped Sinclair in his career in British government service. In information, as in lexis, he knew how to manipulate the discourse of power.

Sinclair's work was far from alone. Another purger of Scots speech and Scotticisms was the spelling reformer, James Elphinston, whose *Propriety Ascertained in Her Picture* (1786–7) devoted almost all its second volume to examining the 'Scottish Dialect' and seeing how to get rid of such 'deviacions from dhe purity ov LONDON'. Elphinston held up Sinclair's work as a glorious example of how 'Dhe patriot wished indeed, not onely by discriminating dhe dialects to' reunite dhe kingdoms; but to' render INGLISH PROPRIETY, first accessibel in evvery part of dhe iland, and dhen in evvery quarter ov dhe globe.'[34] If Elphinston was attracted to world-dominating regularity, numerous other writers—like the Aberdeen poet, philosopher, and academic James Beattie, like James Bannantine, and like Hugh Mitchell—all had the more modest aim of removing from the writing of Scottish authors those linguistic indicators of cultural difference, Scotticisms.[35]

Beattie had corresponded with Sinclair over the matter of Scotticisms, and the two utilized each other's work. Beattie's list includes Scotticisms drawn from the work of Scottish writers such as Smollett (who is accused of writing 'to *take on* for a soldier' instead of to 'inlist' [*sic*]). Beattie ends his work with 'An Exercise', a passage composed so as to contain a large quantity of italicized Scotticisms. Of this he writes defensively:

These idioms are thus huddled together, by way of exercise, to young Scotch people, who may have been reading this pamphlet. But the English

[34] James Elphinston, *Propriety Ascertained in her Picture* (2 vols.; London: Jon Walter, 1786–7), ii, pp. i and ii.

[35] [James Beattie], *Scoticisms Arranged in Alphabetical Order, Designed to Correct Improprieties of Speech and Writing* (Edinburgh: William Creech, 1787); James Bannantine, Letter and list of Scotticisms in the *Monthly Magazine*, Dec. 1798, 434–9; Hugh Mitchell, *Scotticisms, Vulgar Anglicisms, and Grammatical Improprieties Corrected . . .* (Glasgow: Falconer and Willison, 1799).

reader will not suppose, that people of education in North Britain speak so uncouth a dialect. Many of them use a correct phraseology. Yet I fear there may be some (in the lower and middle ranks there are many thousands) who would read this exercise, without suspecting that there is any thing exceptionable in the style of it.[36]

Scots seeking to progress in British institutions were genuinely troubled by the way in which their language threatened to hinder them; their position resembles in many ways that of many present-day 'provincial' speakers. A London Scottish correspondent in the *Monthly Magazine* for 1798 contributes a list of Scotticisms and anecdotes, warning that Scotticisms can divert attention from a speaker's subject-matter, leading him to lose political and other arguments. So, for example,

Mr MONTGOMERY, now chief baron of the court of Exchequer, in Scotland, when lord advocate and member for Peeblesshire, made a speech on some important question, in the house of commons, where he mentioned his having made a note of something or other with a *keeliveyne pen*—the members, puzzled to discover the meaning of this outlandish word, and amused with the ridiculousness of it, had their attention altogether diverted from the argument of the speech—The right honourable orator meant a pencil.[37]

Here, picking a linguistic curio from Montgomery's speech, his largely English auditors perform an action as subtly patronizing and damaging as that of James Joyce's English Jesuit dean in the fifth chapter of *A Portrait of the Artist as a Young Man*, who ignores Stephen Dedalus's arguments, choosing to seize instead on his use of the word 'tundish'. The questions of attitude to language which preoccupied eighteenth-century Scots have far from vanished from the English-speaking world.

All climates are in some way prejudicial, but it is in the context of this particular eighteenth-century Scottish cultural climate that the development of the university teaching of Rhetoric and Belles Lettres, the forerunner of English Literature, should be viewed. If John Stevenson was the herald of the new subject, it really expanded in the mid-eighteenth century under the guidance of the man who, at Glasgow University in 1751, became the most important person since Stevenson to offer an official university course in English which covered the technique and appreciation of modern writers in that

[36] [Beattie], *Scoticisms*, 2, 97, 121.   [37] Bannantine, Letter and list, 434–5.

language as well as writers in the classical tongues. This man was Adam Smith.

It may seem strange that Smith is a progenitor of English studies, but his linguistic and literary work, like his work on economics, jurisprudence, and proto-anthropology, is part of the project of 'improvement'. If aspiring business men would profit from the ideas of the *Wealth of Nations*, there was also profit to be made from studying literary texts that, as models of English propriety, would improve and correct the language and taste of men ambitious to hold public office. In the development of English studies, as in Smith's other concerns, economics mattered.

Smith had been a student at Glasgow under Francis Hutcheson ('father' of the Scottish Enlightenment); afterwards he had spent several years at Balliol College, Oxford, holding the prestigious Snell Exhibition. Then, in 1746, before his Exhibition had come to its end, he had done a surprising thing. He had gone home to Kirkcaldy in Fife. It was Lord Kames who realized that the Smith who had returned after six years in Oxford might be something of a national asset. Encouraged by Kames, Smith 'was induced to turn his early studies to the benefit of the public, by reading a course of Lectures on Rhetoric and the *Belles Lettres* . . . to a respectable auditory, chiefly composed of students in law and theology'. Among Smith's audience was Alexander Wedderburn, who published Smith's work in the *Edinburgh Review* of 1755–6, where the editorial concerned itself with 'the difficulty of a proper expression in a country where there is no standard of language, or at least one very remote'. Also in Smith's Edinburgh audience was Hugh Blair, the future Professor of Rhetoric and Belles Lettres at Edinburgh University. Smith's lectures appealed to Scots who were upwardly mobile: he offered the chance to hear 'proper English' spoken, as well as advice about what constituted good style and examples of great English stylists.[38]

In Edinburgh Smith was lecturing to an audience of ambitious citizens, but in 1751 the lectures themselves became part of an official university curriculum when Smith was appointed a professor at Glasgow University. After he moved there to take up the Chair of Logic, he continued to teach Rhetoric and Belles Lettres to his 'private' class (the class to which the professor taught his own subject of special interest). Notes on these lectures were discovered in 1958 by John Lothian; the most complete scholarly edition, edited by John

---

[38] For these passages and for fuller information, see Smith, *Lectures*, 7–23.

Bryce, appeared in 1983 as part of the Glasgow edition of the *Works and Correspondence* of Adam Smith. Despite Professor Bryce's exemplary edition, the full significance of Smith's work and its impact has yet to be assessed.

W. S. Howell has set Smith's work against the framework of traditional Rhetoric—a necessary approach, but one which distorts what was particularly important about Smith's work in a mid-eighteenth-century Scotland confused about the proper language of Scottish culture.[39] In this context, Smith's work is seen as being at the root of what constitutes the university canon of English Literature, and also at the root of the struggle between vernaculars and the standard English language, both issues which were central not just to the Scottish tradition, but also to the international English-speaking world in the centuries which followed.

The Smith who had decided to give up Oxford did not berate his Glasgow students for being Scottish, but he did subtly stress the need to conform to southern English linguistic models.

Our words must not only be English and agreeable to the custom of the country but likewise to the custom of some particular part of the nation . . . It is the custom of the people that forms what we call propri[e]ty, and the custom of the better sort from whence the rules of purity of stile are to be drawn. As those of the higher rank generally frequent the court, the standard of our language is therefore chiefly to be met with there . . .[40]

Such a point would have had a particular significance for Smith's Scottish audience. 'We in this country are most of us very sensible

---

[39] Wilbur Samuel Howell, *Eighteenth-Century British Logic and Rhetoric* (Princeton, NJ: Princeton University Press, 1971), 536–76. It should be said that, though Howell's interest in the emergence of modern from Classical models of Rhetoric prevents him discussing the significantly 'provincial' importance of writers such as Smith, Blair, Barron, and Witherspoon, he does acknowledge that the development of the eighteenth-century elocutionary movement associated with Rhetoric was significantly provincial: 'During the eighteenth century the British became aware that the many forms which their language had in various districts of England, and the many differences which existed between London English, on the one hand, and Scottish or Irish or American English, on the other, were no longer to be regarded merely as attractive and desirable variations from locality to locality, but as positive hindrances to the cultural, political, commercial, and occupational welfare of a growing and dynamic world empire.' (p. 156.) Howell's book does not aim to chronicle the history of English Literature as a university discipline, but it is valuable to all interested in the subject. He is also unusual in mentioning in his concluding paragraphs 'the capacity of the new rhetoric of eighteenth-century Scotland to bring about the development of English Literature as a field of study in nineteenth-century American and British universities' (p. 716).   [40] Smith, *Lectures*, 4–5.

that the perfection of language is very different from that we commonly speak in.'[41] Having made this point, Smith spends no time in lambasting his audience for its Scotticisms; the records we have of eighteenth-century university lectures on Rhetoric and Belles Lettres contain very little of that. But they do operate subtly, using significant exclusion, insinuation, and implicit guidance to point their audiences clearly in the direction of Anglocentric (though not solely English) Britishness.

What Smith offers his audience is not only an education in Rhetoric as it was conventionally understood, but also the opportunity as Scots to confront a pure English style. The medium of his own lectures— English, not Latin—was an important part of this, as early testimonials to his pure, un-Scottish style demonstrate. The literary material presented was also important. In his particular praise of Swift, whose 'language is more English than any other writer that we have', Smith directs attention to a model of linguistic purity.[42] Again, one of the most important points is implicit. Like Smith's teacher, Francis Hutcheson, Swift was an Irishman with an admired English style, proof that it was possible for the provincial un-Englishman to upstage the London aristocrat if only he tried hard enough.

Another subtle point, left entirely implicit, is that, when exemplifying the sounds to be avoided when trying 'to speak English with the proper accent', Smith begins with the [x] sound.[43] This attack on the most distinctive Scots linguistic shibboleth, the [x] sound of words like 'loch', went hand in hand with a devaluing of writing in Scots, which is remarked on either not at all or else in terms which present it as barbarous or primitive. As an early nineteenth-century Scotsman recognized, the task of the successful eighteenth-century Scottish writer involved 'guarding his diction against the words and phrases which he naturally spoke'.[44] Such guarding, necessary for the promotion of a British ethos, often seemed to involve a turning-away from the distinctively Scottish. Indeed, Smith's lectures contain so few Scottish references that, particularly in view of the circumstances in which they were delivered, this must have been deliberate policy. There are just sufficient quirks to pinpoint Smith's country of birth; only a Scotsman would have been likely to yoke together Pompey and Robert the Bruce, and one may detect a hint of pique

---

[41] Ibid. 42.   [42] Ibid. 4.   [43] Ibid. 15.
[44] Sir James Mackintosh, *Miscellaneous Works* (3 vols.; London: Longman, Brown, Green, and Longmans, 1846), ii. 486.

when, classifying Clarendon as a partial historian, Smith points out how he spends considerable time on minor affairs in the English Court, while 'he discusses in two or three sentences all the actions of Montrose in Scotland tho' of the Greatest importance'.[45] But no Scottish poetry is mentioned in Smith's lecture on poetry, though recent collections of Scots verse, such as Allan Ramsay's *Ever Green* (1724) and *Tea-Table Miscellany* (1724–37), had fought to stress that poetry in the Scots tongue was not only a hardy perennial but also fit to be admitted to polite society. The 1781 catalogue of Smith's library (now in Tokyo) shows not only that he maintained an interest in the development of Rhetoric teaching throughout his life, owning such works as Campbell's *Philosophy of Rhetoric*, his pupil William Richardson's pioneering study of *Shakespear's Remarkable Characters*, and a synopsis of lectures on Belles Lettres at St Andrews University, but also that he owned copies of several important works of eighteenth-century and earlier Scottish Literature, including the works of Drummond of Hawthornden, Macpherson's *Fingal* and *Temora*, Ramsay's *Poems* and *Ever Green*, Bannatyne's *Ancient Scottish Poems*, and a collection of *Scottish Tragic Ballads*. Yet one of the most interesting features of Smith's lectures is that Allan Ramsay is not mentioned. A Scottish acquaintance of Smith's recalled that Smith disliked Ramsay because that poet (who used distinctively Scots forms) did not write like a gentleman.[46]

The only lecture in which Smith does briefly demonstrate his familiarity with the Scottish poetic tradition is not his lecture on poetry; rather, he mentions Scots poetry as part of his explanation of how poetry precedes prose in society's progression from a primitive to a refined state. Detailed examination of this passage shows the workings of Smith's mind—how he skilfully associates Scots writing with the widely scorned Gaelic (the supposed primitive 'Erse' of the recently published Ossianic poems), and then moves swiftly on, in a discrediting anthropological glide, to 'the most Rude and Barbarous nations' and African 'savages', which are contrasted with a culture of commerce, modern security, and urban refinement—exactly the sort of society that the study of proper English models would further in Glasgow and Edinburgh. Smith's subject here was, appropriately,

---

[45] Smith, *Lectures*, 129–30, 115.

[46] Tadao Yanaihara, *A Full and Detailed Catalogue of Books which Belonged to Adam Smith* (Tokyo: Iwanami Shoten, 1951), 75, 89, 91–2, 107, 119; anon., 'Anecdotes . . . of the Late Adam Smith' (1791), repr. in Smith, *Lectures*, 230.

'the Language of Business', which he showed as more highly developed than the earlier, apparently more 'difficult', language of poetry.

It will no doubt seem at first sight very surprising that a species of writin[g] so vastly more difficult should be in all countries prior to that in which men naturally express themselves. Thus in Greece Poetry was arrived to its greatest Perfection before the beauties of Prose were at all studied. At Rome there had lived severall poets of considerable merit before Eloquen[ce] was cultivated in any tollerable degree. There were English poets of very great reputation before [before] any tollerable prose had made its appearance. We have also severall poeticall works in the old Scots Language, as Hardyknute, Cherry and the Slae, Tweedside, Lochaber, and Wallace Wight in the originall Scotts but not one bit of tollerable prose. The Erse poetry as appears from the translations lately published have very great merit but we never heard of any Erse prose. This indeed may appear very unnatural that what is most difficult[y] should be that in which the Barbarous least civilized nations most excell in; but it will not be very difficult to account for it. The most barbarous and rude nations after the labors of the day are over have their hours of merryment and Recreation; and enjoyment with one another; dancing and Gambolling naturally make a part of these dive[r]sions; and this dancing must be attended with music. The Savage nations on the coast of Africa, after they have sheltered themselves thro the whole day in caves and grottos from the scorching heat of the Sun come out in the evening and dance and sing together. Poetry is a necessary attendant on musick, especially on vocall musick the most naturall and simple of any. They naturally express some thoughts along with their musick and these must of consequence be formed into verse to suit with the music. Thus it is that Poetry is cultivated in the most Rude and Barbarous nations, often to a considerable perfection, whereas they make no attempts towards the improvement of Prose. Tis the Introduction of Commerce or at least of opulence which is commonly the attendent of Commerce which first brings on the improvement of Prose.— Opulence and Commerce commonly precede the improvement of arts, and refinement of every Sort. I do not mean that the improvement of arts and refinement of manners are the necessary consequence of Commerce, the Dutch and the Venetians bear testimony against me, but only that [it] is a necessary requisite. Wherever the Inhabitants of a city are rich and opulent, where they enjoy the necessaries and conveniences of life in ease and Security, there the arts will be cultivated and refinement of manners a neverfailing attendent.[47]

A phrase such as 'the old Scots language' is a way of writing off Scots as obsolete; work in it is seen as paralleling the 'merryment' of 'least

[47] Smith, *Lectures*, 136–7.

civilized nations'. The models for university study which Smith puts
forward in Glasgow are entirely non-Scottish. His lectures are a
means of translating his audience, subtly alienating it from the
language of its own culture. In his work of translation, Smith was
hoping to benefit his Scottish auditors. Yet any culture separated
from the language of its heritage faces enormous problems.

At Glasgow University the subject which Smith had introduced
was continued by George Jardine from 1774 to 1824 in his 'system of
lectures on general grammar, rhetoric, and belles lettres', which
included 'the elements of taste and criticism, and . . . the rules of
composition, with a view to the promotion of a correct style, illustrated by examples'. Some of his students, including Francis Jeffrey,
formed an Elocution Society.[48] The teaching of Belles Lettres was
also developing at Aberdeen and St Andrews. In the former, George
Campbell, a friend of Hugh Blair, was developing material for his
Philosophy of Rhetoric in the late 1750s. At St Andrews Robert Watson,
who had secured the approbation of David Hume and Lord Kames
for his public lectures in Edinburgh on language, style, and Rhetoric,
was appointed Professor of Logic, Rhetoric, and Metaphysics at
St Salvator's College. But the teaching of Belles Lettres was most
celebrated at Edinburgh University, where a member of Smith's
earlier Edinburgh audience became a lecturer in Rhetoric and Belles
Lettres in 1759 and Professor in 1762.

This man was Hugh Blair, whose bestselling *Lectures on Rhetoric and
Belles Lettres* (published in 1783) both acknowledge a debt to Smith's
work and take it further. Blair's first biographer saw the Professorship
as entailing 'the task of forming the taste of the rising generation'.
Blair was one of those, we are told, who worked 'with a generous view
to improve literature in Scotland, where, though it had begun to
dawn, it was but little advanced'. Blair stresses his lectures' unique
potential to 'improve' their audience and 'cultivate their Taste'.
Linguistic care was seen as the distinguishing feature of behaviour.
'Among nations in a civilized state, no art has been cultivated with
more care, than that of language, style, and composition.' 'Taste is a
most improveable [*sic*] faculty.'[49] Rather than being satisfied with the

---

[48] George Jardine, *Outlines of Philosophical Education* (Glasgow: no publisher given,
1818), title-page; see also entry on Jardine in Robert Chambers (ed.), *A Biographical
Dictionary of Eminent Scotsmen*, rev. Thomas Thomson (London: Blackie, 1875); for the
Elocution Society, see McElroy, *Scotland's Age of Improvement*, 123.

[49] John Hill, *Life of Hugh Blair* (Edinburgh: T. Cadell and W. Davies, 1807), 24–5;
Blair, *Lectures*, i, pp. iv, 2, 19.

brief examples of good English given by Smith and Jardine, Blair
gives numerous extended quotations, intended as practical guides to
eloquence. A series of 'Critical Examinations' subjects essays from
the *Spectator* and passages of Swift to minute scrutiny. One senses
that, from the given examples, a canon is developing of writers who
are approved for their proper English style. Like Smith, Blair avoids
labouring any hostility towards writing in Scots; he simply concentrates on English models, taking it for granted that 'pure' style must
be 'without Scotticisms or Gallicisms'.[50] Composing his lectures
slightly later than Smith, Blair is occasionally able to introduce the
work of Scottish writers as exemplifying such pure English. So, for
instance, to illustrate the correct use of the words 'with' and 'by',
Blair is able to point out how 'The proper distinction in the use of
these particles, is elegantly marked in a passage of Dr. Robertson's
History of Scotland.' The Edinburgh friends Blair and Robertson
revised each other's writings prior to publication.[51] In his literary
orientation Blair was a defender of what he saw as Scottish excellence,
but that excellence had to be of the 'proper English' variety. He was a
conscious Briton, writing of 'Our Island', at the same time as lacking
no alacrity in championing such respectable Scottish eminences as the
Latinist George Buchanan.[52] As a Scot, Blair manifested peculiar
daring in his excursions into detailed criticism of the language of
Addison (a favourite subject of the *Lectures*). What is particularly
interesting is that, where Smith had almost entirely avoided mention
of literature produced in Scotland, Blair's cultural patriotism leads
him to devote rather more attention to Scottish works, and even to
discuss Ramsay's pastoral, *The Gentle Shepherd*.

Here an awkwardness manifests itself: on the one hand, Blair is
anxious to claim merit for the Scottish writer and to encourage
'improved' writing in Scotland; but, on the other, he is uneasy about
the way in which *The Gentle Shepherd* employed Scots as well as English.
Like Smith, Blair tries to historicize Scots, seeing it as simply 'the old
rustic dialect' and so fatally limited, yet at the same time clearly
wishing to inscribe this Scottish presence into his approved canon of
'Pastoral Poetry'.

I must not omit the mention of another Pastoral Drama, which will bear
being brought into comparison with any composition of this kind, in any

---

[50] Blair, *Lectures*, i. 187.   [51] Ibid. 201; Hill, *Hugh Blair*, 179.
[52] Blair, *Lectures*, ii. 284.

language; that is, Allan Ramsay's Gentle Shepherd. It is a great disadvantage to this beautiful Poem, that it is written in the old rustic dialect of Scotland, which, in a short time, will probably be entirely obsolete, and not intelligible; and it is a farther disadvantage, that it is so entirely formed on the rural manners of Scotland, that none but a native of that country can thoroughly understand, or relish it.[53]

Like Smith, Blair engages not only with Classical writing and with the accepted English classic writers (such as Shakespeare and Addison), but also with modern writing. In the latter area there are several slight, but important, attempts to make the new English voice of Scotland (the voice, usually, of Blair's own friends) speak its part in correct literature. So, for instance, the 'Scottish Homer', William Wilkie, author of *The Epigoniad* (1757), is included in the consideration of epic, while John Home's 1756 Scottish drama *Douglas* (which David Hume at one time thought superior to Shakespeare) is regarded as worthy of consideration alongside the work of Euripides, Shakespeare, and Racine.[54] Knowledge of Home's play is taken for granted by Blair, as is knowledge of the Scottish text with which his name has come to be most intimately associated: the poems of Ossian.

'The Substance' of Blair's *Critical Dissertation on the Poems of Ossian, the Son of Fingal* 'was delivered by the Author in the Course of his Lectures on Rhetoric and Belles-Lettres, in the University of Edinburgh', but the *Dissertation* was published in 1763 and is not contained in the published *Lectures* of twenty years later.[55] What is conspicuous there, though, is Blair's assumption that the reader's knowledge of Ossian's work can be taken for granted. Its popularity came about partly because the Ossianic poems represented a valorization not of those contemporary Highlanders so despised by Sir John Sinclair, but of the antique Gael, and because the poems appeared not in the barbarous language of Gaelic, but in a decorous 'translatorese' which harmonized beautifully with eighteenth-century standards of propriety. Blair delighted in the way in which the Ossianic poems combined 'the fire and the enthusiasm of the most early times' with 'an amazing degree of regularity and art', so that compared, for instance, with early Scandinavian poems, these Highland artworks seemed 'a fertile and cultivated country'; but his *Lectures* presented Gaelic as clearly both ancient and obsolete.[56] Like the *Lectures on Rhetoric and Belles*

[53] Ibid. 351-3.
[55] Blair, *Dissertation*, p. ii.
[54] Ibid. 410, 506.
[56] Ibid. 11; Blair, *Lectures*, i. 169-70.

Lettres themselves, the Ossianic poems (whose collection by James Macpherson had received the sponsorship of Blair and his dramatist friend John Home) are a skilled effort at cultural translation, turning Scottish material of an unacceptable kind into a form acceptable to a new British audience.

In some ways the aims of these early university teachers of Rhetoric and Belles Lettres are seen most clearly in the career of William Barron, who succeeded Robert Watson, Adam Smith's follower, at St Andrews. In looking at his work and its links with that of lecturers like Adam Smith and Hugh Blair, we can see some common features in the operation of their literary teaching. Like Blair, Barron began as a Church of Scotland minister, with several improving interests. One of these was in agriculture. Barron's first book, the 1774 *An Essay on the Mechanical Principles of the Plough*, is an attempt to have the heavy old Scots plough, designed for uncultivated soils, replaced by a more modern model, suitable for a soil which is now more cultivated.[57] Barron's lectures at St Andrews, where he was Professor of Belles Lettres and Logic from 1778 to 1803, were designed to help eradicate older Scottish literary and linguistic forms which were now considered uncultivated, and to replace them with more acceptably refined southern forms. Barron, like the author of the *Wealth of Nations*, was interested in the 'industry' that would, by hard work, promote cultural and commercial success; his *Lectures on Belles Lettres and Logic* demonstrate clearly the way in which the study of Belles Lettres was to enable Scots to 'progress', through communication that would let them 'attain in society a greater portion of influence and fame.' He points out that

> The beneficial influence of assiduous culture is not restricted to individuals; even communities partake something of the same distinction. It is industry that constitutes the internal strength and opulence of a nation. It improves her arts, enriches her subjects, enlightens her philosophers, and extends her reputation. Here also, as in the case of individuals, it surmounts the defects of nature, and the inconveniences of climate and situation.[58]

The relevance of such arguments to eighteenth-century Scotland would not have been lost on Barron's St Andrews audience; the examples which he cites as success stories—'the diminutive republics

[57] William Barron, *An Essay on the Mechanical Principles of the Plough* (Edinburgh: J. Balfour, 1774), 27–8.
[58] William Barron, *Lectures on Belles Lettres and Logic* (2 vols.; London: Longman, Hurst, Rees, and Orme, 1806), i. 11, 12–13.

of ancient Greece' as well as the modern Swiss cantons and Holland—are all geographically small units which have achieved high eminence despite their size. Sometimes Barron is quite explicit about how Scotland is to progress: 'Her intercourse with England, now so common and easy, the introduction of many of the arts with which that rich and industrious country abounds, the refinement of manners and taste, the extension of science and literature, all have concurred to cultivate the public ear, and to polish her language and her pronunciation.'[59] This 'polishing' meant that a levelling-down of Scotticisms and accent, the most obvious markers of cultural difference, was being encouraged among the ambitious intelligentsia through the university system. But the impact of the subject went much further than that. For the effect of the Scottish universities' concentration only on those literary models which conformed to proper English norms represented an official attack on the traditional, vernacular Scottish culture, with the result that only groups educated 'properly' in the new way were to be taken seriously in questions of taste. Barron's *Lectures* show this most explicitly. He writes of how the acquisition of a thorough knowledge of the principles and structure of a language will allow his audience to relish refined composition, and so 'ensure success in circumstances the most disadvantageous'.

These remarks have been verified by experience, founded on the literary history of Scotland. While she remained a separate kingdom, the remoteness of her situation from the seat of politeness and power, was accounted an insuperable obstruction against all attempts to compose in the language of England. Her writers, of course, either confined their compositions to the imperfect dialect of their own country, or undertook to express their sentiments in the language of ancient Rome. Even posterior to the union, when intercourse between the kingdoms became easy and frequent, it was long thought impossible that a North-Briton should produce any work, which could be relished or applauded by an Englishman of taste. A few eminent examples have invalidated every opinion of this sort, and have demonstrated that Scotland, by proper culture, is qualified to furnish English compositions, not less pure and correct, perhaps, than England herself. Let not, then, our situation discourage us in the study either of speaking or writing. Let us remember, that industry is certainly sufficient to surmount every obstacle or inconvenience. A consciousness of difficulty, and the fear of incurring censure, may, probably, excite a spirit of attention and perseverance, which may surpass any merit that has yet appeared; and which may convey to our language an energy and an elegance, which have not hitherto distinguished

[59] Ibid. 13, 36.

the most perfect of her writers. The ambition to excel by honourable means, is at least laudable; and though it should not be crowned with immediate success, yet it may excite a spirit of emulation, which may be finally triumphant.[60]

This, I believe, is the clearest statement of the motivation which underlies the teaching of all the eighteenth-century Scottish university rhetoricians; it harmonizes, too, with the work of the many independent teachers, and it represents nothing less than a huge attempt at cultural restructuring. It would be too easy to label this 'southern English cultural imperialism', because what we have to remember is that, while the metropolis and English Court may have taken it for granted that their own standards should be adopted as universal, it was none the less Scots, and generally 'provincials', who encouraged other Scots and provincials to adopt these standards so as to improve and, ultimately, to benefit themselves and their community. The attack on the distinctive Scottish cultural tradition was one mounted by Scots themselves, and what made the attack all the easier was the increasingly widespread dissemination of English texts regarded as canonical. The works which became canonical were those affording examples of the 'proper English' which would permit speakers of provincial dialect, even if they were unable to master correct southern pronunciation, to write a uniform, standard English purged of cultural peculiarity. Barron's students studied models which would allow them to realize that 'Persons may write well a language they cannot speak', and to improve until they could 'write the language of the purest authors of their age. There is, therefore, nothing in the nature of the thing, that should hinder the language of England from being written well in India or America. The authors of the latter country will not, I suspect, readily admit, that their situation is in any respect detrimental to the merit of their compositions.'[61] The provincials could win respect, partake of power, and compete with the men of the capital on the capital's own ground; but only by paying stylistic homage to the capital's standards. However they might continue to *speak*, as *writers* they had to exist in translation, more muzzled precursors of the young Irishman Stephen Dedalus, whose soul frets

---

[60] Ibid. 118–19.

[61] William Barron, *Synopsis of Lectures on Belles Lettres and Logic Read in the University of St Andrews* (Edinburgh: John Balfour, 1781), 47; Barron, *Lectures*, i. 131.

under the shadow of the English Jesuit dean's language: 'The language in which we are speaking is his before it is mine.'[62]

Yet the fact that access to proper English Literature gave access to 'proper' language—and so to power—explains the popularity of the subject of Rhetoric and Belles Lettres in North America. William Small, Jefferson's Scottish tutor at the College of William and Mary in Virginia, taught the subject to his young pupil, who went on to include Rhetoric and Belles Lettres in his own subsequent plans for institutions of higher education, though he also wondered whether Americans should not have their own dialect similar to that of Robert Burns.[63] At Princeton the Scot John Witherspoon became President of the College of New Jersey in 1768 and, during his first year as President, inaugurated the first formal course in Rhetoric in North America. Though Witherspoon in the 1780s was to wonder 'whether we shall continue to consider the language of Great Britain as the pattern upon which we are to form ours: or whether, in this new empire, some center of learning and politeness will not be found, which shall obtain influence and prescribe the rules of speech and writing to every other part', his Princeton lectures on eloquence were to follow the work of Blair, and where the Scots had produced lists of *Scotticisms* in an effort to curb the native tendencies of their speech when they came to write, so Witherspoon coined the word 'Americanism' and published a list of these. Blair's own *Lectures* were very popular in late eighteenth-century America (being published in Philadelphia in 1784, very soon after their first appearance in Edinburgh) and featured in many college courses. The *Edinburgh Review* (edited by George Jardine's old student William Jeffrey) went on to achieve a cultural power in America comparable with that of the *Spectator* in eighteenth-century Scotland, though there were also native productions such as the *Belles-Lettres Repository*. Harvard established a Chair of Rhetoric and Oratory in 1806, while Yale appointed the editor of

[62] James Joyce, *A Portrait of the Artist as a Young Man* (1916), repr. in *The Essential James Joyce*, ed. Harry Levin (1948; repr. Frogmore, St Albans: Triad/Panther, 1977), 316.

[63] Thomas Jefferson, *Writings*, ed. Merrill D. Peterson (Cambridge: Cambridge University Press, Library of America Series, 1984), 463, 1296; for detailed considerations of the interaction of Scottish rhetorical teaching and American political life, see David Daiches, '*Style Périodique* and *Style Coupé*: Hugh Blair and the Scottish Rhetoric of American Independence', and Thomas P. Miller, 'Witherspoon, Blair and the Rhetoric of Civic Humanism', in Richard B. Sher and Jeffrey Smitten (eds.), *Scotland and America in the Age of the Enlightenment* (Edinburgh: Edinburgh University Press, 1990), 209–26, 110–14 respectively.

Select British Eloquence to be its Professor of Rhetoric in 1817.[64] Benjamin Franklin had told Hume that the best British English would always be the American standard, and Americanisms continued to be viewed as 'barbarous phraseology' in early nineteenth-century America.[65] Though we may at times think of Noah Webster and his *American Dictionary of the English Language* in terms of an assertion of cultural difference, Webster's greatest pride was that 'the genuine English idiom is as well preserved by the unmixed English of this country, as it is by the best English writers'.[66] There is certainly American pride here, but it is a pride like that of many an eighteenth-century Scot who has managed to rid himself of any features of diction that would mark him out as 'impurely' un-English. According to A. W. Read, no one in England could recognize an American accent until the nineteenth century; the wide variety of English domestic dialect meant that Americans were simply taken as provincials.[67]

As Andrew Hook has shown, Scottish textbooks like those of Blair and Campbell continued to be used in American universities well into the mid-nineteenth century, so that a classical British propriety remained the standard model. Even as late as 1911 an American entrepreneur of correspondence courses was republishing condensed versions of Blair's *Lectures*, claiming they were 'too well known to require an extended introduction'. Earlier Americans took stands against Americanisms, but also against importing recent British 'vulgarisms and provincialisms'. If John Neal (who wrote in *Blackwood's Magazine* in 1824–5 championing the American use of '*low* words, unless they are wholly beneath us') refused to write '*classical* English', and if the young Whittier in 1829, began writing a small number of poems in Scots as the Robert Burns of Massachusetts ('thinking of the temperance lyrics the great poet of Scotland might have written had he put his name to a pledge of abstinence'), then

[64] L. T. Chapin, 'American Interest in the Chair of Rhetoric and English Literature in the University of Edinburgh', *University of Edinburgh Journal*, Autumn 1961, 119; see also Andrew Hook, *Scotland and America, 1750–1835* (Glasgow: Blackie, 1975), 81–2.

[65] Richard Bridgman, *The Colloquial Style in America* (New York: Oxford University Press, 1966), 40.

[66] Webster, in his 1828 *Dictionary*, quoted in Harry R. Warfel, *Noah Webster, Schoolmaster to America* (New York: Macmillan, 1936), 362.

[67] Charles Bremner, 'U.S. Boffin Acts as a Guide through Twists of the English Tongue', *The Times* (London), 25 June 1988, 7. This article draws on and outlines material forthcoming in the dictionary of 'Briticisms' being compiled by Allen Walker Read and John Alego for the Clarendon Press.

these examples only emphasize the similarity between the predicaments of Scottish and American writers, who wished to assert their cultural difference in the face of educational norms aimed principally at the eradication of that difference.[68] Later, Scott and his follower, Fenimore Cooper, 'the American Scott', would find their own compromises and solutions, but these solutions were arrived at in a climate in which the higher education system continued to echo the eighteenth-century call of the teachers of Rhetoric and Belles Lettres, who sought to improve Scots out of their Scotticisms, Americans out of their Americanisms, and to promote the study of a canon whose works were, by and large, the literary embodiment of English metropolitan taste.

The teaching of Rhetoric and Belles Lettres in Scotland in the early nineteenth century does seem to be largely made up of echoes. After the pioneering work of Smith, and the impact of that long-reigning triumvirate of Blair, Barron, and Jardine, their successors seem to have taken the subject little further, simply retracing their masters' footsteps. Jardine's successor, Robert Buchanan, was celebrated for 'the correct and chaste style of his lectures', and, as author of various Scottish patriotic dramas, his publications and teaching manifested more enthusiasm for Rhetoric and creative writing than for philosophy. By the time Buchanan retired as Professor of Logic at Glasgow in 1864, the University already had its first specifically designated Professor of English Language and Literature, John Nichol, appointed in 1862. At St Andrews the Chair of Logic, Rhetoric, and Metaphysics continued to be held by men of notably literary interests, such as the popular William Spalding, author of a *History of English Literature* and Professor of Rhetoric and Belles Lettres at Edinburgh (1840–5), who held the St Andrews Chair from 1845 until 1859. The present Chair of English at St Andrews dates from 1896. At Edinburgh Spalding was followed by the poet W. E. Aytoun, then by David Masson, who was still correcting Scotticisms when, in the 1860s, the Chair was renamed 'Rhetoric and English Literature'.[69]

---

[68] Hook, *Scotland and America*, 81–2; Hugh Blair, *Lectures on Rhetoric*, ed. Grenville Kleiser (New York: Funk & Wagnalls Co., 1911), p. v; Bridgman, *Colloquial Style*, 22, 41–3, 48; John Greenleaf Whittier, *Poetical Works*, ed. W. Garrett Horder (London: Oxford University Press, 1904), 530 (head-note to 'The Drunkard to his Bottle').

[69] 'Buchanan, Robert' and 'Jardine, George', *DNB*; Philip Hobsbaum, 'A Guid Chepe Mercat of Languages: The Origins of English Teaching at Glasgow University', *College Courant*, Sept. 1983, 10–17; 'Spalding, William', in M. F. Connolly, *Biographical Dictionary of Eminent Men of Fife* (Cupar, Fife: John C. Orr; and Edinburgh:

But, rather than taking up further space here in reciting the lines of succession that connected the teaching of Rhetoric and Belles Lettres to what, in the mid-nineteenth-century Scottish universities, became explicitly the teaching of 'English Literature', it seems best to conclude this chapter by indicating the result of the impact of the Scottish developments on what most commentators have seen as the launch site of the discipline: the early London Chairs of English.

When we look at these, we see that there was a continuity between them and the earlier growth of the subject under the title of 'Rhetoric and Belles Lettres' in Scotland. In University College, London, there was, from the beginning, a strong Scottish influence. Twelve of the twenty-eight early professors had been Scottish graduates, only six having come from Oxford (none from Cambridge); and both University College (founded in 1828) and King's College (founded in 1831) were, like the Scottish universities, institutions which could be seen as challenging the monopoly of Oxbridge.[70] Thomas Dale, an evangelical clergyman, achieved the remarkable feat of becoming the first Professor of English at both University and at King's. D. J. Palmer writes that Dale's plan for English study 'does not differ essentially from the rhetorical studies of the old dissenting academies and of the Scottish universities'.[71] We might emphasize more strongly the similarities between Dale's work in London and Hugh Blair's work in Edinburgh by noting Dale's regular lectures on the principles of composition, and the fact that, in 1845, he produced the last influential edition of Hugh Blair's *Lectures on Rhetoric and Belles Lettres*, supplementing it with a brief account of English literary history. By this time Dale had retired from his Chair of English Literature and History, but the ease with which Blair's work could be adapted to such an English context is salutary. Because in important ways Blair's work had been geared to a task of cultural conversion, of Anglicizing upwardly mobile Scots to make them acceptable Britons, it easily pointed forward to the development of English studies in England. There the emphasis, as Baldick, Palmer, and others have indicated, came to fall on 'English Literature' as bound up with the culture of England; the emphasis also fell on the teaching of English Literature as a useful implement in the attempted conversion of provincials or

Inglis and Jack, 1866); Henry W. Meikle, 'The Chair of Rhetoric and Belles-Lettres in the University of Edinburgh', *University of Edinburgh Journal*, Autumn 1945, 89–103.

[70] McMurtry, *English Language, English Literature*, 119.
[71] Palmer, *The Rise of English Studies*, 16–24.

the lower classes to an acceptable metropolitan standard. This can be seen, for instance, in the work of F. D. Maurice, the founder of working men's colleges. D. J. Palmer's valuable, though sometimes timid, book gives a perfect example of the way in which the cultural conversion practised by these working men's colleges (and encouraged by Matthew Arnold) paralleled the effect which Hugh Blair's *Lectures* had on a lower-class reader. Palmer quotes from the autobiography of the Chartist Thomas Cooper, who recalls his early discovery of what was then being institutionalized as 'English Literature': 'Blair's "Lectures on Rhetoric and Belles Lettres" was another book that I analysed very closely and laboriously, being determined on acquiring a thorough judgment of style and literary excellence. All this practice seemed to destroy the desire of composing poetry of my own.' Cooper goes on to describe how great poetry transported him out of his immediate 'vulgar world of circumstances'.[72] This curbing of vulgarity, silencing or showing great condescension to provincial and lower-class expression, and the heavy stress on models of English propriety (which only the educated might possess) were just what the eighteenth-century Scottish fathers of the university teaching of English Literature had in common with those who developed the subject in nineteenth-century England.

When the study of English Literature expanded in English universities, there were no voices to question its association with Englishness of a metropolitan variety. The provincial origins of the subject might have been expected to counterbalance the narrowly upper-class, metropolitan Anglocentric emphases pointed out by Baldick and others. Instead, they encouraged just those emphases. In their own anxiety to adopt non-native standards, the Scottish—like the American, Canadian, Irish, Indian, and English—provincial inventors of English Literature were responsible for developing, through the educational system, problems which these cultures would have to face over the next century and a half. The university subject of 'English Literature' was born among conditions of 'provincial' socio-economic ambition and anti-provincial linguistic prejudice. It found its voice not just in singing the merits of one dominant cultural tradition, but also in its silencing of others. First and emblematic among those silenced others were the other tongues of Scottish culture—Scots, Gaelic, and the Scotticisms which impregnated English with signs of Scottish difference. 'English Literature' promoted an Anglocentric,

[72] Ibid. 33–4.

even a London-centred, cultural hegemony even as it championed Britishness. If this Britishness was to allow Scotland to flourish in various ways, it seemed that she could do so only if she spoke and wrote as she was told to do. Any deviation from the model of instruction would be patronized as ethnically curious or quaint. The literary consequences of the Scottish invention of 'English Literature' are present in the writing of Burns and Boswell, just as they are evident in Scott, Carlyle, Joyce, MacDiarmid, Seamus Heaney, and Tony Harrison. It was a particular idea of 'improvement' which gave eighteenth-century Scotland commercial and intellectual eminence. It was that same idea of improvement which gave rise to, and encouraged, the development of the university subject under discussion, and stimulated its institutionalization of forces geared towards the suppression of cultural difference. Scotland can claim much of the credit for inventing 'English Literature'; it can also claim, in common with all other societies which are sometimes labelled 'provincial', to have felt the need to try and escape from that invention.

2

British Literature

IF there is an important sense in which 'English Literature', the university subject, was a Scottish invention, then there was a parallel eighteenth-century Scottish invention in the arena of creative writing. That invention is British Literature. The Union of Parliaments in 1707, and its consequences, had very little effect on literature written in England. That simply went on being English Literature. Certainly, the Englishness of 1789 was scarcely identical to that of 1688. England's own identity altered, reshaped by a Hanoverian monarchy and an American colony in revolt as well as by other factors, but none of this seems to have involved any profound confrontation with the issue of being British as opposed to English. The wish of some Scottish thinkers and teachers to acquire fundamentally Anglocentric mores may have contributed to the shaping of eighteenth-century English-ness in ways that remain to be investigated, but as regards the development of a literature that was in any meaningful sense British, the English showed little or no interest. Panegyrics to 'Britain' were common in early eighteenth-century England, but 'Britain' was usually seen as a synonym for 'England'. Matthew Prior prefaces his *Solomon* by explaining that his 'digressive *Panegyric* upon Great Britain' is part of his desire to be 'thought a good *English-man*', and the panegyric sees Britain as England. When Pope in 'Windsor Forest' (1713) hymns 'British blood' (l. 367) and 'BRITISH QUEEN' (l. 384), there is no hint that the word 'British' means anything other than 'English'. Bonamy Dobrée, writing in 1949 on 'The Theme of Patriotism in the Poetry of the Early Eighteenth Century', takes it for granted that Britannia and Britain should be equated with England in this way, but a more sceptical critic, John Lucas in *England and Englishness* (1990), points out that 'when English poets speak of Britain as a nation of the free they usually mean England'.[1]

[1] Matthew Prior, cited in Bonamy Dobrée, 'The Theme of Patriotism in the Poetry of the Early Eighteenth Century', *Proceedings of the British Academy*, 35 (1949), 52; John Lucas, *England and Englishness: Ideas of Nationhood in English Poetry, 1688–1900* (London: Hogarth Press, 1990), 3.

If English poets after 1707 changed little in their Anglocentric outlook, the same is surely true of English novelists. This is not to say that English novelists totally avoid characters and attitudes from the areas of Britain beyond England, but these are, at best, small side-issues. The books often seen as the great English novels of the eighteenth-century, such as *Tom Jones*, *Clarissa*, and *Tristram Shandy*, are just that—great *English* novels. They offer no interrogation of the nature of Britishness, and show no interest in the new Britain as opposed to England. This was not a subject that (despite the Act of Union and the Jacobite Rebellions) particularly impinged on English authors. Though one can find a good supply of insults directed against the Scots in eighteenth-century English writing, few of the major English writers paid much attention to the revised relationship between the united kingdoms.

Through the eighteenth century, though, in response to the cultural and political pressures outlined in the previous chapter, and sometimes in a direct response to the teachings of Rhetoric and Belles Lettres, Scottish Literature involved a continuing examination of, and response to, the strains and possibilities of Britishness. Insightfully, awkwardly, entrepreneurially, Scottish writing entered its British phase. It is this Britishness which, more than anything else, distinguishes Scottish from English Literature in the eighteenth century. The Scots' concern with identity, discrimination, and the possibilities of 'improvement' or advancement makes prejudice one of the main themes of Scottish books in this period. If we wish to see how a society may attempt to articulate a non-English cultural identity while using a (sometimes modified) form of the English language, it is to eighteenth-century Scotland that we must turn for the first full-scale example.

This chapter does not offer a history of eighteenth-century Scottish writing. It largely ignores, for instance, any examination of the bonding properties of writing geared to form a more cohesive Scottish society at a time when social division was being accentuated— bonding properties examined by John Dwyer in his study of *Virtuous Discourse*. Again, except for gesturing towards Adam Smith, it avoids discussion of the Scottish contribution towards sentimental writing treated by John Mullan in *Sentiment and Sociability*.[2] Instead, the

[2] John Dwyer, *Virtuous Discourse: Sensibility and Community in Late Eighteenth-Century Scotland* (Edinburgh: John Donald, 1987); John Mullan, *Sentiment and Sociability: The Language of Feeling in the Eighteenth Century* (Oxford: Clarendon Press, 1988).

present discussion is structured round consideration of three writers crucial to eighteenth-century Scottish writing: Smollett, Boswell, and Burns. It is with regard to their work that the concept of Britishness seems most important. It is these three writers who leave the greatest legacy to the figures and movements considered later in this book. In the broad field of eighteenth-century Scottish literature it is these three writers who are most exciting to us today, partly because they sceptically experimented with the new possibilities offered by being British.

By way of prologue, though, there is the case of James Thomson, in whose work we see fleetingly topics which will possess the later Scottish developers of British Literature. Distinctively Scottish subject-matter and vocabulary mark out Thomson's work: so claims Mary Jane Scott, but the very title of her study, *James Thomson, Anglo-Scot*, shows that Thomson's Scottishness has to be qualified.[3] Rather than joining M. J. Scott and the line of earlier critics who lay stress on such slippery, unquantifiable factors as the influence on Thomson's *The Seasons* of the Border landscape of his youth, it seems more fruitful to look at the ways in which Thomson asserts a Britishness which is not (as it is in Pope) the equivalent of Englishness. It is here that Thomson is most clearly, problematically, and often covertly, Scottish.

Thomson's early patrons, such as Sir William Bennet and Robert Riccaltoun, were interested in cultural and personal improvement. Born in Roxburghshire in 1700, though he 'spoke broad Scots all his life', Thomson grew up attempting to perfect his English according to Addisonian models, while he also admired Allan Ramsay's 'Tartana', a wide-ranging poem in English which celebrated 'Caledonian beauties' and valorized tartan as well as dealing with changing weathers and geographies, alluding sophisticatedly to the Classics and to Newtonian science, and embedding in its course substantial narrative passages.[4] Thomson, who imitated in Scots one of Ramsay's 1718 elegies, was clearly aware of the linguistic possibilities open to him, but, as an Edinburgh divinity student, he chose to

[3] Mary Jane Scott, *James Thomson, Anglo-Scot* (Athens, Ga.: University of Georgia Press, 1988).

[4] James Sambrook, *James Thomson, 1700–1748: A Life* (Oxford: Clarendon Press, 1991), 16; see James Thomson, 'Upon Beauty', in *Liberty, The Castle of Indolence, and Other Poems*, ed. James Sambrook (Oxford: Clarendon Press, 1986), 231; see also Allan Ramsay, 'Tartana, or The Plaid', in *Works*, i, ed. Burns Martin and John W. Oliver (Edinburgh: William Blackwood for the Scottish Text Society, n.d.), 27–37.

develop an English so ornate that his professor warned him he would need to 'express himself in language more intelligible to an ordinary congregation'.[5] Such criticism, blended with hopes of a better reception in the south, led to Thomson's decision to try his fortune in London in 1725. This move to London did not represent the abandonment of a Scottish milieu. It was advocated by his Scottish kinswoman, Lady Grizel Baillie (some of whose verse was anthologized by Allan Ramsay), and it was in the household of her son-in-law, the Scottish nobleman Lord Binning, that he found initial employment as a tutor. Thomson immediately contacted his fellow Edinburgh alumnus, the literary opportunist David Mallet, who was a tutor in the Duke of Montrose's household.[6] After moving south early in the 1720s, Mallet had changed his surname from the distinctively Scottish 'Malloch', apparently after the English critic John Dennis had called him 'Moloch'—though Mallet told a Scottish friend that the reason was that the English mispronounced his name. Certainly, his Scottish background marked him out as a target. Dr Johnson described him as 'the only Scot whom Scotchmen did not commend'.[7]

In London, helped by Mallet, Thomson formed acquaintances with a wide Scottish circle. He cultivated patrons such as Duncan Forbes of Culloden, as well as the patriotically Scottish painter William Aikman, who became Thomson's close friend and was a friend of Allan Ramsay. Thomson, like a typical immigrant, sought out his fellow-countrymen and was homesick. A letter to William Cranstoun, a friend left behind in Scotland, in October 1725 mentions the poem *Winter* for the first time, and sets its composition beside Thomson's longing for the more rugged Scottish landscape, and his imagining Cranstoun 'in the well known Cleugh' back in Scotland.[8] Yet, for all its mentions of wild terrain, if the poem has a specifically Scottish aspect, it is a totally private one, and the modern reader will be unaware of it unless *Winter* is set beside the letter to Cranstoun. *Winter* was hard to publish. Mallet and Thomson 'walked one November day to all the Booksellers in the Strand, and Fleet-

[5] Patrick Murdoch, 'Life', in *The Works of James Thomson* (2 vols.; London: A. Millar, 1762), i, p. vi.

[6] James Thomson, *The Seasons*, ed. James Sambrook (Oxford: Clarendon Press, 1981), p. xxxv. Line numbers in the present text refer to this edition's main text of *The Seasons* or (if so indicated) to the texts of Appendix A.

[7] Samuel Johnson, 'Mallet', in *Lives of the English Poets* (1779 and 1781; 2 vols.; repr. London: Oxford University Press, 1961), ii. 441.

[8] James Thomson, *Letters and Documents*, ed. Alan Dugald McKillop (Lawrence, Kan.: University of Kansas Press, 1958), 16.

Street, to sell the copy', without luck. Eventually another Scotsman, John Millan, agreed to publish the poem, using Archibald Campbell, surely a further Scot, as his printer.[9] Developed and manufactured in this London Scottish circle, *Winter* had to reach a wider audience if it was to be commercially successful. Thomson's English adviser, Aaron Hill, helped by writing a commendatory poem for a second edition. Mallet (signing himself D. Malloch) supplied another, and was in turn puffed as 'a *Gentleman* of . . . fine, and exact Taste' by Thomson in his Preface.[10]

These two Scottish poets in London regularly exchanged ideas, aiding and abetting one another. In 1725 Thomson was residing with Mallet in the house of Mallet's then employer, the Duke of Montrose. In August 1726 Thomson sent drafts of *Summer* to Mallet, hinting that the use of British patriotic materials is partly a flattering and opportunistic marketing ploy, but also making it clear that Thomson is aware of being a Scottish poet who sees Britain as a cultural amalgam comprising more than just England. He envisages 'a Panegyric on Brittain, which may perhaps contribute to make my Poem popular. The English People are not a little vain of Themselves, and their Country. Brittania too includes our native Country, Scotland.'[11] Those pronouns ('*their* Country', '*our* native Country') reveal Thomson's sense of cultural differences within the United Kingdom of 'Britannia'. In the first (1727) edition of *Summer* the 'Panegyric on Brittain' lasts a little over a hundred lines, beginning 'HAPPY BRITANNIA!' (l. 498), and describes a very English-seeming landscape of 'Guardian-Oaks', 'Meadows', and '*Villas*'. Among British patriots and sages, Thomson lists only English names. But the ensuing line, 'And should I northward turn my filial eye' (l. 558), alerts us to Thomson's northern loyalties and makes it clear that he has so far regarded only South Britain. When he writes of 'thy *Caledonian* sons' (l. 562), the pronoun 'thy' refers back to 'Britannia', reminding readers that Scotland too is part of the British amalgam. Thomson draws attention to the ancient Scots' independent spirit ('A gallant, warlike, unsubmitting Race! | Nor less in *Learning* vers'd': ll. 563–4) as well as their wisdom and honour. Though this first version of *Summer* is generally geared to appeal to those English who are 'not a little vain of Themselves, and their Country', it also contains a subtly

9 Thomson, *The Seasons*, pp. xxxvi–xxxvii.   10 Ibid. 307.
11 Thomson, *Letters*, 48.

embedded reminder that Britain must take account of other—Scottish
—cultural sensitivities and pride.

This furthering of the Scottish cultural presence under the aegis
of Britishness found support among some of Thomson's literary
countrymen. Allan Ramsay's patron, the polymathic improver Sir
John Clerk of Penicuik (1676–1755), who was both pro-Union and
pro-Scotland, worried about a southward drift of Scottish talent, yet
recorded his pleasure at meeting in London on 30 April 1727 'Mr
Maloch and Mr Thomsone two young Lads of my Country & justly
esteam'd at present the best poets in Britain'.[12] Clerk's partisan
remark hints at the continuing rivalries between the English and the
Scots within the cultural as well as the political arena of 'Britain'. In
1730 the London success of Thomson's first play, *Sophonisba*, was
attributed by a hostile critic to 'Scotchmen with tuneful hands and
many feet' who had packed the theatre.[13] Too pronounced a Scottishness could be a potent disadvantage to the writer who, rather than
being content to be a provincial curio, wished to operate on the central
ground of British culture, as did Thomson in his play *Agamemnon*.
Reading the play to the actors at a rehearsal, Thomson was greeted
with 'a loud laugh' because he 'pronounced every line with such a
broad Scotch accent'. He retorted that 'though I can write a tragedy,
I find I cannot read one'.[14]

Ridiculed for their pronunciation, some aspiring Scottish writers
saw in the written word the chance of concealing their origin, retaining their sense of Scottishness only as a private possession for exposure
to a trusted circle. Success in the English-dominated British market
would come more easily if Scottishness were concealed or elided, as is
surely the case with *The Masque of Alfred* (1740), co-authored by
Thomson and Mallet, and containing Thomson's best-known piece,
'Rule, Britannia!'. This masque celebrates the England of Alfred the
Great, so mention of Scotland would scarcely be appropriate. But
what is significant is the ease with which 'Rule, Britannia!' was
removed from its specific, purely Old English context, and came to be
adopted as an anthem for contemporary Britain. One suspects that
Thomson planned it opportunistically, knowing well that it might
appeal to the pride of Englishmen, for whom Britain and England
were equivalent.

---

[12] Quoted in Douglas Grant, *James Thomson, Poet of 'The Seasons'* (London: Cresset
Press, 1951), 73.   [13] Ibid. 91–2.   [14] Ibid. 181.

Given the ease with which 'Britain' and 'Britannia' could come to
mean simply 'England', and the temptation for the Scottish writer to
submerge his national identity under a purely Anglocentric notion of
Britishness, what is particularly noteworthy in Thomson's work is the
way in which he several times points to Scotland not merely for the
sake of adding attractive landscape descriptions to a pre-Romantic
palette, but in order to remind readers that Scotland too is a part of
Britain and should be seen as such. It is this which gives Thomson's
work a distinctive British accent, produced as a result of his Scottish
cultural loyalties. That these loyalties persisted is clear from his
continuing association with his Scottish intellectual and social circle in
the south of England. Thomson's Scottish friend and first biographer,
Patrick Murdoch, became tutor in 1729 to John Forbes, another close
friend of the poet and the son of Thomson's patron, Duncan Forbes.
A glance at Thomson's strong interest in Newtonianism again leads
us to the poet's Scottish network. Patrick Murdoch edited and introduced the posthumous *Account of Sir Isaac Newton's Philosophical Discoveries* (1748) by his Scottish friend and former teacher, Colin
Maclaurin. Thomson and several of his circle, including John Armstrong, Duncan Forbes, and Mallet, were among the subscribers, as
were Allan Ramsay and Andrew Mitchell, a close mutual friend of
Thomson and Maclaurin, and Andrew Reid, another literary Scot in
London, whose periodical, the *Present State of the Republick of Letters*,
disseminated Newtonian ideas and carried an enthusiastic notice of
Thomson's *Spring* in its May 1728 issue. That poem was published by
yet another Scotsman, Andrew Millar, who had established himself
in the Strand and who would later publish the histories of Hume and
Robertson, becoming 'the generous patron of Scotch authors'. In
London 'the majority of Thomson's close friends were expatriate
Scotsmen'.[15] With the London Scots George Strahan and John
Millan, and with Thomson himself and Allan Ramsay in Edinburgh,
Millar took subscriptions in 1728 for a proposed edition of four
'Seasons', about which Thomson wrote to Clerk of Penicuik on
18 January 1728.[16]

The genesis of much of Thomson's early work, culminating in the
eventual publication of *The Seasons* in 1730, was a Scottish enterprise
involving a group of Scottish literary men intent on succeeding in the
new British market. Such ambitions are also encoded in the poem.

[15] *DNB* entry for Andrew Millar; Sambrook, *James Thomson*, 154.
[16] Thomson, *The Seasons*, p. xliii.

Though Thomson transferred his treatment of Caledonia from
Summer to *Autumn* ('his favourite season for poetical composition'),
leaving the panegyric to Britannia in *Summer* to refer to England
alone, the revised and extended Caledonian material speaks specifically of Scotland's aspirations within Britain.[17] The poet approaches
Scotland via its roughest landscape—the 'stormy Hebrides' (*Autumn*,
l. 865)—but Caledonia is seen as potent with romantic beauty and
natural resources. It is hinted that these are not being used to the full
by the modern Scots. Caledonia may be naturally rich, but she is
'incult' (uncultivated), though this may be part of her attraction.
Thomson translates Scotland to win the sympathy of the dominant
majority of the British readership. He gives us lakes not lochs, and
vales not glens. One need only think of Johnson's vision of a treeless
Scotland to realize how different a picture might have been presented
from that of *Autumn*:

> And here a while the Muse,
> High-hovering o'er the broad cerulean Scene,
> Sees CALEDONIA, in romantic View:
> Her airy Mountains, from the wavering Main,
> Invested with a keen diffusive Sky,
> Breathing the Soul acute; her Forests huge,
> Incult, robust, and tall, by Nature's Hand
> Planted of old; her azure lakes between,
> Pour'd out extensive, and of watry Wealth
> Full; winding deep, and green, her fertile Vales;
> With many a cool translucent brimming Flood
> Wash'd lovely . . .   (ll. 878–89.)

It is at this point that Thomson again reveals his own Scottish origin,
going on to ask if there is not 'some Patriot' to bless Scotland's
fortune by helping 'To chear dejected Industry'. Thomson wishes to
improve particular industries—weaving, linen, and fishing—which
were of great contemporary importance in Scotland.[18] Partly, this
passage is designed to flatter the Duke of Argyll and Thomson's
friend Duncan Forbes of Culloden, introduced as the wished-for
patriots, patrons of Scottish industry and of her parallel 'reviving
Arts' (l. 946). It is also designed to emphasize Thomson's clear wish

[17] Murdoch, 'Life', p. xvii; Sambrook's edn. of *The Seasons* allows the reader to
study Thomson's revisions with ease.
[18] Alan Dugald McKillop, *The Background of Thomson's Seasons* (Minneapolis:
University of Minnesota Press, 1942), 136.

that Scottish interests be given their full share in Britain. He uses the only distinctly Scottish word in the passage when he writes of how 'our Friths' (l. 923) were raided by Dutch fishermen, and contends that, if such a threat is not countered, the Scots are not receiving fair treatment as part of the United Kingdom.

In 1730 Thomson arrests his readers by highlighting the word 'Britain', forcing a consideration of just what that word means, as he wishes for a new energy, for

> all-enlivening Trade to rouse, and wing
> The prosperous Sail, from every growing Port,
> Uninjur'd, round the sea-incircled Globe;
> And thus united BRITAIN BRITAIN make
> Intire, th'imperial MISTRESS of the day.   (ll. 924–8.)

Thomson sees an imperial role for Britain, and wants Scotland to be able to take advantage of it by being fully part of Britain. The line 'And thus united BRITAIN BRITAIN make' is striking, but possibly a little obscure. This section is the only part of *Autumn* to be significantly revised during the 1730s. James Sambrook, who demonstrates that 'the newly united kingdom provides a unifying idea for much of the *Seasons*', suggests that these lines 'are rewritten to eliminate a jingle'—'BRITAIN BRITAIN'.[19] This is probable, but it is also clear that Thomson rewrites the lines to make his political point absolutely unmistakable:

> And thus, in Soul united as in Name,
> Bid BRITAIN reign the Mistress of the Deep.

Thomson requires greater efforts from both Scots and English people to make Britain truly united in spirit. His worry is that Britain may be only a 'Name'.

Such a concern with a view of Britishness that makes full room for Scotland surfaces only occasionally in Thomson's work, but it is present. *Liberty* (1734–6) is a poetic essay on a topic which was to fascinate Scottish Enlightenment writers, the history of civil society. For most of the poem Britannia seems synonymous with England, but eventually the goddess Liberty makes it clear that Britain is a cultural amalgam:

---

[19] Sambrook, *James Thomson*, 101; Thomson, *The Seasons*, p. lix.

'She rears to *Freedom* an undaunted Race:
'Compatriot zealous, hospitable, kind,
'Her's the warm CAMBRIAN: Her's the lofty Scot,
'Fir'd with a restless an impatient flame,
'That leads him raptur'd where Ambition calls:
'And ENGLISH MERIT'S Her's . . .[20]

If Thomson here portrays the Scot as a follower of 'Ambition', he gives us something of a self-portrait. He was ambitious that both he and his native country might exploit the concept of Britain to the full. In personal terms, he did this often by his use of his network of Scottish contacts and friends in London and beyond. His public voice appears English, except for those significant moments when the idea of Britain is deployed in such a way that Scotland is seen to play her part. These moments are sufficient to show him as a Scottish pioneer of British Literature.

The Castle of Indolence (1748) reveals only a covert Scottish presence. Several of Thomson's London Scottish circle, including his old college friend William Paterson, his literary protégé John Armstrong, John Forbes, and Patrick Murdoch, have been identified in James Sambrook's 1986 edition as portrayed among the inhabitants of the castle. Yet none has his own name, nor, without scholarly and biographical information, would we know them as Scottish. Interestingly, Thomson's plays deal repeatedly with the themes of intolerance and the strains which arise between private and patriotic loyalties, so that one might speculatively suggest a 'Scottish' interpretation, subsidiary to their immediate political resonances. Thomson has a clear interest in the well-motivated individual caught between opposing cultural codes—whether the Arab Selim and his sister, victims of English prejudice in *Edward and Eleonora* (1739), or the Coriolanus whose self-interest ran counter to his ties to the motherland in Thomson's last play, an adaptation of Shakespeare.

In *The Castle of Indolence* and much of his other work Thomson indicated one course open to Scottish writers who wished to succeed in the British, predominantly English, market: to internalize and conceal any Scottish component so that it remained present only as a secret trace, identifiable to those in the know and largely unnecessary to the functioning of the text. Yet, occasionally in his earlier work, when, unlike his English contemporaries, Thomson carefully accented Scotland's place in Britishness, he exhibited a seed that would grow in

[20] Thomson, *Liberty*, 129.

the works of his friend Smollett and in those of his admirer James Boswell. Smollett would write what is arguably the first fully *British* novel—*Humphry Clinker*. Boswell's project would examine the John Bull figure of Johnson from the point of view of a Scottish author. While Boswell would anxiously and strongly accent his Scottish position, Smollett, like Thomson, would sometimes accent his much more subtly, and sometimes not at all. Robert Burns, another eulogist of Thomson, would adopt all these strategies. It is the nature of their Britishness which marks out these writers as peculiarly Scottish. Thomas Blackwell gave Thomson's work a position of honour in '*British* Poetry'.[21] So should we. Fleetingly but perceptibly, Thomson is (like Allan Ramsay) one of the inventors of British Literature precisely *because* he is a Scot.

SMOLLETT AND PREJUDICE

The British novel, as pioneered by Tobias Smollett, is a novel about prejudice, a subject with which, as an ambitious eighteenth-century Scot with his eye on the British capital, Smollet was quite familiar. He went to school in Dumbarton on the Firth of Clyde in the 1720s. His close Scottish friend and first biographer, John Moore, pointed out that Dumbarton Rock had been a stronghold of William Wallace, whose exploits in Scotland were 'repeated by every school-boy'.[22] Smollett's earliest verses were addressed to the anti-English Wallace, and were encouraged by a 1722 popular version of Blind Harry's *Wallace* rewritten by Burns's future correspondent, William Hamilton of Gilbertfield, as part of the post-Union resurgence of Scottish cultural nationalism led by Allan Ramsay. At Dumbarton's grammar school Smollett also enjoyed *Rerum Scoticarum Historia*, the history of Scotland by the great Scottish Latinist George Buchanan, supposed a former pupil of the school. Moore points out that Smollett's first full-length literary work, the orotund, sub-Shakespearian tragedy *The Regicide* (written in Glasgow when Smollett was an 18-year-old medical student), is based on the passage from Buchanan's *Historia* which deals with the assassination of James I of Scotland. If the young

---

[21] See Ralph Cohen, *The Art of Discrimination: Thomson's 'The Seasons' and the Language of Criticism* (London: Routledge and Kegan Paul, 1964), 152.

[22] John Moore, 'Memoir', prefaced to *The Works of Tobias Smollett* (8 vols.; London: Law *et al.*, 1797), i, pp. cv–cvi; information about Smollett's life is drawn from this work and the studies noted below.

Smollett grew up very much aware of the long historical rivalry between the now united kingdoms of Scotland and England, his adult career would keep him aware that such rivalries had not vanished.

To many Englishmen of the later seventeenth century, Scotland was a 'scabbie Land', verminous, exporting mangy itches; by the mid-eighteenth century Scots were spreaders of 'Caledonian Poison'; they were 'bare ars'd *Caledonian* Rogues', whose literature was the worthless rant of 'Caledonian pedlars'; on stage and in hostile pamphlets they were stereotyped as 'Sawney', the indigent immigrant; their presence in England and in the British government was 'the Scots Scourge'. By 1773 Horace Walpole had come to scorn the word 'Briton' (so celebrated by Thomson) because it included the Scots.[23]

In such a climate, the fortune-seeking Scottish writers in London (whose number Smollett joined in 1740, *The Regicide* in his luggage) tended to band together, utilizing—as Thomson had done—a Caledonian network. This in turn bred charges of Scottish cabals, cliques, and conspiracies. Despite the Union, Scots tended to be viewed in London as an immigrant community. If London was now the political and administrative centre of *Britain*, it remained the capital of an England often hostile to Scots who sought to play their part in British affairs. Foremost among Smollett's supporters in the sustaining community of London Scots was his relative Sir Andrew Mitchell, friend of Thomson and a member of the Scots circle which included Armstrong, Mallet, and Millar. Though Smollett soon embarked on a seagoing career, there is plenty of evidence that he made strong Scottish contacts in London, meeting Thomson, befriending Armstrong, and having some of his verses set to music in 1744–5 by James Oswald, the London Scottish composer and the first man in England to make Aeolian harps. Early English reactions to Smollett's first novel, *Roderick Random* (1748), show that readers noted his national origins as well as his profession. The book was published anonymously, but before long one reader was remarking that it was written by 'a Scotch sea surgeon', and another that its author was 'Mr Smollett, a Scotch surgeon'.[24] The word 'Scotch' here may seem

[23] See Eric Rothstein, 'Scotophilia and *Humphry Clinker*: The Politics of Beggary, Bugs, and Buttocks', *University of Toronto Quarterly*, 52/1 (Fall 1982), 63–4; George M. Kahrl, *Tobias Smollett, Traveler-Novelist* (Chicago: University of Chicago Press, 1945), 67–9.

[24] Quoted in Lewis Mansfield Knapp, *Tobias Smollett, Doctor of Men and Manners* (Princeton, NJ: Princeton University Press, 1949), 95–6.

neutral, but, as Smollett's career developed, he was to be marked out by English literary enemies as a 'Sawney', presiding (as editor of the *Critical Review* in 1756) over a tribunal of interloping 'Scotch gentlemen critics'. Sneered at as the 'Scottish critic' by Cuthbert Shaw (an English poetaster who had tried to imitate Thomson), Smollett featured in caricatures directed against Scots seeking power in Britain. He was a 'Scotch adventurer', a 'Caledonian Quack', a 'vagabond Scot'.[25] After he had saved Doctor Johnson's black servant, Francis Barber, from the press-gang, Johnson granted that Smollett was 'a scholarly man, sir, though a Scot'.[26] Schooled in Scotophobia, Smollett's triumph was to write the eighteenth century's greatest novel on the theme of prejudice. In order to understand *Humphry Clinker* in these terms, and to see for the first time its seminal place in Scottish literary tradition, it is helpful to review some aspects of his first novel and of his wider career.

'I was born in the northern part of this united kingdom'—the opening words of the first chapter of *Roderick Random* (1748)—signal both Scottishness and Britishness, and the book goes on to examine various ways in which the two interact. Red-haired Roderick's poor, harsh upbringing in Scotland none the less gives him a broad classical education and leaves him particularly valuing his 'taste in the *Belles Lettres*' (VI. 27).[27] After some brief satire on Rory's people's mores, both commercial and sexual, Random, as his creator had done, sets off for London.

At once British and alien, Roderick is laughably naïve, yet he is used in this novel of immigrant experience to complicate and defuse crude anti-Scottish prejudice. Random and his friend Hugh Strap remain visible in the text as Scots, constantly recognized as such by the other characters. Yet, in a novel linguistically alert to speech mannerisms and to French, Welsh, and Irish accents among the minor characters, Roderick's Scottishness of speech is never represented on the page. The result is that the reader, though kept aware that the protagonist is a Scot, tends to sense his voice as that of standard

25 See Paul-Gabriel Boucé, *The Novels of Tobias Smollett*, trans. Antonia White (London: Longman, 1976), 23; 'Smollett, Tobias', *DNB*; Knapp, *Tobias Smollett*, 177, 149, 167.   26 Quoted in Kahrl, *Smollett*, 68.
27 Since there is as yet no complete standard edition of Smollett's works, quotations from *Roderick Random* will be keyed to the chapters of the novel, with chapter numbers followed by a page reference to the 1973 reprint of the Everyman edition of the novel (Letchworth: Dent, 1973), from which widely available text quotations are taken.

English. This is surely a deliberate ploy on the part of Smollett to win over non-Scottish-speaking readers to Roderick's side by preventing the prejudice which would greet the realistic transcription of the voice of a protagonist whose very name, when pronounced with rolled Scots 'r's', would constitute a Scottish linguistic shibboleth. Other characters in the book hear Roderick's Scottish accent. We are made aware of it, but do not hear it directly. This strategy is part of Smollett's countering of expected prejudices in his English-dominated British audience.

That audience is humoured with some predictable mockery of Scottish innocents abroad, which also hints at the prejudice which such innocents encounter. One instance among many may suffice to demonstrate this. Cursed as 'a lousy Scotch guard', Random and Strap are immediately mocked when 'a wag, who sat in a box, smoking his pipe, understanding by our dialect that we were from Scotland, came up to me, and, with a grave countenance, asked how long I had been caught?' (XIII. 68–9). This needling almost exactly anticipates the celebrated anti-Scots gibe of Johnson, recorded later by Boswell as 'Much . . . may be made of a Scotchman, if he be *caught* young', while the ensuing, insulting question about the contents of Strap's knapsack also anticipates the lexicographer's Scotophobic wit: 'Is it oatmeal, or brimstone, Sawney?' (XIII. 69).[28]

Predictably, on arrival in London, Strap takes Roderick to seek out an established Scottish relative, who may be able to aid the young hopefuls. Mr Concordance is a mockery not only of Scottish linguistic preoccupations in the eighteenth century, but also of the Scot clumsily longing for a particular kind of upward mobility in London. In dress, as in language, where Smollett's transcription shows him to be affectedly English, this 'gentleman' only recently come from Scotland is desperate to dissociate himself from the habits of his countrymen and to conform absolutely to metropolitan fashion. Though he does not turn the young Scots away, this fop distances himself from them (he says 'your country', not 'our country'), and sees their Scottish dress as marking them out as distinctly primitive. Smollett adds a nice irony by having him wear 'a nightgown of plaid' (XIV. 73).

The meeting with Mr Concordance is only one of a number of encounters to exhibit Roderick and Strap as Scots in the context of the British capital. *Roderick Random* offers an example of the sort of network of mutual assistance which (however disrupted) seems to have

[28] *Life*, ii. 194.

operated among eighteenth-century Scots in London; it also furnishes plenty of examples of the prejudice to which such networks were opposed. Roderick is told by the Board of Surgeons that 'you Scotchmen have overspread us of late as the locusts did Egypt' (XVII. 91). His supposed meanness as a true 'Scotchman' is mocked (XVIII. 99). Though Mr Concordance is a caricature, his presence does indicate some of the difficulties which Scots perceived themselves as facing in the south. Smollett himself was very sensitive about Scotticisms.[29] Strap tells Random that 'the schoolmaster . . . has undertaken to teach you the pronunciation of the English tongue, without which, he says, you will be unfit for business in this country' (XVIII. 101).

Gradually, the story exposes Roderick's naïve 'prejudice in favour of my country' (XXI. 114), at the same time as showing the hostility he encounters as a 'lousy Scotch son of a whore' (XXIV. 145). As the book develops, though, we become much less aware of Roderick and Hugh Strap as two young Scots against the rest of Britain, and more aware of them as existing among various outsiders and victims of prejudice in the society. Miss Williams, despised as a fallen woman, is revealed (like the young Scots) as highly educated, if unfortunate. Eventually this young Englishwoman marries Strap. In his portrayal of life at sea, Smollett is introducing into English-language fiction details of a life lived largely outside that of respectable British society. In the navy Roderick is befriended by a fellow Celt, Morgan, a 'Cambro-Briton' (XXV. 152) who observes that 'in all likelihood, the ancient Scots and Britons were the same people' (XXVII. 159). With his Welsh accent ('praise Got' : XXVII. 159), Morgan may be a figure of fun, but his kindness to Roderick prevents the reader from dismissing him as such. On board ship, Roderick becomes involved in quarrels—linguistic and otherwise—with some Irish sailors, and gets into trouble. Yet no sooner does he seem to be alone again than he meets up with his countryman Thomson, who had been presumed drowned but has in fact found his way to Jamaica and the inevitable Scottish network—'the master's name was Robertson, by birth a North Briton, whom I knew at first sight to be an old schoolfellow of mine' (XXXVI. 204).

Roderick Random, with its absurd 'random' coincidences and caricatures, takes us inside the eighteenth-century Scottish network, and, to a degree, sends it up. At the same time, though, it pushes strongly

---

[29] See James G. Basker, *Tobias Smollett, Critic and Journalist* (Newark, Del.: University of Delaware Press, 1988), 82–4.

to win sympathy for the young Scot and for other social outsiders. Smollett produced the first important novel in the English language to have a Scot as hero. For all Roderick's naïvety, and for all the comic portrayal of some of his fellow Scots, the hero of the novel is as sympathetic as Tom Jones. Smollett's success with *Roderick Random* is all the more striking when we remember that the book appeared only two years after the defeat at Culloden, at a time when prejudice against Scots was strong, and when (unfairly) all Scots were often seen as tainted with Jacobitism. Rather than skirting round the question of anti-Scots prejudice, Smollett prefers to explore it and confront it directly. He does this most obviously in his Preface to the novel, where he apologizes for his use of a Scottish protagonist, yet makes this apology pointedly double-edged.

Smollett's contrast in the Preface of 'simplicity of manners in a remote part of the kingdom' with the implied sophistication of 'the capital' and its environs may sound like polite flattery of the English reader. However, Smollett has just come close to suggesting that the average 'North Briton' may be better educated than the average Englishman; perhaps 'simplicity of manners' may have more to do with innocence and virtue than with ignorance. Several of the characters we meet in the opening Scottish scenes of the novel are far from idyllic, yet the central Scottish character, Roderick, does exhibit a simplicity that is certainly not ignorance, but is innocence. What appears to constitute an apology in the Preface may in fact merely lay the foundations of a critique of supposed metropolitan sophistication. Such a critique was not new in the English novel—it is present in *Joseph Andrews* (1742), and clearly in *Tom Jones* (1749). But by choosing a Scot as his hero, and making the reader aware of the way in which Scottishness is treated, Smollett is beginning to construct a fiction that is not English (in the national sense) but both Scottish and truly British.

Attention is drawn less frequently to Roderick's Scottishness in the novel's middle sections. Continually the victim of his perceived origins in the earlier part of the book, he wins through to play his accepted part in British life, without being a marked man. The conclusion of the novel makes clear that this does not represent a diminution of Roderick's Scottishness, but is a sign of his full acceptance into the United Kingdom. Important also as pioneering naval fiction, a book in which most of the action takes place outside the bounds of normal English society, and as a forerunner of the great naval novels by those un-English writers, Fenimore Cooper, Stevenson, and Conrad, *Roderick Random* has not been appreciated as a pioneer novel of British prejudices. Fielding and Smollett both presented innocents abroad, and looked at education while exposing prejudice. But Smollett's concern with British identity is his own.

The acceptance of Roderick's continuing Scottishness within a British union is emblematized in the novel's final chapter in a way that anticipates the Scott of *Waverley*. Roderick the traveller Scot has found his long-lost Scottish father in the West Indies, as well as an English bride in England. The whole party travels north, Scots and English in happy accord: 'Everything being thus settled, we took leave of all our friends in London, and set out for Scotland.' (LXIX. 425.) Arrival there does not bring a clear-cut Arcadia. Roderick's father is immediately besieged by a scrounging former enemy. Yet Scottish society is portrayed as both conservative and welcoming: 'As there is no part of the world in which the peasants are more attached to their lords than in Scotland, we were almost devoured by their affection.' (LXIX. 426.) Not only Roderick and his father are welcomed—there is also Narcissa, Roderick's English bride. The union of Roderick and Narcissa signifies on a personal level a uniting of kingdoms; its particular nature also has very different overtones from the early attitude of the English towards Roderick, for Narcissa in Roderick's Scottish home is 'so well pleased with the situation of the place, and the company round, that she has not as yet discovered the least desire of changing her habitation' (LXIX. 427). That 'as yet' might promise something to the cynical English reader, but, on balance, *Roderick Random* concludes with a vindication of the Scotland so battered in the novel's early section, and a vindication of the young immigrant with whom, in the face of prejudice, the reader has been brought to sympathize and who has established his place in British society.

The Smollett who wrote *Roderick Random* was aware of Scottishness under pressure. He was the pro-British Smollett who in 1746 reworked a line from *The Regicide* as the refrain of his Culloden poem, 'The Tears of Scotland', which voiced 'Resentment of my country's fate' as it urged 'Mourn Caledonia, mourn'.[30] Alexander Carlyle recalled being with Smollett in London at the time of Culloden. The two men had to walk with drawn swords, for fear of attack, and

---

[30] Howard Swazey Buck, *Smollett as Poet* (New Haven, Conn.: Yale University Press, 1927), 32.

Smollett cautioned Carlyle 'against speaking a word, lest the mob should discover my country and become insolent, "for John Bull", says he, "is as haughty and violent to-night as he was abject and cowardly on the Black Wednesday when the Highlanders were at Derby" '; Damian Grant points out the strong terms in which Smollett returns to the attack on Cumberland, the Butcher of Culloden, in *The History and Adventures of an Atom* (1769).[31] Though consideration of what it meant to be a Scot in Britain was largely absent from Smollett's middle novels, he retained strong Scottish cultural loyalties. His literary and medical livelihoods depended on the rich London market, but in 1754, after a recent visit to Scotland, he wrote to Alexander Carlyle: 'I do not think I could enjoy Life with greater Relish in any part of the world than in Scotland among you and your Friends.' English prejudice and ignorance continued to annoy him: 'One of our Chelsea Club asked me if the weather was good when I crossed the Sea from Scotland . . .'.[32]

In London the favourite coffee-house of Smollett and his London Scottish community was *The British*. Writing to a Scottish kinsman in 1756, he points out proudly that the etymology of Dumbarton is 'Dunbritton', since it is capital of an ancient kingdom of 'Britons', from whom 'I would fain derive myself'.[33] The word 'Briton' was important to Smollett. When anti-Scottish feeling reached a climax during the administration of Lord Bute, Smollett was invited to edit a magazine to defend Bute's policies (which often favoured Scots). Smollett called his magazine the *Briton*, a title emphasizing that Britishness went beyond Englishness. Well aware that, for the English public, the word 'Scotchman' had become 'a term which implies everything that is vile and detestable', he tried to mock such facile prejudices, attempting to win sympathy for the Scots by imaginatively crossing into the English camp and ironically voicing their bigotry.[34] As 'Winifred Bullcalf', he caricatured John Bull's prejudice in terms that recall an anti-Scottish remark in *Roderick Random* cited above, as well as Smollett's complaint about the Londoner who thought Scotland was cut off from Britain by sea. 'Althof my neighbour Firkin says you can't rite English, therefore must be a Scotchman; and being a Scotchman, you have no right to

[31] Quoted in Knapp, *Tobias Smollett*, 58; Damian Grant, *Tobias Smollett: A Study in Style* (Manchester: Manchester University Press, 1977), 206.

[32] Tobias Smollett, *Letters*, ed. Lewis M. Knapp (Oxford: Oxford University Press, 1970), 33.　　　　　　　　　　[33] Ibid. 43–4.

[34] Quoted in Boucé, *The Novels of Smollett*, 32.

call yourself a Brittin; and as how you are a vagabond people, that come over in shoals with every fair wind, like locusts to devour us; yet I knows what's what.'[35]

It would seem, though, that such sentiment, rather than provoking amusement and sympathy for the Scots, did indeed come close to actual English views and anti-Scottish strategies. John Wilkes was hired to combat the *Briton*. Previously friendly with Smollett, as with Armstrong, Wilkes's exploitation of anti-Scottish feeling in his attacks on the *Briton* led to the breaking-off of cordial relations. Smollett's friend and biographer Moore recorded that:

> It was then proposed, that Mr. Wilkes should publish a paper in answer to the *Briton*, and call it the *Englishman*. He agreed to the proposal, but chose another name for the paper. This was the origin, as Mr. Wilkes himself related to me, of the famous *North Briton*.
>
> Mr. Wilkes was sensible that nothing that could be urged against Lord Bute would have so much weight with the multitude, at that period, as his being a Scotchman; this accusation therefore was often repeated in the *North Briton* . . . much concomitant abuse was extended to the whole Scottish nation . . .[36]

The popular strategy of the *North Briton* (which far outsold the *Briton*) was to stress that Scots, however much they might profess themselves to be a part of Britain, had no right to play a part in any affairs that affected England. The paper was, in effect, pro-English and anti-British, seeing Scots as 'aliens', and making clear that 'In spite of all their specious arguments . . . reason could never believe that a Scot was fit to have the management of *English* affairs.'[37] Seeking to see Britain as the prerogative of England, the paper was dedicated to 'the English Nation . . . by Englishmen'.[38] The magazine's first issue may have begun by paying lip-service to the ideal of freedom of expression for all Britons—'*The Liberty of the press* is the birth-right of a BRITON'—but the last concluded with a quotation from Dryden which was clearly deployed to limit this freedom to England— 'Freedom is the *English Subject's* Prerogative.'[39] Emphasizing that the Scots' unruliness before the Union, the Jacobite Rebellions, and their 'declared enmity to England' had made 'the very name of *Scot* hateful

[35] Ibid. 198.
[36] Moore, 'Memoir', p. clxv.
[37] *North Briton*, 44, 2 Apr. 1763, 223; also 34, 23 Jan. 1763, 111.
[38] Dedication in the *North Briton*, 1 (London: J. Williams, 1763).
[39] Boucé, *The Novels of Smollett*, 56 and 32.

to every true Englishman', the *North Briton* mounted a full-scale attack on Scottish culture.[40]

One of its tactics was to regard as synonymous the words 'Scot', 'Tory', and 'Jacobite'. Though this was absurdly unfair, it worked as propaganda that played to popular prejudices. It was also useful to deploy against the author of 'The Tears of Scotland', whose outrage at the Culloden massacre was mocked. All Scottish literary pretensions and anxieties about linguistic identity were ridiculed.

> Mr. MACPHERSON's fifteenth Course of Lectures on Oratory began yesternight, and will be continued *timeously* every evening, the Sabbath only excepted. Select passages out of *Allan Ramsay*, and other celebrated writers, will be read for the better illustration of the precepts. At the conclusion of the course, Mr. MacPherson purposes a general exercitation of all his pupils, as formerly; but as many of them have on foregoing occasions, through the want of a proper command of voice, run into discordant notes, to the great annoyance of the delicate ears of the North British nobility, who have attended to mark the progress of the young gentlemen, it is expected that for the future they will submit it to have their voices properly pitched by the drone of a bagpipe . . .[41]

Such mockery by national stereotypes was sharp and effective. The *Briton* ceased publication. Soon, on 8 April 1763, Bute fell from office, and Smollett had to contend in the same period with a much more powerful blow—the death of his daughter. His own health failing, he was forced, after a final visit to Scotland, to spend his last years in exile.

Out of that exile came his final version of Britishness, *Humphry Clinker*, a novel which embodies a cultural anthology. Louis Martz has seen Smollett as an important figure in the eighteenth century's 'Age of Synthesis', an age that produced such eclectic projects as dictionaries, encyclopaedias, universal histories, and compendious works like the *Wealth of Nations*, but Martz fails to note the disproportionately strong Scottish contribution to this synthesizing urge.[42] The first chapter of this present book has suggested that the eighteenth-century Scottish collection of social and anthropological data by such figures as Ferguson, Smith, and Millar arose in part, like the teaching of Rhetoric and Belles Lettres, from concern about Scottish 'improve-

[40] *North Briton*, 44, 2 Apr. 1763, 215.

[41] *North Briton*, 6, 10 July 1762, 47–8.

[42] Louis L. Martz, *The Later Career of Tobias Smollett* (New Haven, Conn.: Yale University Press, 1942), 5.

ment'. We might see Smollett's work on such projects as *A Compendium of Authentic and Entertaining Voyages* (1756), *The Modern Part of the Universal History* (which appeared between 1759 and 1765), and *The Present State of All Nations* (1768–9), with its detailed presentation of Britain and substantially unbiased treatment of Scotland, as relating to the activities of Sinclair and the *Statistical Account of Scotland* or to the later formation of the *Encyclopaedia Britannica*—all part of a potent, eclectic, anthological as well as anthropological drive in eighteenth-century Scottish Literature.

Editorial projects also kept Smollett aware of anti-Scottish prejudice. Though he edited the *Critical Review* (1756–63) with an Englishman, and had been helped by an Irishman and by John Armstrong, Smollett was accused of running a '*Scotch Tribunal*' and producing a magazine that was like 'an Edinburgh mobile privy'.[43] Publishing, for instance, a hostile review of work by Smollett's friend, the Scottish dramatist John Home, the *Critical* was no simple Scottish publicity machine, but it did help to keep Smollett abreast of intellectual developments, not least Scottish ones.[44] Earlier attempts to relate Smollett's work to Scottish Enlightenment thinking have been so general as to be unconvincing, but recent scholarship can serve as a base for linking Smollett with Adam Smith. P. J. Klukoff has argued convincingly for Smollett's authorship of a long review of Smith's *Theory of Moral Sentiments* in the May 1759 issue of the *Critical Review*. Comparing a sentence early in *Ferdinand Count Fathom* (1753) with the opening paragraph of the Adam Smith review, Klukoff fails to point out that both deal with Smollett's favoured theme of prejudice:

For not withstanding that deference and regard which we mutually pay each other, certain it is, we have often differed, according to the predominancy of those different passions, which frequently warp the opinion, and perplex the understanding of the most judicious. (*Ferdinand Count Fathom*.)

Even the few who are entitled to judge of their merit, have often their sentiments warped by innocent, because unavoidable prejudices; and having previously embraced some system of their own, with regard to these objects of enquiry, receive with reluctance, if not with aversion, any attempt to overturn those opinions, which they have been accustomed to look upon as certain or indisputable. (*Critical Review*.)[45]

[43] Boucé, *The Novels of Smollett*, 22–3.

[44] Basker, Smollett, *Critic and Journalist*, 84 and *passim*.

[45] Philip J. Klukoff, 'New Smollett Attributions in the "Critical Review"', *Notes and Queries*, Nov. 1967, 418–19. Such detailed work, reinforced by that of Basker, allows us to move to specifics and away from the generalities of M. A. Goldberg,

After considering the 'numerous prejudices' which must greet a new
work of philosophy, the review in the *Critical* moves on to the way in
which Smith seeks to follow 'the practice of our modern naturalists,
and make an appeal every moment to fact and experience'.

> He begins with observing, that, however selfish men may sometimes be
> supposed, there is a principle in their nature which interests them in the
> fortunes of others, and gives them a sympathy with the movements and
> affections of their fellow-creatures. This sympathy he endeavours to account
> for, by supposing, that while we survey the pains or pleasures of others, we
> enter into them by the force of imagination, and form so lively an idea of these
> feelings, that it approaches by degrees to the feelings themselves.[46]

The reviewer then quotes a passage from Smith's book which begins:
'That this is the source of our fellow-feeling for the misery of others,
that it is by changing places in fancy with the sufferer, that we come
either to conceive or to be affected by what he feels, may be demonstrated by many obvious observations . . .'.[47]

In a sense, Smollett would have learned little that was new to him
from his reading of Smith; the philosopher's account conveys essentially what Smollett had attempted in *Roderick Random* by winning the
reader's sympathy for a character whom many might have tended to
view through the warped perceptions of prejudice. The sympathetic
reader is then brought to undergo with Roderick some of the hostility
which prejudice engenders. Also relevant to Smollett's fiction are
other passages from Smith quoted in the *Critical Review* for May 1759,
such as those which discuss the distorting perceptions of travellers, or
the way in which, though absolute virtue is not relative, differences
in manners can provoke hostile reactions when one culture meets
another. 'The modes of furniture or dress which seem ridiculous to
strangers, give no offence to the people who are used to them.'[48]
Observations like this, springing from individual and cultural comparison, are very pertinent to the treatment of prejudice throughout
Smollett's writing, but nowhere more so than in *Humphry Clinker*.

This novel, claimed Horace Walpole, was written by 'the profligate
hireling Smollett . . . to vindicate the Scots'.[49] The book provoked and

Smollett and the Scottish School (Albuquerque, NM: University of New Mexico Press,
1959).
    [46] Review of Adam Smith, *The Theory of Moral Sentiments*, *Critical Review*, 7, May
1759, 384–5.   [47] Ibid. 385.   [48] Ibid. 395.
    [49] Quoted in Wolfgang Franke, 'Smollett's *Humphry Clinker* as a "Party Novel"',
Studies in Scottish Literature, 9/2–3 (Oct.–Jan. 1971–2), 97.

contains prejudice; it is about prejudice. *Humphry Clinker*, Frederick M. Keener has remarked, brings together 'extremists from the extremities of Britain'.[50] In its presentation of a party from rural Wales who undertake a tour of England and Scotland which involves meetings with natives of England, Scotland, and Ireland, *Humphry Clinker* is a novel of cultural cross-overs, with a particular interest in the cross-over between England and Scotland. It has a comparative perspective, both at a cultural and personal level. Its epistolary presentation of characters' views of each other and of their surroundings floods the text with a variety of points of view, revealing not only individuals' self-perceptions but also what Smith had thought important in *The Theory of Moral Sentiments*—the way in which each individual is perceived by others. In the Scottish–English context, as in other contexts, the comparative perspective exposes prejudices between the countries. The epistolary technique makes prominent the novel's constant crossing of language boundaries—from the supposedly sophisticated standard English reportage of the Oxonian Melford, to the Malapropic dialect of Win Jenkins. Such lively cross-overs show how misunderstandings and prejudices develop as characters 'translate' each other into their own perceptions and idiolect. For the Welshwoman Tabitha Bramble, 'Mr Quin' becomes 'Mr Gwynn' (30 April: 51).[51]

As Welsh people, Matt Bramble and his 'Cambrian companions' (2 April: 8) provide an un-English point of view largely exempt from the sort of English prejudice which greeted Scots. The travellers, though, are not just national stereotypes, but distinct, and distinctly comic, individuals undertaking a journey which is also a process of education. Jery Melford, the Oxford student, is 'tutored' (24 April: 29) by what he experiences with his uncle. The heady Lydia, Jery's sister, shows a Miranda-like innocence when, fresh from Wales, she sees Bath as 'to me a new world' (26 April: 37). As in the case of Roderick Random, Lydia's remote simplicity of manners is both made fun of and celebrated. She sees what meets her in terms of her own culture ('the Pump-room . . . is crowded like a Welsh fair':

[50] Frederick M. Keener, 'Transitions in *Humphry Clinker*', in O. M. Brack, jun. (ed.), *Studies in Eighteenth-Century Literature*, xvi (Madison, Wis.: University of Wisconsin Press, 1986), 155.

[51] In the absence of a standard modern edition of *Humphry Clinker*, references in the present text are keyed to the dates of the letters in this epistolary novel, each date being followed by the page number in the widely available Everyman reprint (London: Dent, 1968).

26 April: 37), though her aunt tells her that such a reaction 'is the effect of a vulgar constitution, reared among woods and mountains; and, that as I become accustomed to genteel company, it will wear off' (26 April: 40).

This represents a very simple comic example of differing cultural perspectives. More complexly ridiculous is Melford's report of a Bath doctor's learned, quasi-anthropological disquisition upon 'stinks', and the way in which 'individuals differed *toto coelo* in their opinion of smells, which, indeed, was altogether as arbitrary as the opinion of beauty' (18 April: 17). The doctor goes on to illustrate how various societies ('the French . . . the Hottentots . . . the Savages in Green-land . . . the Negroes on the Coast of Senegal . . . the inhabitants of Madrid and Edinburgh') reacted totally differently to similar stenches (18 April: 17). This is like a burlesque version of the question of the relativity of taste which interested Smollett in his review of *The Theory of Moral Sentiments*, where he quoted with approval Smith's view that 'There is, perhaps, no form of external objects, how absurd and fantastical soever, to which custom will not reconcile us, or which fashion will not render even agreeable.'[52] The Smith review had also focused on the unreliability of travellers' tales. Though the Bath doctor makes Edinburgh sound like a sewer, later accounts of the place will give us a more favourable and accurate view. Indeed, Bramble's long passages on the effluvia on the waters in Bath might urge us to compare Edinburgh and Bath to the latter's disadvantage.

Early in the novel Jery Melford remarks in a letter: 'You know we are the fools of prejudice.' (30 April: 48.) Melford is discussing being taken in by appearances and presuppositions—a theme that relates to the character of Young Wilson, Lydia's suitor, who has to court her in constant disguise in order to counter the prejudices of her family, who are ignorant of his real nature. Similar ignorance and prejudice have to be dispelled in the case of Humphry Clinker, the Methodist servant, whom a change of clothes renders unrecognizable and whose virtue is eventually clear enough to overcome even Tabitha Bramble's prejudice. Again, in the case of Scotland, the ignorant prejudice which has disguised the true image has to be dispelled. Matt Bramble at least realizes that such prejudices should be confronted. One might suspect that he will find in the Scots' favour, since his initial decision to visit Scotland is expressed as part of an admission that the Scots (like the Welsh) are a part of Britain, and that 'I think, it is a reproach

[52] Review of Smith, *The Theory of Moral Sentiments*, 395.

upon me, as a British freeholder, to have lived so long without making an excursion to the other side of the Tweed.' (8 May: 63.)

Scotland and Clinker are introduced early in the narrative, but the revelation of each takes considerable time. First, the party goes to London, which the sceptical provincial Bramble can view with fresh eyes ('London is literally new to me': 29 May: 82). Considering it 'Bedlam' (29 May: 85), middle-aged valetudinarian Bramble decides to shorten his stay there, annoying his niece—'People of experience and infirmity, my dear Letty, see with very different eyes from those such as you and I make use of.' (31 May: 89.) We are constantly aware of viewpoints being prejudiced. Jery remarks that 'fumes of faction not only disturb the faculty of reason, but also pervert the organs of sense' (2 June: 91). Matt discusses 'prejudices' in the context of 'the liberty of the press' (2 June: 99) and the falseness of literary judgements, themes dear to the former editor of the *Briton*, who caricatures himself as the benign manager of a literary factory which is itself a model of the varieties of Britishness presented in this novel. 'Not only their talents, but also their nations and dialects were so various, that our conversation resembled the confusion of tongues at Babel. We had the Irish brogue, the Scots accent, and foreign idiom, twanged off by the most discordant vociferation . . .' (10 June: 120). In the midst of this we find a 'Scotchman [who] gives lectures on the pronunciation of the English language' (10 June: 120). Smollett is inviting his readers to laugh in the usual way at the Scots. He is exploiting national caricatures and inviting the stock reactions of prejudice—'The Irishman . . . Lord Potatoe' (10 June: 120). Yet, as *Humphry Clinker* develops, the reader may also be provoked to question such automatic prejudices as that of Matt Bramble against Clinker's Methodist preaching.

Characters are seen both exhibiting prejudice and countering it. Where others dismiss the supposedly corrupt Edward Martin at once, Matthew Bramble gives him fairer and closer scrutiny. Tabitha emerges as one of the main purveyors of prejudice (particularly against Humphry), yet her own biases run up against those of other people and other regions, as when she discusses the Countess de Melvil with a Yorkshire landlord: ' "Handsome! (cried Tabby) she has indeed a pair of black eyes without any meaning; but then there is not a good feature in her face." "I know not what you call good features in Wales (replied our landlord); but they'll pass in Yorkshire." ' (26 June: 160.) It seems barely relevant at this point that the

Countess against whom Tabitha is so prejudiced is Scottish, but, increasingly, Scottish characters join the narrative. At this same inn is 'The Scotch Lawyer, Mr. Micklewhimmen' (26 June: 163), who recommends the Harrogate waters to Matt Bramble. Taking waters, Bramble points out 'that my organs were formed on this side of the Tweed' (26 June: 164), and we seem firmly in the territory of anti-Scottish jokes as Micklewhimmen is revealed as laying court to ugly Tabitha in the dialect, whining selfishly, and dosing himself with medicinal liquor. Micklewhimmen emerges as wily. He outwits a practical joke against himself ('He that would cozen a Scot, mun get oop betimes': 1 July: 166), but he is hardly attractive. When the house seems to be burning down, he is interested only in saving his own skin. We are back in the territory of the Scot as enemy.

Matt Bramble's brooding on the way in which foreigners use the phrase '*old English hospitality*' (26 June: 155) only ironically, and his memories of an Englishman who, when abroad, made no concessions to his host country, continue the themes of individual and national bias. The ensuing focus on Lieutenant Lismahago concentrates attention on anti-Scottish prejudice. Matt Bramble sympathizes with Lismahago's hard life, but Jery sees him essentially as 'this Caledonian . . . a self-conceited pedant, aukward, rude, and disputacious' (10 July: 181). When the Scot discusses 'war, policy, the belles lettres, law, and metaphysics' (13 July: 182) with Matt before recounting his North American adventures, he demonstrates a learning beyond that of the Oxford student, as well as an unusual ability to cross boundaries which others cannot cross. Listening to Lismahago's American tale, Tabitha can imagine the Indian bride's wedding-dress only in terms of European high fashion, despite Lismahago's explanations 'in general, that Indians were too tenacious of their own customs to adopt the modes of any nation whatsoever', and his stress on their 'simplicity of . . . manners' (13 July: 185). Significantly, this wandering Scot's account of his American adventures comes just before the Bramble party's arrival in Scotland. The English tended to view Scottish society as primitive in comparison with their own, but the Scotland to be presented by these Welsh visitors appears all the more refined by contrast with the account of Amerindian 'simplicity of . . . manners', where fingers are sawn off with rusty knives and eyes are scooped out to be replaced with burning coals.

It is Jery Melford's letter which juxtaposes Lismahago's adventures with crossing into Scotland. Because Jery has manifested anti-Scots sentiments, it is all the more effective that it should be he who records Lismahago's politeness in the face of the fact that 'From Doncaster northwards, all the windows of all the inns are scrawled with doggerel rhymes, in abuse of the Scotch nation.' (13 July: 188.) Similarly, Jery is impressed by Lismahago's arguments that 'a North-Briton is seen to a disadvantage in an English company, because he speaks in a dialect that they can't relish, and in a phraseology which they don't understand' (13 July: 189–90). To counter English presuppositions about the Scots as a dourly humourless people, Lismahago recommends to Jery that he read Allan Ramsay's works, which Jery resolves to do. This is the sign that Jery is crossing not only the literal border, but also the border of intercultural prejudice.

The erosion of anti-Scottish prejudice deepens in the ensuing Scottish tour. Eric Rothstein, the critic most alert to this theme, points out that Lismahago does not accompany the Welsh travellers in Scotland because he would be too crudely pro-Scottish a witness. Smollett wishes to champion Scotland a little more subtly, and to champion the full sense of the word 'Britain'.[53] So Matt Bramble's discussion of Scots and English advancement in the Bute administration is paralleled by Jery's citing of the Edinburgh toast 'may a' unkindness cease between John Bull and his sister Moggy' (8 Aug.: 216). Smollett draws on his own experience as a British Scot when, for instance, he attributes to Tabitha the misconception held by one of the members of his Chelsea club:

She was so little acquainted with the geography of the island, that she imagined we could not go to Scotland but by sea . . . If the truth must be told, the South Britons in general are woefully ignorant in this particular. What, between want of curiosity, and traditional sarcasms, the effect of ancient animosity, the people at the other end of the island know as little of Scotland as of Japan (18 July: 203–4.)

Jery Melford (the writer here) feels Scottish cultural differences acutely: 'their looks, their language, and their customs, are so different from ours, that I can hardly believe myself in Great-Britain' (18 July: 204). The point is that Jery *is* in Great Britain, and must be educated to revise his ideas of what Britain is by revising his ideas about Scotland. This process of education makes the book an informational text at times, drawing on Smollett's writing in *The Present*

53 Rothstein, 'Scotophilia and *Humphry Clinker*', 74.

State of All Nations in the ways in which Martz had indicated.[54] But
the information is all geared to the central theme of the crossing of the
barriers of prejudice.

As Scotland is described and compared with other parts of Britain,
the difficulties and possibilities of understanding it emerge. 'I don't
speak their lingo' (18 July: 210), says Win Jenkins, but Jery confesses
to undergoing a degree of cultural conversion: 'If I stay much longer
at Edinburgh, I shall be changed into a down-right Caledonian.'
(8 Aug.: 211.) Interpreting Scotland to his Oxford friend 'Sir Watkin
Phillips, Bart.', Jery becomes what Scott's Edward Waverley will
become, a broker between cultures, a figure able to partake of
cultural territory on one side of a border of prejudice and interpret it
to the other. Prejudice may be dissipated as Jery explains Scotland to
'The English, who have never crossed the Tweed . . .' (8 Aug.: 213),
but it does not vanish. If Jery comes to enjoy hearing Scots, Matt
Bramble maintains that it is 'clownish', and he adopts a position
which is very much that of the teachers of Rhetoric and Belles Lettres
and their supporters:

I think the Scots would do well, for their own sakes, to adopt the English
idioms and pronunciation; those of them especially, who are resolved to push
their fortunes in South-Britain. I know, by experience, how easily an Englishman is influenced by the ear, and how apt he is to laugh, when he hears his
own language spoken with a foreign or provincial accent. I have known a
member of the house of commons speak with great energy and precision,
without being able to engage attention, because his observations were made
in the Scottish dialect, which (no offence to lieutenant Lismahago) certainly
gives a clownish air even to sentiments of the greatest dignity and decorum.
I have declared my opinion on this head to some of the most sensible men of
this country, observing, at the same time, that if they would employ a few
natives of England to teach the pronunciation of our vernacular tongue, in
twenty years there would be no difference, in point of dialect, between the
youth of Edinburgh and London. (8 Aug.: 220.)

Bramble, too, becomes engaged in questions of cultural 'comparison'
(8 Aug.: 220), and sometimes favours the southern customs. Yet he
also exposes the different strengths of Scotland, particularly its legal
and educational systems. The Scots literati and Smollett's own
friends (such as John Moore) pass before the reader, who is also made
aware that all Scotland cannot be lumped together. The travellers are
conscious of cultural divisions within Scotland, particularly between

54 Martz, *The Later Career of Smollett*, 147–62.

Lowlanders and Highlanders. These differences are translated into familiar terms by being compared with non-Scottish parallels. So, north of Loch Lomond, 'This country appears more and more wild and savage the further we advance; and the people are as different from the Lowland Scots, in their looks, garb, and language, as the mountaineers of Brecknock are from the inhabitants of Hertford-shire.' (3 Sept.: 227.) Jery grows able to see 'Caledonia in romantic view' using an attractive lens polished less by James Thomson than by James Macpherson: 'I feel an enthusiastic pleasure when I survey the brown heath that Ossian was wont to tread.' (3 Sept.: 228.) Yet the natives are not stereotyped. Jery meets Dougal Campbell, a Highlander who detests bagpipes. Much of the comedy deployed here at least turns the Highlanders from objects of scorn to subjects of amusing scrutiny.

Concern with Highland–Lowland division accompanies an interest in Scottish 'improvement' (28 Aug.: 223) and in the differences between Scotland's conservatively patriarchal culture and her in-creasingly mercantile present. The reader learns of the ancient *British* territory round Glasgow, particularly Dumbarton—'Dunbritton', as Bramble, like Smollett, discovers it to be. Loch Lomond, 'romantic beyond imagination', is 'the Arcadia of Scotland' (28 Aug.: 235–7). Yet in the same letter we are conscious of the Glaswegian 'spirit of enterprise' in manufacturing and shipping. The Highland landscape so close to 'Dunbritton' is paradisal yet menacing. Bramble perceives that 'The chieftainship of the Highlanders is a very dangerous influ-ence operating at the extremity of the island' (6 Sept.: 242), and sees this Scotland as territory to be colonized by commercial concerns, with the Highlanders, for a moment, glorified as Red Indians:

It cannot be expected, that the gentlemen of this country should execute commercial schemes to render their vassals independent; nor, indeed, are such schemes suited to their way of life and inclination; but a company of merchants might, with proper management, turn to good account a fishery established in this part of Scotland. Our people have a strange itch to colonize America, when the uncultivated parts of our own island might be settled to greater advantage. (6 Sept.: 244.)

This subject of 'the natural progress of improvement' and the divided society within Scotland is returned to in Bramble's letter of 20 September, which also contains his argument with Lismahago over the Union. This represents the most passionate discussion in the book of the Scottish, English, and British portions. Lismahago puts the anti-Union case forcibly, pointing out the small number (forty-five) of Members of Parliament at Westminster. He argues that all the advantages from the Union have been on the English side, all the disadvantages on the Scottish. Bramble is initially nettled by being 'schooled in so many particulars' he had not known, but he remains less than completely won over. We cannot assume that Lismahago's voice, as it argues in the terminology of Adam Ferguson about 'the distinctions of civil society', is that of Smollett, but we can see how intently the author wishes his readers to reflect on the nature of Britishness.

Eric Rothstein has demonstrated spendidly how the theme of anti-Scottish prejudice is allied to the prejudice which greets the Humphry Clinker who, early in the novel, appears with his backside showing through his clothing. Rothstein marshals an array of eighteenth-century taunts of the Caledonian (imaged as a kilted Highlander) for being 'bare ars'd'.[55] To these we might add John Armstrong's letter to Smollett in March 1769, pointing out to him that a number of sympathetic readers felt that the commentary on Scotland in *The Present State of All Nations* had 'too much exposed the posteriors of our brothers in the north'.[56] Surely Tabitha's horror of the bare-buttocked Clinker is analogically related to the novel's concern with anti-Scottish prejudice. Rothstein's article points out that, because Clinker is not a Scot, the prejudice against him is more clearly seen as wrong. Perception of this paves the way for examination of prejudices against the Scots. Both sets of biases are part of the novel's crucial focus on prejudice and its tempering. As Win Jenkins states: ''Tis a true saying, *live and learn*' ([3 Oct.]: 291).

Having enjoyed Scottish hospitality, the travellers learn that hospitality and domestic economy south of the Border can be poor by comparison when they encounter the uncaring Mrs Baynard. Again Matt hears how the virtuous Mr Dennison encounters prejudice, 'despised among the fashionable company, as a low fellow, both in breeding and circumstances' (11 Oct.: 312). As readers we have learned Mr Dennison's true worth, and can dismiss such prejudice against him, but its presence is a significant part of the novel, which, as it draws to its conclusion, maintains its stress on the theme of prejudice which goes beyond the mere valorizing of a despised Caledonia.

---

[55] Rothstein, 'Scotophilia and *Humphry Clinker*', 63–6.
[56] Quoted in Martz, *The Later Career of Smollett*, 130.

Discovering the suspicious Wilson to be, in fact, the virtuous George Dennison, Jery Melford exclaims (perhaps too blatantly) that he is 'mortified to reflect what flagrant injustice we every day commit, and what absurd judgment we form, in viewing objects through the falsifying medium of prejudice and passion' (14 Oct.: 316). The marriage of Lydia to young Dennison signals the overcoming of prejudice and the perception of true worth; so, in its way, does the comical marriage of two of the most prejudiced individuals in the book, Tabitha and Lismahago.

In the novel's concluding alliances the despised—Methodist Humphry, absurd Tabitha and Lismahago, persecuted 'Wilson'— find the prejudices against them dispelled and new unions made possible. Such personal unions parallel the treatment of the United Kingdom and, in particular, the Union of Scotland and England. The way to proper unions in the book is through 'expedition'—in the sense not only of travelling (through which sympathetic encounters with other cultures modify earlier presuppositions) but also of freeing, of liberation from literal imprisonment (when Humphry finds himself gaoled) and, as importantly, from the prison-house of prejudice. In its juxtapositions of societies and individuals within a comparative perspective, this Scottish writer's last and investigatively British novel achieves a comparative and anthropological framework which was to feed into the writing of Scott. More immediately, it looks toward the concerns of James Boswell, who, like Smollett and Scott, would be fascinated by the place of the Scot within Britain, as well as with the prejudices of the English, and who would take not a Welsh country squire but the embodiment of John Bull himself upon a more adventurous, yet in some ways less rewardingly conclusive, northern expedition.

BOSWELL'S BRITAIN

Perhaps no writer was more uneasy about his Scottishness than the James Boswell who, as a Glasgow student, had listened attentively to Adam Smith's lectures on Rhetoric and Belles Lettres in 1759, and who had attended and then subscribed to the publication of Thomas Sheridan's Edinburgh lectures at the start of the following decade.[57]

---

[57] Adam Smith, *Lectures on Rhetoric and Belles Lettres*, ed. J. C. Bryce (Oxford: Clarendon Press, 1983), *17–18*; Thomas Sheridan, 'Subscribers', in *A Course of Lectures on Elocution* (London: W. Strahan, 1762), p. xv.

The title of the book which made his name as a writer, the 1768 *Account of Corsica*, suggests a Boswell eager to flee Scottish problems. He boasts that the book's orthography is that of the most proper English, as sanctioned by 'the illustrious Mr. Samuel Johnson'.[58]

Corsica, however, was a focus of eighteenth-century Scottish interest. Rousseau pointed out to Boswell on 30 May 1765 that 'My Lord Marischal of Scotland . . . is one of the most zealous partisans of the Corsican nation.' (*Corsica*, 265.) The *Account*'s acknowledgements mention Smollett's interest in Corsica, and culminate with tributes to the Scottish proto-anthropologist, Lord Monboddo (who had commented on the book), and to Lord Hailes, in whose Scottish library Boswell had studied Corsican documents (*Corsica*, pp. xii, xvi--xvii). Many details in the *Account* show Boswell bringing Scottish experience to bear. Corsican horses are likened to the 'shelties' (*Corsica*, 39) of the Scottish Highlands, while Boswell notes animals like stags in the Corsican mountains, through which few roads run. On this last point, he cites the opinions of his Scottish countryman, Sir Alexander Dick, celebrated for his 'public spirit in promoting good roads in an improved age', and goes on to reflect that 'It was in a good measure owing to her rugged hills, that ancient Scotland preserved her independency.' (*Corsica*, 39.) Boswell's Corsicans, like ancient Scotland, represent liberty under threat. Boswell acknowledges Smollett's 'warmth in favour of the Corsicans' (*Corsica*, 124–5), to whom Smollett referred in his 'Ode to Independence'.[59] James Thomson had hymned *Liberty* in 1734, and James Boswell opens his *Account of Corsica* with a similarly directed panegyric, one which demonstrates that the Scots' heritage is fully part of the British love of liberty. Valorizing the Corsicans' fierce independence, Boswell quotes from Tacitus's account of 'Galgacus, the ancient Scottish chief' (*Corsica*, 117). He also links the Corsicans with contemporary Highlanders (so recently kept in check by military road-building) and with the Highlanders 'in former lawless times' (*Corsica*, 204–5). The Corsicans' small stature makes them 'much like the Scottish Highlanders' (*Corsica*, 226), as does the fact that 'The women do the greatest part of the drudgery work.' (*Corsica*, 218.) Peter France highlights the *Account*'s interest in a society poised between

---

[58] James Boswell, *An Account of Corsica* (Glasgow: Robert and Andrew Foulis, 1768), p. xviii. Since there is no standard modern edition of this work, references are keyed to the original edition and are given in the body of the text, using the abbreviation *Corsica*.   [59] See Buck, *Smollett as Poet*, 77.

the 'ancient simplicity of freedom-loving mountain people' and the cultivation of 'the arts of civilisation'.[60]

Boswell's Corsica, then, resembles the environment that fascinated so many eighteenth-century Scots. Celebrating rough, picturesque Corsican mountain dwellings, Boswell is already starting to play his part in the Romanticization of the Highlands. Sometimes the Corsicans in the *Account* see Boswell as Scottish, sometimes as English, and sometimes as British. Boswell does little to sort out these labels, but the epigraph on the title-page of his book asserts the connection implied elsewhere between Corsica and Scottish tradition. That epigraph comes from the 1320 Scottish 'Declaration of Independence', the Declaration of Arbroath:

Non enim propter gloriam, divitias aut honores pugnamus, sed propter libertatem solummodo, quam nemo bonus nisi simul cum vita amittit.

We fight not for the sake of glory, riches, or honours, but for the sake of liberty alone, which no good man relinquishes except with his life. (My translation.)

In its implicit and explicit juxtapositions of Scotland with Corsica, Boswell's first book anticipates the comparative perspective which clearly characterizes his other, more famous account—that of the journey to the Western Isles—and also his *Life of Samuel Johnson*.

That *Life* is not just a classic biography. It is also a Scot's subtle work of cultural comparison, furthering the process of British investigation developed in *Humphry Clinker*. For Boswell perceives Johnson as '*Jean Bull philosophe*', and as Boswell provokes Johnsonian pronouncements or stage-manages his subject's actions, we are constantly made aware of an English mind confronting a Scottish one. In his endeavours, Boswell is very much continuing eighteenth-century Scottish preoccupations. It was, after all, the Scot John Arbuthnot who had established the figure of John Bull as the representative Englishman.[61] This confrontation of English and Scottish sensibilities is clearest in Boswell's motivation, organizing, and recording of the visit to the Hebrides in 1773, the subject of Johnson's *A Journey to the Western Islands of Scotland* (1774) and of Boswell's first account of his relations with Johnson, *The Journal of a Tour to the Hebrides* (1785). Though Johnson had been intrigued by the Hebrides

---

[60] Peter France, 'Western European Civilisation and its Mountain Frontiers (1750–1850)', *History of European Ideas*, 6/3 (1985), 303.

[61] *Life*, i. 467–8; see entry for John Arbuthnot in *DNB*.

since childhood, it was Boswell's motivating power which made the expedition a reality.

Johnson's *Journey* plunges into narrative almost immediately, but Boswell's *Journal* takes longer to set the expedition in a framework of cultural comparison, locating in Scotland some of the features which he had celebrated in Corsica, and projecting the journey as one which will investigate another culture, one whose description will demonstrate the great contrasts between the societies which are united as 'Britain'.

Martin's Account of those islands had impressed us with a notion that we might there contemplate a system of life almost totally different from what we had been accustomed to see; and, to find simplicity and wildness, and all the circumstances of remote time or place, so near to our native great island, was an object within the reach of reasonable curiosity. (*Journal*, 167.)

Boswell's use of the word 'us' here is not simply his way of ingratiating himself with Johnson; it emphasizes that both men are Britons, citizens of 'our native great island', with a good deal in common as they set off for those smaller British islands so remote from London that Voltaire thought the proposed journey like a voyage to the North Pole (*Journal*, 167). But Boswell's introductory remarks also place a certain distance between himself and Johnson, consciously marking a cultural difference between Scot and Englishman:

He was indeed, if I may be allowed the phrase, at bottom much of a *John Bull*, much of a blunt *true-born Englishman* . . . I am, I flatter myself, completely a citizen of the world.—In my travels through Holland, Germany, Switzerland, Italy, Corsica, France, I never felt myself from home; and I sincerely love 'every kindred and tongue and people and nation.' I subscribe to what my late truly learned and philosophical friend Mr. Crosbie said, that the English are better animals than the Scots; they are nearer the sun; their blood is richer, and more mellow: but when I humour any of them in an outrageous contempt of Scotland, I fairly own I treat them as children. And thus I have, at some moments, found myself obliged to treat even Dr. Johnson. (*Journal*, 172.)

Here, characteristically, Boswell both grovels and struts, is both child and adult at once. If he often appears to be Johnson's pupil, in Scotland it is Boswell who guides his mentor through a process of education, an attempted releasing from prejudice which would itself provoke further prejudice:

To Scotland however he ventured; and he returned from it in great good humour, with his prejudices much lessened, and with very grateful feelings of the hospitality with which he was treated; as is evident from that admirable work, his 'Journey to the Western Islands of Scotland,' which, to my utter astonishment, has been misapprehended, even to rancour, by many of my countrymen. (*Journal*, 172–3.)

Demanding comparison with both *Humphry Clinker* and *Waverley*, Boswell's arranging and recording of Johnson's journey juxtaposes this quintessential John Bull figure not only with the alluring simplicities of the Gaelic-speaking Hebrides, but also with the manners and people of the aspiring commercial Scottish Lowlands of the Enlightenment. Extremes of Britain meet. By stressing his own amazement, Boswell maintains an emphasis on Johnson's otherness: 'I now had him actually in Caledonia'; 'I was much pleased to see Dr. Johnson actually in St. Andrews . . .'; 'I had a romantick satisfaction in seeing Dr. Johnson actually in [Macbeth's castle] . . .' (*Journal*, 173, 200, 242). At the same time, Boswell shows how the two travellers can be united in the pleasures of the journey. Boswell's 'us' subtly appropriates Johnson for the proto-Romantic sensibility, as well as functioning as a subliminally emphatic British pronoun. Yet it is constantly splitting into the 'I' of the Scottish observer and the 'he' of the observed Englishman:

It was a delightful day. Lochness, and the road upon the side of it, shaded with birch trees, and the hills above it, pleased us much. The scene was as sequestered and agreeably wild as could be desired, and for a time engrossed all our attention.

To see Dr. Johnson in any new situation is always an interesting object to me; and, as I saw him now for the first time on horseback, jaunting about at his ease in quest of pleasure and novelty, the very different occupations of his former laborious life, his admirable productions, his *London*, his *Rambler*, &c. &c. immediately presented themselves to my mind, and the contrast made a strong impression on my imagination. (*Journal*, 243.)

The great contrast here is not just between work and leisure, but between the metropolitan world of *London* and the *Rambler* and that of the Highlands. The book's structure presses this home: in the ensuing episode the great writer enters a Highland woman's hut, but cannot even speak to its occupant. Johnson must speak to Boswell, who speaks to a guide, who in turn communicates with the woman in 'Erse' (Gaelic) (*Journal*, 243). Boswell the Lowlander acts as cultural

broker here, and is careful to point out that Johnson is 'pleased at
seeing, for the first time, such a state of human life' (*Journal*, 244).

Like Adam Ferguson's *Essay on the History of Civil Society*, to which
Boswell pays tribute, his own Hebridean *Journal* is an exploration and
juxtaposition of different types of social organization and development (*Journal*, 184). Isolated among Gaelic-speakers, he finds them
'black and wild in their appearance as any American savages', and
suggests to Johnson that 'it was much the same as being with a tribe
of Indians' (*Journal*, 250–1). Johnson is recorded as pointing out that
the latter would be much more terrifying. Boswell is constantly
anxious to deal with Johnson's notorious antipathy to Scotland.
When it cannot be qualified or skirted, Boswell at least achieves a
comic *rapprochement*:

> One night, in Col, he strutted about the room with a broadsword and
> target, and made a formidable appearance; and, another night, I took the
> liberty to put a large blue bonnet on his head. His age, his size, and his bushy
> grey wig, with this covering on it, presented the image of a veritable *Senachi*:
> and, however unfavourable to the Lowland Scots, he seemed much pleased to
> assume the appearance of an ancient Caledonian. (*Journal*, 379.)

Carefully, Boswell brings out Johnson's sympathies with the old, patriarchal way of life in the Highlands and his liking for the Highlanders.
Boswell does not disguise, but sometimes pointedly apologizes for,
Johnson's bursts of prejudice against Lowlanders. In this book about
Britishness Boswell the Scot is constantly alert to prejudice. Not the
least of his successes was that, by organizing the Highland expedition,
he stimulated Johnson to write one of the rare English books in which
the concept of Britishness is important. For, even when Johnson is
attacking aspects of Scotland, he at least shows himself as the only
major eighteenth-century English writer after the spy Defoe to devote
detailed attention to the Scottish part of Britain. In simply securing
Johnson's provocative comments on Scotland, Boswell has ensured a
raising of the Scottish profile in the British literary world. Later,
Boswell takes care to record in the *Life* a remark by the biographer
which states clearly his achievement in arranging the Scottish trip:
'You and I, Sir, have, I think, seen together the extremes of what can
be seen in Britain—the wild rough island of Mull, and Blenheim
park.' (*Life*, ii. 451.)

Boswell's examination of Britishness in his private diaries, letters,
and papers has been acutely examined by Gordon Turnbull, who

rightly sees it as part of the Scottish Enlightenment's 'great revision-
ary interrogation of the British identity and its making from the
perspective of the post-Union Scot'.[62] Though Turnbull pays hardly
any attention to the *Life of Johnson*, that work reveals a similar preoccupation, and, like much of Boswell's compendious writing, relies
on a minutely detailed eclecticism as Boswell orders vast data banks
about his subject. This eclecticism appears of a piece with that of
Smollett, the compiler of reference books, and with the other vast
eclectic projects, from the *Encyclopaedia Britannica* to the *Wealth of
Nations*, which have already been cited. If Adam Smith's review of the
French *Encyclopédie* in the 1755 *Edinburgh Review* may have prepared
the way for the *Britannica*, we can see that Smith's lectures also
provided justification for Boswell's copious and eclectic biographical
method.[63] Near the start of the Highland *Journal*, his first published
work on Johnson, Boswell writes: 'Let me not be censured for mentioning such minute particulars. Every thing relative to so great
a man is worth observing. I remember Dr. Adam Smith, in his
rhetorical lectures at Glasgow, told us he was glad to know that
Milton wore latchets in his shoes, instead of buckles.' (*Journal*, 171.)
Boswell is recalling this point after about a quarter of a century.
Though, partly under Johnson's influence, he grew less devoted to
Smith in time, there is every reason to believe that Smith's teaching
deeply impressed him. In 1763, fourteen years after attending
Smith's class, Boswell points out to Johnson how highly Smith had
regarded rhyme (*Life*, i. 427). Paying tribute to Johnson's literary
judgement, Boswell can think of no corroboration better than that of
his old teacher: 'Dr. Adam Smith, than whom few were better judges
on this subject, once observed to me, that "Johnson knew more books
than any man alive."' (*Life*, i. 71.) Boswell's loyalty and debt to his
Scottish teachers, and, more importantly, his constant preoccupation
with the Scottish–English cultural negotiations of Britishness, remind
us that, if the massively impressive John Bull figure of Johnson has
so weighted the *Life* that it is firmly lodged in an honoured, if little-
visited, hall of English Literature, it might also benefit from being

---

[62] Gordon Turnbull, 'James Boswell: Biography and the Union', in Andrew Hook
(ed.), *The History of Scottish Literature*, ii *1660–1800* (Aberdeen, Aberdeen University
Press, 1987), 157.

[63] Adam Smith, 'Letter to the *Edinburgh Review*' (1755), repr. in *Essays on Philosophical Subjects*, ed. W. P. D. Wightman and J. C. Bryce (Oxford: Clarendon Press,
1980), 242–54.

decentred and viewed cheekily as part of that line of Scottish writings
so obsessed with Britishness.

No part of the *Life* more immediately demands to be read from this
angle than the set piece where the 22-year-old Boswell first meets
Johnson. The account of the meeting is heavily prefaced by reminders
of 'provincial' Scotland's attempts to become more fully British
through linguistic 'improvement'. Boswell recalls reading Johnson
'with delight and instruction' as a student. It was from a linguistic
'improver' that Boswell first heard of Johnson: 'Mr. Gentleman, a
native of Ireland, who passed some years in Scotland as a player, and
as an instructor in the English language, had given me a representation of the figure and manner of DICTIONARY JOHNSON! as he was
then called . . .' (*Life*, i. 384–5). Boswell then recalls how he was
often in the company of Thomas Sheridan, who was in Edinburgh in
1761 delivering his lectures on the English language and public
speaking, and heard more about Johnson from him. An apparent
digression follows, in which Boswell describes Sheridan as a 'man of
literature', and his teaching of 'the arts of reading and speaking with
distinctness and propriety'. In particular, Boswell details the attempt
to teach Alexander Wedderburn, who had attended Smith's Edinburgh lectures and who edited the *Edinburgh Review*:

> though it was too late in life for a Caledonian to acquire the genuine English
> cadence, yet so successful were Mr. Wedderburn's instructors, and his own
> unabating endeavours, that he got rid of the coarse part of his Scottish accent,
> retaining only as much of the 'native woodnote wild', as to mark his country;
> which, if any Scotchman should affect to forget, I should heartily despise him.
> Notwithstanding the difficulties which are to be encountered by those who
> have not had the advantages of an English education, he by degrees formed a
> mode of speaking, to which Englishmen do not deny the praise of eloquence.
> (*Life*, i. 386–7.)

Wedderburn stands here in Boswell's text not simply as a man who
has assiduously striven for linguistic conversion, but as someone
who has done this and, without abandoning all signs of Scottishness,
has proceeded to rise spectacularly in public life. Wedderburn
represents an ideal of the teaching of Rhetoric—he went on to translate himself to London as Lord Loughborough.

Thematically, this detailing of Wedderburn's career just before
Boswell's account of his own first meeting with Johnson is no
digression. It emblematizes the efforts and aspirations of the bio-

grapher, who had himself modified his Scottish speech through rhetorical study, and come to London seeking fame.

I have dwelt the longer upon this remarkable instance of successful parts and assiduity, because it affords animating encouragement to other gentlemen of North Britain to try their fortunes in the southern part of the island, where they may hope to gratify their utmost ambition; and now that we are one people by the Union, it would surely be illiberal to maintain, that they have not an equal title with the nations of any other part of his Majesty's dominions. (*Life*, i. 387.)

Yet, blocking the way of such aspirants, and fascinating them, were those Englishmen for whom England, not Britain, represented the significant cultural and political environment. The Johnson whom Boswell first met appeared to be just such a man. Discussing *London* (1738) in the *Life*, Boswell has already mentioned the Doctor's 'prejudices as a "true-born Englishman"', not only against foreign countries but against 'Ireland and Scotland'. Boswell, anxious that the true meaning of the word 'British' should not be forgotten, reminds us in a footnote that this epithet 'ought to denominate the natives of both parts of our island' (*Life*, i. 129–30). He delights in indicating that five of Johnson's six *Dictionary* amanuenses were 'natives of North-Britain, to whom he is supposed to have been so hostile' (*Life*, i. 187). Such anxious attempts to counter English biases are summed up in Boswell's account of his first meeting with Johnson in a London bookshop:

I said to Davies, 'Don't tell where I come from,'—'From Scotland,' cried Davies, roguishly. 'Mr. Johnson, (said I) I do indeed come from Scotland, but I cannot help it.' I am willing to flatter myself that I meant this as light pleasantry to soothe and conciliate him, and not as an humiliating abasement at the expence of my country. (*Life*, i. 392.)

Boswell squirms, but he does so in such a way that he attempts to defuse Johnson's anti-Scottish prejudice by highlighting how absurd it would be to blame anyone for being born in a particular place. Significantly, Johnson's ensuing conversation, as recorded by Boswell, soon proceeds to a discussion of barbarous and polished societies. A moment later Johnson recommends Kames's *Elements of Criticism* (*Life*, i. 393).

Detailing their second meeting, Boswell departs from documentary chronology to describe Johnson's disagreement with Hugh Blair over the Ossianic poems (*Life*, i. 396). Though not the book's ostensible subject, concern with Scottishness constantly affects Boswell's structuring of his *Life of Johnson*. No other biography of the Doctor devotes nearly as much attention to that Scotophobia with which Boswell constantly negotiates. Boswell the Scottish Briton attempts to discredit anti-Scottish prejudice. Goldsmith's contradiction of praises of Scotland is delivered, Boswell assures us, 'very untruly, with a sneering laugh' (*Life*, i. 425). Johnson's anti-Scottish remarks are recorded, but they are also commented on, sometimes feebly. When the Doctor declares that 'the noblest prospect which a Scotchman ever sees, is the high road that leads him to England!' (*Life*, i. 425), Boswell writes lamely: 'After all, however, those who admire the rude grandeur of Nature, cannot deny it to Caledonia.' (*Life*, i. 426.) In the structure of the *Life*, Boswell's account of the genesis of the Hebridean visit is preceded by mention of Sir James Macdonald, 'a young man of most distinguished merit, who united the highest reputation at Eton and Oxford, with the patriarchal spirit of a great Highland Chieftain' (*Life*, i. 449). This translated Highlander, the synthesis of British extremes, prefaces the idea of the Hebridean journey just as the emblematic Wedderburn, translated Lowlander, prefaces Boswell's first meeting with Johnson.

Carefully, Boswell charts and annotates Johnson's fluctuating attitudes towards Scotland. In 1768, when Johnson mocks the Scot's wish to be out of Scotland, Boswell's text also cites the Doctor's animadversions on Scottish writers such as Hume and Kames, remarking pointedly how 'His prejudice against Scotland appeared remarkably strong at this time.' (*Life*, ii. 53.) Alerting readers to Johnson's Scotophobic bigotry, Boswell uses him elsewhere to accent Scottish cultural differences—for example, when he quotes all of Johnson's 1766 letter to the Society in Scotland for Propagating Christian Knowledge, which opposed the translation of the Bible into Gaelic 'from political considerations of the disadvantage of keeping up the distinction between the Highlanders and the other inhabitants of North-Britain' (*Life*, ii. 27). In writing 'North-Britain', Boswell stresses Scotland's loyalty to Britain; simultaneously, he reminds readers of its distinctive cultural patterning, using Johnson's letter to emphasize this point. When Johnson is John Bullish towards the Scottish literati, Boswell upbraids him for it ('Yes, Sir, you tossed and gored several persons'), unobtrusively drawing attention to his subject's prejudice (*Life*, ii. 66).

The most celebrated Johnsonian statements in Boswell's *Life* are so wittily anti-Scottish that readers too easily ignore the way in which Boswell contains them, modifies them, or places them in the wider context of the Johnson who advises his biographer to collect Scottish antiquities and compile, as a significant contribution to the history of the language, a Scottish dictionary (*Life*, ii. 91). Boswell continually brings Johnson back to pronouncing and opining on Scottish matters —law cases, literature, manners, speech traits—so that Scottishness/ Englishness/Britishness is maintained as a thread through the book. Nor is Boswell content simply to limit himself to being Johnson's prompter. He intervenes, analyses, and comments on what is said. Extracting from Johnson the retort, 'Sir, your pronunciation is not offensive', Boswell goes on to advise 'my countrymen of North-Britain' to preserve (as Wedderburn had done) 'a small intermixture of provincial peculiarities' (*Life*, ii. 160). He tellingly records Johnson's confession that his own speech sometimes exhibited provincial traces. Boswell desires a reconstructed, modified Scottishness which may be an asset to British life. Repeatedly in the *Life* he continues his Hebridean project to convert John Bull to a Britishness which acknowledges the Scottish heritage.

So, in the literary arena, Boswell, like Blair and Barron, contends that Ramsay's *Gentle Shepherd* be classed among the finest pastorals:

> I spoke of Allan Ramsay's 'Gentle Shepherd,' in the Scottish dialect, as the best pastoral that had ever been written; not only abounding with beautiful rural imagery, and just and pleasing sentiments, but being a real picture of manners; and I offered to teach Dr. Johnson to understand it. 'No, Sir, (said he,) I won't learn it. You shall retain your superiority by my not knowing it.' (*Life*, ii. 220.)

Johnson is portrayed here as jamming out Scottish culture in a patronizing way. If Boswell learns from, and records, the words of John Bull, John Bull here appears determined to learn neither about nor from Scottish materials. This Johnson, prepared to accept Boswell only as 'the most *unscottified* of Scots', is subtly shown up as prejudiced in his bouts of Scotophobia (*Life*, ii. 242). His celebrated remark, 'Much may be made of a Scotchman, if he be *caught* young', follows a mention of Lord Mansfield, held up by Boswell as a creditable Scot: [Johnson] 'he was educated in England' (*Life*, ii. 194).

Oat-defining Johnson is notorious, but the theme of Scottishness–Englishness in Boswell's writings is constructed as a cultural give and take in which the placing of Johnson's remarks is significant. For instance, Boswell sabotages the gibe about catching a Scotsman by preceding it by Johnson's pronouncement that 'there is no permanent national character'—which suggests Englishness as well as Scottishness is in the melting-pot (*Life*, ii. 194). Again, very soon after this, Boswell shows Johnson closely considering 'a question purely of Scotch law' (*Life*, ii. 196). As elsewhere in the book, Johnson's attentions reflect well on Scotland's distinctive legal system. Only Boswell would have given such detailed consideration to the Great Cham's pronouncements on often obscure Scottish legal matters. These act as a subtle counterbalance to Johnson's better-known Scotophobic wit, demonstrating that here was a Scotland Johnson took seriously. So Boswell juxtaposes Johnson's praise of Scottish law reports with his patronizing dismissal of Allan Ramsay (*Life*, ii. 220).

Insufficient attention has been paid to the way in which Boswell orders, provokes, and juxtaposes his materials in constructing the *Life*. More than any other theme of the biography, the Johnsonian pronouncements on Scottish matters are carefully stimulated and arranged. The *Life* becomes cultural biography—and cultural autobiography, since the biographer features so prominently as part of the matter recorded. This is true not only at the level of the large set pieces, such as Boswell's first meeting with Johnson, but also at the level of detail—for example, when Boswell seeks 'to adopt one of the finest images in Mr. Home's Douglas' to describe a Johnsonian fit of temper (*Life*, iii. 80). Boswell, who, like the Scottish literati, thought *Douglas* 'beautiful and pathetick', was well aware of how cheeky he was being in explicitly applying to Johnson a phrase from what the Doctor elsewhere in the *Life* provocatively denounces to Thomas Sheridan as 'a foolish play' (*Life*, i. 486; and ii. 320). The *Life* offers clear evidence that, in highlighting Johnson's cultural prejudice, Boswell knew well what he was about:

> I ventured to mention a person who was as violent a Scotchman as he was an Englishman; and literally had the same contempt for an Englishman compared with a Scotchman, that he had for a Scotchman compared with an Englishman; and that he would say of Dr. Johnson, 'Damned rascal! to talk as he does of the Scotch.' This seemed, for a moment, 'to give him pause.' It, perhaps, presented his extreme prejudice against the Scotch in a point of view somewhat new to him, by the effect of *contrast*. (*Life*, iii. 170.)

To maintain the Scottishness–Englishness theme through *contrast*, through the maintenance of a comparative cultural perspective, is a preoccupation which permeates Boswell's writings, which seek a balance of Scottishness and Englishness in Britishness.

Yet Johnson is much less Boswell's opponent than his guru, and Boswell's worries about Scotland being 'too narrow a sphere' form part of his attraction to Johnson and his world (*Life*, iii. 176). Balancing on a cultural frontier, Boswell is continually pulled in two directions, as the discussions in the *Life* about whether or not he should stay in Scotland show. Anticipating Scott's cultural invest-igations and complicating the Scottishness–Englishness topic, Boswell's treatment of Ossian's poetry (about which he agrees with Johnson) is of great interest here. In the *Life* Boswell drew parallels between the Highland journey and the Ossianic controversy. This reminds us that, for eighteenth-century readers, Johnson's *Journey* mattered not least for its powerful denunciation of the Ossianic poems —a product, the Doctor contended, 'of Caledonian bigotry'. For Johnson, the poems 'never existed in any other form than that which we have seen' (*Journey*, 107). Their 'translator', Macpherson, defi-antly tried to convince Johnson that originals of the poems did exist, and seems to have threatened Johnson; the Doctor's strong, insulting response appeared in contemporary accounts, and underpins Boswell's account of the controversy in the *Life*.[64]

Boswell's *Journal* presents Johnson's most outspokenly anti-Ossianic remarks—'as gross an imposition as ever the world was troubled with' (*Journal*, 320). Yet, later, Boswell, who diligently inquired about the Ossianic poems during his Highland tour, and who records a discussion of them by himself, Johnson, Blair, and other Scottish literati, protests in a footnote that 'I desire not to be understood as agreeing *entirely* with the opinions of Dr. Johnson, which I relate without any remark.' (*Journal*, 422.) Typically, though, he does offer some comment, attempting to mollify Scottish readers by asserting that Johnson would have been equally scathing had Ossian been an English epic. He tries to calm matters by saying, not entirely truth-fully, that 'The subject appears to have now become very uninterest-ing to the public', yet he cannot resist returning to the topic, suggesting that parts of the poems may be authentic (*Journal*, 423). Boswell similarly hedges his bets in the *Life*, recalling his early en-thusiasm for Ossian's 'wild peculiarity', yet confessing to later doubts

---

[64] The fullest modern account of the Ossianic poems and the controversy they aroused is Fiona J. Stafford, *The Sublime Savage: James Macpherson and the Poems of Ossian* (Edinburgh: Edinburgh University Press, 1988).

about the materials to whose collection he had subscribed (*Life*, ii. 302). Wanting to have things both ways, Boswell defends Johnson's right to his strictures upon the Ossianic poems, however offensive to Scots, yet he also points to Johnson's prejudice: 'That he was to some degree of excess a *true-born Englishman*, so as to have entertained an undue prejudice against both the country and the people of Scotland, must be allowed.' (*Life*, ii. 300.)

Boswell's uneasy oscillation over the Ossianic question typifies his uneasy exploration of Scottishness/Englishness/Britishness. He remains anxious to avoid Johnson's charge that Scots tend to 'love *Scotland* better than truth' (*Journey*, 108), and he is equally anxious to encourage Johnson to praise as many aspects of Scottish life as possible. Sometimes Boswell seems to be contending high-mindedly that the truth is neither Anglo- nor Scotocentric; but at other times he seems to be toadying. Together the anxious Boswell and dominant Johnson are like a scale model of the United Kingdom. Boswell's anxieties as an eighteenth-century Scot, a part of that culture so preoccupied with Rhetoric and Belles Lettres, propelled him towards an investigation of Britishness and a continuing concern with associated prejudices which clearly positions his work in a literary genealogy that includes *Humphry Clinker* and *Waverley*. Yet such chameleon-like shifts of attitude as Boswell reveals over the Ossian controversy show him as wishing to inhabit a wide and sometimes contradictory variety of postures within that expanded cultural arena of Britishness. Boswell can act the metropolitan sophisticate and the local Scottish patriot. The British Boswell is a figure who realizes that access to Britain offers him a wider variety of roles than he could play in Scotland alone. The Laird of Auchinleck had native access to a Scottish cultural heritage and an awareness unknown to most Englishmen, but he wished to have access to what England might offer as well. Boswell wished for the scope of full Britishness. It is this desire for a wide cultural spectrum which, in unexpected ways, links him to the poet whose name has become one of the richest of cultural shibboleths: Robert Burns.

BRITISH BURNS

Burns was bicultural—another cultural broker. He drew on, and negotiated between, the Scots folk-culture of his family and background, and the officially dominant Anglicized culture of his formal education and of most of the metropolitan values with which he came into contact. He moved between the plough and Hugh Blair. Yet, for all its occasional deference, his work contains a critique of those forces behind the Scottish development of Rhetoric and Belles Lettres.

Insisting on the importance of his local vernacular, writing a deliberately impure language, deeply inscribing himself in a culture outside the prevailing metropolitan one, Burns marginalized himself in a way that was to make him fruitful for a wide spectrum of later writers—not only Wordsworth and the Romantics, but also Whitman, MacDiarmid, and other Modernists. Burns has been seen too much only in terms of what preceded him, as 'a decadent representative of a great alien tradition', in T. S. Eliot's words.[65] But, as is suggested later in this book, even Eliot learned from brooding on Burns.

Burns shared a country with Boswell, but not a social class. Like Smollett, he was weaned on tales of Wallace (*Letters*, i. 62) and Scottish independence. He admired Smollett's 'Ode to Independence' (*Letters*, ii. 45), and would himself celebrate independence as a cardinal virtue. Admiring the 'incomparable humour' (*Letters*, i. 296) of Smollett's novels, Burns corresponded revealingly with the novelist's close friend Dr John Moore, who encouraged him, and whose works Burns valued. 'Vive l'amour et vive la bagatelle were my sole principles of action' (*Letters*, i. 141), he claimed rakishly to Moore, who treated just such attitudes in his 1786 novel *Zeluco*, greeted by Burns as 'A glorious story' (*Letters*, ii. 469) and by Byron as a template for Childe Harold's career.[66] For Moore, the Scot in London, Burns mixes deference with self-assertion: 'I have taken a whim to give you a history of MYSELF'—the capitals have an engaging and self-deflating mock-grandiosity, at the same time as a self-assurance (*Letters*, i. 133). In verse Burns frequently presents himself in a guise both self-deprecating and confidently assured, constantly implying that the manifestly little may stand confidently beside the mighty.

The effect of this technique is to upset established categories, raising questions about the way in which we casually assign cultural value. Burns loves taking the small and the great and making them

---

[65] T. S. Eliot, *The Use of Poetry and the Use of Criticism* (1933; repr. London: Faber and Faber, 1970), 106.

[66] Lord Byron, 'Addition to the Preface to the First and Second Cantos of *Childe Harold's Pilgrimage*' (1813), in *Poetical Works* (Oxford: Oxford University Press, 1904), 175.

rhyme incongruously, as is evident in the early song 'Mary Morison', where the individual Scottish girl of the title is made to rhyme with 'sun to sun', which means simply 'sunrise to sunset' but suggests the huge workings of the universe (*PS*, 31–2). Mary's transitory 'smiles and glances' are seen to 'make the miser's treasure poor', the 'rich reward' of lyrical eroticism destabilizing established values of commerce. Similar ideas motivate Burns's cancelling-out of a bank-note by writing his poet's oath over its promise of value ('Lines written on a Banknote': *PS*, 202), or the individual heart's over-balancing of widely validated worldly 'riches' in 'Green grow the Rashes' (*PS*, 44), where the heart's non-monetary spending power seems to triumph through being linked to things proverbially worth-less ('not worth a rush'). Placing the 'rashes' (rushes) in the poem's refrain, the poet ensures that they reproduce themselves before the audience, while the other 'riches' remain unmultiplied.

> The warly race may riches chase,
>   An' riches still may fly them, O;
> An' tho' at last they catch them fast,
>   Their hearts can ne'er enjoy them, O.
> *Green grow the rashes, O;*
> *Green grow the rashes, O;*
> *The sweetest hours that e'er I spend,*
> *Are spent amang the lasses, O.*

In the next stanza 'canny', which can mean both 'frugal, careful with expense', and 'lucky *or* pleasant', makes a cheekily ambivalent introduction to an up-ending, a turning 'tapsalteerie', of conventional wisdom as the little is asserted as superior to the large.

> But gie me a canny hour at e'en,
>   My arms about my Dearie, O;
> An' warly cares, an' warly men,
>   May a' goe tapsalteerie, O!

Such upsetting of the conventionally great is akin to some of the tricks of *Gulliver's Travels*, or to Hugh MacDiarmid's favouring of the 'bonnie broukit bairn' over all the splendid planets in his poem of that title.[67] Burns goes on to destabilize conventionally prudent ('douse') ideas of wisdom by a daring stroke.

---

[67] *MCP*, 17; Burns's familiarity with *Gulliver's Travels* is pointed out in John S. Robotham, 'The Reading of Robert Burns', *Bulletin of the New York Public Library*, 74/9 (November 1970), 575.

> For you sae douse, ye sneer at this,
>> Ye're nought but senseless asses, O:
> The wisest man the warl' saw,
>> He dearly lov'd the lasses, O.

To Dr Moore Burns wrote: 'I assure you, Sir, I have, like Solomon whose character, excepting the trifling affair of WISDOM, I sometimes think I resemble, I have, I say, like him "Turned my eyes to behold Madness and Folly;" and like him too, frequently shaken hands with their intoxicating friendship.' (*Letters*, i. 133–4.) The song, though, is more daring than the letter, because it explicitly asserts (in a way that parallels Blake) that true wisdom may be what the 'douse' would account madness and folly. Through the provocative juxtaposition of little and large, it moves to an overturning and questioning of accepted values.

Effects like this are ubiquitous in Burns's songs. 'My bony Mary' (*PS*, 354) holds more power over the poet than the world of armies and the elements, while in 'Louis what reck I by thee' (*PS*, 358) the setting of little against large mixes the 'high' and the 'low' so that the 'high' are not only rudely translated into the low (King George becoming 'Geordie'), but actually have their sovereignty stolen while being themselves accused of rudeness ('Reif randies' = thieving, rude people):

> Louis, what reck I by thee,
>> Or Geordie on his ocean:
> Dyvor, beggar louns to me,
> I reign in Jeanie's bosom.

> Let her crown my love her law,
>> And in her breast enthrone me:
> Kings and nations, swith awa!
> Reif randies I disown ye!—

The small triumphantly trumps the large in the elegant lyric 'A red red Rose' (*PS*, 582), where the beloved, likened to small, vulnerable things—a young rose or a well-played melody—is juxtaposed with vast expanses of geological time. Repeated terms of endearment ('My Luve . . . my Luve . . . my bonie lass . . . my Dear . . . my Dear . . . my only Luve . . . my Luve') assert the continuity of a particular human affection throughout the greatest geological changes revealed by contemporary Scottish scientists such as James Hutton—'Till a'

the seas gang dry . . . | And the rocks melt wi' the sun.' The small
and particular outwits the universe.

In 'To a Louse' (*PS*, 156) the tiny beast reaches 'The vera tapmost,
towrin height | O' *Miss's bonnet*', making all her grandeur ridiculous;
in 'To a Mouse' (*PS*, 101) the 'WEE, sleeket, cowran, tim'rous *beastie*'
prompts the thought that mouse and man are equivalent; in 'To a
Mountain Daisy' (*PS*, 182) the daisy too is 'wee', but its fate is also
that of man. Familiar with the story of David and Goliath and with
Gulliver's Travels, Burns loves to depict the small unsettling the great,
and can turn this technique against himself. At a verbal level he yokes
great to small when, in 'On Willie Chalmers' (*PS*, 211), he rhymes
'holy Palmers' with 'Willie Chalmers', or, in 'The Author's Earnest
Cry and Prayer, to the . . . Scotch Representatives in the House of
Commons' (*PS*, 149), he complains that '*Scotland* and *me*'s in great
affliction'. As its title suggests, 'Death and Dr Hornbook' (*PS*, 60) is
another meeting of large and little, as is 'Tam o' Shanter' (*PS*, 443)—
essentially the tale of how an Ayrshire peasant beats hell's legions. In
'Holy Willie's Prayer' (*PS*, 56) Holy Willie's denunciation of a minor
miscreant's peccadilloes blasts a hole in his own great unco guidness.
Juxtaposition of major and minor to unsettle established values—this
is the technique Burns used not only in his verse, but also in his
negotiations with the wider world of literature and, in particular,
with its custodians, those Edinburgh literati and teachers of Rhetoric
and Belles Lettres. The mere Ayrshire peasant outwitted the great
professors.

After reviews of the Kilmarnock edition of his *Poems* in late 1786,
the literary world began to accord Burns star status; he responded by
referring to himself as 'my Bardship' (*Letters*, i. 119, 125; *PS*, 256).
Gray's *The Bard* and Macpherson's Ossian (much admired by Burns)
had furthered an enthusiasm for ancient, dignified, vatic bards. In
fun, Burns donned regal bardic robes in a letter to William Chalmers
and John M'Adam in November 1786, in which he simultaneously
sets out his great poetic aspirations and belittles them through hyper-
inflation:

In the Name of the NINE. *Amen.*

We, ROBERT BURNS, by virtue of a Warrant from NATURE, bearing date
the Twenty-fifth day of Janueary, Anno Domini one thousand seven hundred
and fifty-nine, POET-LAUREAT and BARD IN CHIEF in and over the Districts
and Countries of KYLE, CUNNINGHAM, and CARRICK, of old extent . . .
(*Letters*, i. 65).

A similar spirit of self-mocking exaggeration led Burns in 1787 to coin the word 'Bardship' (*Letters*, i. 119, 125, 128, 133, 160, 164, 308). He describes a fall with his horse Jenny: 'and down came the Highlandman, horse & all, and down came Jenny and my Bardship' (*Letters*, i. 128). Particularly during 1787, when he was coming to terms with the fame accorded him by the literati, Burns deployed this inflated term of greatness which he had invented.

Characteristically, Burns's grand term 'bardship' was a linguistic counterbalance to the diminutive 'bardie' which Burns found in Robert Fergusson's work, and which he first used in 'Poor Mailie's Elegy' (*PS*, 26) when mock-heroically but affectionately lamenting the death of a pet sheep. In this piece Burns affords other poets the title '*Bards*', but calls himself '*Our Bardie*' and '*Robin*'. This act of familiar naming and use of the term 'bardie', the Scots diminutive of the literati's English word 'bard', allows him to remain at one with his local audience. Only once does he use the pronoun 'I' here; all the other times he is 'Robin' or 'Our Bardie'. It is as if Burns is voicing the poem more through his community (which perceives him familiarly and in the third person) than through the 'I' of the poet. Such acts of inscribing himself in a particular vernacular community take various forms in Burns's work.

Probably his letter-poem of May 1785, 'To W. S——n, Ochiltree' (*PS*, 72), was actually sent, making it a concrete act of poetic intercourse with his community. In it he uses another self-deprecating diminutive when he writes of 'my poor Musie', who is Coila (Burns's coinage—the Muse of Kyle), a specifically *local* deity whose '*Bardies*' seem to be set against the grand '*Poet*' who would view Burns's haunts with disdain:

> Auld COILA, now, may fidge fu' fain,
> She's gotten *Bardies* o' her ain,
> Chiels wha their chanters winna hain,
>   But tune their lays,
> Till echoes a' resound again
>   Her weel-sung praise.
>
> Nae *Poet* thought her worth his while,
> To set her name in measur'd style;
> She lay like some unkend-of isle
>   Beside *New Holland*,
> Or whare wild-meeting oceans *boil*
>   Besouth *Magellan*.

Burns asserts himself here as 'Bardie', not 'Poet', stressing his determination to be part of a community that celebrates not the grand,
cosmopolitan, great rivers '*Thames* an' *Seine*' mentioned later in the
poem, but the local '*Irwin*' and '*Lugar*' which 'Naebody sings'. His
confident deployment of such local names, like his use of dialect
words, asserts the importance of the supposedly 'provincial', defiantly
setting the little beside the great. The use of place-names and dialect
forces readers to consider the text's local origin as part of the poem's
meaning, its assertion that the bardie's apparently obscure, small
culture may be valued at least as much as the poet's grand, celebrated
one. Seamus Heaney, Douglas Dunn, and Les A. Murray were to do
something similar with place-names two centuries later.

In his first 'Epistle to J. L——k, an Old Scotch Bard' (*PS*, 65),
Burns accords his correspondent the title 'bard', but calls himself a
'rhymer'. When he attacks 'Critic-folk' and the institutional literary
culture of his day, with its assumption that classical, generic, and
grammatical knowledge gained in '*Colledge-classes*' was the basis of
literary judgement, he is confronting just the sort of learning promoted by the teachers of Rhetoric and Belles Lettres. Yet Burns also
made use of that culture to which Hugh Blair and his fellows belonged.
His poem 'The Vision' (*PS*, 80) enthusiastically draws on the fashionable discussion of bards, and has Burns's Muse address him as '*my
own* inspired Bard'. By presenting his Muse as 'A tight, outlandish
Hizzie', Burns signals her distance from the Classical Muses, and his
position as more bardie than bard. In a stanza of this poem found in
the Stair manuscript he salutes Boswell's literary achievement, but
clearly separates his own homely Muse from that of Auchinleck and
its pomp:

> [Nearby] arose a Mansion fine,
> The seat of many a Muse divine;
> Not rustic Muses such as mine,
>   With holley crown'd,
> But th' ancient, tunefull, laurell'd Nine,
>   From classic ground.   (*PS*, 84.)

The 'classic ground' here is that of the classics, bound up with a
'high' tradition which Burns pillaged but into which he did not wish
to be assimilated. He would assert to Moore that Scotland had her
own 'classic ground' which he might celebrate, but such celebration
was to be through the inspiration of COILA, not the 'Colledge-class'
Muse (*Letters*, i. 107).

'The Vision' also bows towards the cultural milieu of Dr James Beattie, Professor of Moral Philosophy at Aberdeen and collector of Scotticisms; yet Burns deferentially points out through Coila that Beattie is inspired by a high Muse, while lower Muses are assigned 'The humbler ranks of Human-kind', such as 'The rustic Bard' who is to be placed beside 'The Artisan'. Like his categorization of himself as a bardie, presenting himself as a rustic bard allows Burns to maintain a solidarity with his own community and culture even as he engages with, and steals from, 'high' culture. Using a similar strategy, he seizes upon a term applied to him by the literati, 'the Scotch Bard', and refers to it in several letters of late 1786 and early 1787, writing to Mrs Dunlop in March 1787 that 'The appelation of, a Scotch Bard, is by far my highest pride.' (*Letters*, i. 101; cf. i. 75.) Again, this term 'Scotch bard' is different from 'bard', since the word 'Scotch' acknowledges not only the country, but the language of the author of *Poems, chiefly in the Scottish Dialect*. 'Bard' was primarily used of ancient poets or those in 'primitive' societies such as Gaeldom. It was a term fashionable in the 'high' culture of men like Blair, whose preferred language was decorous formal English. For that culture to recognize a contemporary bard who used Scots was double-edged. On the one hand, it constituted an acknowledgement that Scots, the vestiges of which the teaching of Rhetoric and Belles Lettres was designed to eradicate from polite society, might still have some worth. On the other hand, it served to categorize Burns rather patronizingly as a belated primitive curiosity—as if one were to describe someone today as 'a Zumerzet provincial'.

In adopting the description 'a Scotch bard' and calling it his 'highest pride', Burns attempts to valorize this patronizing term. It allows him again to side firmly with those supposedly rather primitive Ayrshire peasants from whose society he comes and whose language he often uses. As he negotiates with the term 'bard', Burns's invention of the mockingly inflated 'bardship', his deployment of the self-deprecating diminutive 'bardie', and his adoption and celebration of the term 'a Scotch bard' show him well able to cope with the values of the metropolitan world of literature, and able to avoid being trapped in the potentially patronizing or embalming bardolatry of being simply a 'bard'. It is just such tactical canniness which he was to exhibit in his face-to-face dealings with the literati of Edinburgh, subverting and redeploying the power of their culture while remaining on good terms with men whose linguistic and other ideals were very different from his own.

Burns was acceptable and fascinating to the literati of Edinburgh because he never sought to be one of them. Their literary strictures were directed to the world of the upwardly mobile, the aspirants of polite society. Maintaining his ties with his own class and community, Burns was able to remain independent of this world, though involved in it. He informed Sir John Whiteford in Edinburgh in December 1786 that 'learning never elevated my ideas above the peasant's shed, and I have an independent fortune at the plough-tail' (*Letters*, i. 68). No mere self-abasement, this is also a declaration of the independence of mind which Burns so prized and celebrated, most notably in such songs as 'For a' that and a' that—' (*PS*, 602):

> Ye see yon birkie ca'd, a lord,
>   Wha struts, and stares, and a' that,
> Though hundreds worship at his word,
>   He's but a coof for a' that.
> For a' that, and a' that,
>   His ribband, star and a' that,
> The man of independant mind,
>   He looks and laughs at a' that.—

Though Burns did not express himself so forcibly to the literati, it was precisely his combination of independence, solidarity with his own non-metropolitan, vernacular peasant background, and precise awareness of his literary values which allowed him to engage so successfully with the world of metropolitan literature. The otherness which made him so fascinating was also what allowed him to hold his own. In terms of literary, social, and actual geography, Burns knew his place, and manœuvred himself into it with skill. As he told one of Edinburgh's literary teachers: 'I have long studied myself, and I think I know pretty exactly what ground I occupy, both as a Man, & a Poet.' (*Letters*, i. 74.)

Nowadays we can appreciate that Burns was a well-read man. Cataloguing works mentioned in his writings, John Robotham produced a list of well over two hundred items, ranging from Cicero to Goethe, from Adam Smith's *Theory of Moral Sentiments* to William Derham's *Astro-Theology*, from Cervantes to Racine.[68] Burns acknowledged to Moore that he had been 'an excellent English scholar' (*Letters*, i. 135) at school, where he had studied *A Collection of Prose and*

---

[68] Ibid.

Verse, from the Best English Authors by Arthur Masson, who, though he had been a friend of Robert Fergusson, saw Thomas Sheridan as his mentor and was one of those who, as acquaintances of Hugh Blair, cultivated the development of Belles Lettres in Scotland.[69] As one might expect, there is little sense of Scottishness in Masson's anthology, which is geared rather to instilling English models into provincial Scots. Certainly, Burns as a youth had acquired a knowledge of what was proper according to 'correct' English taste—'my knowledge of modern manners, and of literature and criticism, I got from the Spectator' (*Letters*, i. 138). His early (i.e. pre-1787) poems, quite apart from including many quotations, adaptations of, and allusions to a number of writers, explicitly mention Aeschylus (*PS*, 155), Mrs Barbauld (*PS*, 155), Beattie (*PS*, 66), the Bible (*PS*, 37), Robert Blair (*PS*, 143), Boswell (*PS*, 151), Bunyan (*PS*, 180), Calvin (*PS*, 196), Demosthenes (*PS*, 240), Robert Fergusson (*PS*, 73), Gray (*PS*, 89), William Hamilton of Gilbertfield (*PS*, 72), Goldsmith (*PS*, 122), Herd (*PS*, 155), Homer (*PS*, 155), Macpherson's Ossian (*PS*, 80), Milton (*PS*, 155), Pope (*PS*, 66), Ramsay (*PS*, 73), Thomas Reid (*PS*, 180), Sappho (*PS*, 155), Shakespeare (*PS*, 155), Shenstone (*PS*, 89), Adam Smith (*PS*, 180), Steele (*PS*, 66), Dr Taylor of Norwich (*PS*, 75), Theocritus (*PS*, 155), and James Thomson (*PS*, 89), as well as other figures such as Hogarth (*PS*, 240) and Socrates (*PS*, 106). Confident deployment of all these hardly suggests a poet who is a mere child of nature, yet it was essentially in this capacity that Burns was received by the literati of Edinburgh and by critics further afield.

This was no accident. For, stretching the term this way and that, with constant self-mockery, Burns nevertheless encouraged his educated readers to accord him the role of bard. As Blair himself had indicated in his writing on Amerindian bards and in his enthusiasm for Ossian, the bard stood outside the restrictions imposed by the new methods of composition as taught in Rhetoric and Belles Lettres. The bard was an inspired singer, drawing on 'primitive' tradition, rather than a modern, cultured, and correct poet.[70] Bards were thus rude, 'natural' geniuses in a way that modern poets were not. Bards were to be admired, but they were remote from the modern world of literature and literary teaching. With their remoteness went a naturally inspired

[69] Alexander Law, *Education in Edinburgh in the Eighteenth Century* (London: University of London Press, 1965), 151–3; Ian Michael, *The Teaching of English from the Sixteenth Century to 1870* (Cambridge: Cambridge University Press, 1987), 175–6.

[70] See e.g. Hugh Blair, *A Critical Dissertation on the Poems of Ossian, the Son of Fingal* (London: T. Becket and P. A. de Hondt, 1763), 2.

wildness, a *furor poeticus*, which allowed them, unlike the adherents of
Rhetoric and Belles Lettres, to break the rules, to maintain an
'independent mind'.

Burns's readers ignored the poet's clear knowledge of a wide range
of literature, and accepted him on the terms that he invited—as a
'natural' rather than a cultured, literary author; as a rude, modern
Scots version of the Ossianic bard; as someone allowed the liberties of
the genius that lies outside the constraints of polite Belles Lettres. On
such terms he was splendidly acceptable.

Who are you, Mr. Burns? will some surly critic say. At what university have
you been educated? . . . perhaps honest Robert Burns would make no
satisfactory answers. 'My good Sir,' he might say, 'I am a poor country man
. . . I have not looked on mankind *through the spectacle of books*. An ounce of
mother wit, you know, is worth a pound of clergy; and Homer and Ossian,
for any thing that I have heard, could neither write nor read.' The author is
indeed a striking example of native genius . . .[71]

Though this critic (a friend of Blair) recognizes that Burns has read
Ramsay and Fergusson, he treats him as otherwise unlettered. His
slight knowledge of Homer and Ossian is attributed only to 'any
thing that I have heard'. Ignoring the fact that the 1786 *Poems* have a
detailed reference to a footnote in volume 2 of Macpherson's work
(*PS*, 80)—hardly suggestive of only vague contact—this review un-
questioningly accepts Burns's presentation of himself in the poems.
Henry Mackenzie, whose work Burns so admired (*Letters*, ii. 269) and
who was strongly interested in Ossian, similarly went along with
Burns's description of himself as a 'rustic bard'.[72] Mackenzie uses
this term from 'The Vision' without acknowledging it as a borrowing,
but he does speak of 'The Vision' with particular admiration.

As Blair had lamented the provincial quaintness of Allan Ramsay's
language, so Mackenzie laments Burns's use of 'provincial dialect
. . . now read with . . . difficulty', and generally patronizes Burns as
a freak of nature, most clearly in the phrase 'Heaven-taught ploughman'.[73] Nevertheless, his being such an unlikely prodigy grants him
sufficient excuse for transgressing the rules of politeness and decorum
which those in polite society must observe. Carol McGuirk has
stressed the way in which the upper classes liked to emphasize the

[71] Donald A. Low (ed.), *Robert Burns: The Critical Heritage* (London: Routledge and
Kegan Paul, 1974), 63–4 (unsigned notice in *Edinburgh Magazine*, Oct. 1786).
[72] Ibid. 69 (Henry Mackenzie, unsigned essay in the *Lounger*, Dec. 1786).
[73] Ibid. 69–70.

strength of their sensibility by patronizing Burns as 'the humble bard', 'the untutored bard', 'uncultivated genius'.[74] Even the exception among these early reviewers, Blair's protégé John Logan, who noted that Burns was 'better acquainted with the English poets than most English authors that have come under our review', accepts that Burns is essentially 'a *natural*, though not a *legitimate*, son of the muses'.[75] In granting Burns this natural illegitimacy, even Logan grants him the liberty of operating beyond the laws and strictures of cultivated literature.

These early reviews of Burns's *Poems* confirm that there was a place within eighteenth-century poetics for the primitive, wild, bardic, and 'natural'—all that was opposite to the sort of composition taught by the teachers of Rhetoric and Belles Lettres. As 'the Scotch bard', Burns managed to position himself in that space. There he was able to be admired by the literati even as he developed further a style of writing which was both outside and, in many ways, opposed to their strictures.

This is clear in his dealings with the university teachers of Rhetoric and Belles Lettres, Hugh Blair and William Greenfield, both moderate Church of Scotland ministers. Burns called Greenfield 'a steady, most disinterested friend', whose 'good sense . . . joyous hilarity . . . sweetness of manners and modesty, are most engagingly charming'.[76] Yet towards Greenfield the literary professor Burns's attitude is subtly subversive. Sending Greenfield two songs by 'Ayrshire Mechanics', Burns addresses him as 'a Professor of the Belle lettres de la Nature' (*Letters*, i. 73–4). The poet's misspelling of 'Belles Lettres' protests (consciously or not) his ignorance of such matters, but 'de la Nature' is a curious phrase. If it means that Greenfield is a natural professor, then this mocks the whole idea of literary judgement as a matter of learning, of culture rather than nature. If it means that Greenfield is a Professor of the Belles Lettres of Nature, of the uncultivated contemporary world from which Burns (like the 'Ayrshire Mechanics') comes, then the idea that this world might have its own Belles Lettres makes nonsense of the university subject whose ground rule was that it required educated cultivation.

---

[74] See Carol McGuirk, *Robert Burns and the Sentimental Era* (Athens, Ga.: University of Georgia Press, 1985), esp. 59–83; Low, *Robert Burns*, 87, 80.

[75] Low, *Robert Burns*, 77.

[76] Cited in Robert Burns, *Poems and Songs*, ed. James Kinsley (3 vols.; Oxford: Clarendon Press, 1968), 1235 (abbreviated hereafter as Burns, *Poems and Songs*).

Like many of Burns's other linkings of high and low, the phrase 'a Professor of the Belle lettres de la Nature' explodes contemporary assumptions. Comparable is a line from the concluding song of 'The Jolly Beggars', 'Life is all a VARIORUM' (*PS*, 169), where the low uncultivated beggars' 'Life' is yoked to the cultivated, high, and literary 'VARIORUM'. Again, the effect is to assert the possibility of a worthwhile world which runs counter to that of '*Colledge-classes*'. When Burns sends the Ayrshire mechanics' songs to Greenfield as 'Professor of the Belle lettres de la Nature', he says he looks upon this position as 'a kind of bye Professorship, not always to be found among the systematic Fathers and Brothers of scientific Criticism' (*Letters*, i. 73). Burns calls the Ayrshire mechanics 'Bards', so dignifying them and indicating their independence from the world of 'the systematic Fathers and Brothers'. The mechanics are 'Bards such as I lately was; and such as, I believe, I had better still have been'. The implication is that Burns is allying himself with, and continuing to champion, another 'small' tradition beyond the conventionally 'great' world of Belles Lettres, even as he acts as broker between the two. Such a suggestion is reinforced in Burns's letter to the Scottish theorist Archibald Alison in 1791, thanking him for his *Essay on the Nature and Principles of Taste*, which, Burns writes, has stimulated him 'to draw up a deep, learned digest of strictures on a Composition, of which in fact, untill I read the book, I did not even know the first principles' (*Letters*, ii. 71). At first glance flattering, this tribute makes it clear that Burns has succeeded in his own work without any of Alison's advice. Beside the letter to Alison we might set this verse from Burns's epistle 'To W. S——n, Ochiltree':

> In days when mankind were but callans,
> At *Grammar*, *Logic*, an' sic talents,
> They took nae pains their speech to balance,
>   Or rules to gie,
> But spak their thoughts in plain, braid Lallans,
>   Like you or me.   (*PS*, 75.)

These lines are at one with the teachers of Belles Lettres in asserting that primitive communities had a direct eloquence, but glide daringly from such a position to the assertion that this is true of contemporary communities dismissed as uncultivated vernacular users. The Burns of the stanza just quoted looked towards the Wordsworth of *Lyrical Ballads* even as he was politely accepting Hugh Blair's suggestions for the editing of a new edition of his *Poems*.

Interestingly, these suggestions did not involve alteration to the Scots language of the poems. Clearly, for Blair, Burns's position as an outsider licensed his use of the 'dialect', but Blair sought to deny Burns other licences by advising the removal of oaths, indecencies, and 'quite inadmissible' references to the Scriptures. He advises the suppression of the whole of 'The Jolly Beggars' as 'much too licentious'. Blair protests that he is writing as 'one who is a great friend to Burns's poems' and who 'wishes him to preserve the fame of Virtuous Sensibility', yet it was just such unco guidness and correctness which Burns's work was geared to question.[77] Towards Blair Burns remained formally deferential, writing cannily about 'the embarrassment of my very singular situation' (*Letters*, i. 110)—the very situation which caused Burns to appeal to Blair's taste. Writing to Burns on 4 May 1787, Blair mentions how he treasures his own part in bringing Ossian's poems to public notice, and says that Burns 'has stood his charge', in other words, has been someone for whom Blair has felt responsible.[78] Clearly, Blair has seen Burns proprietorially as a novel sort of Ossian, a wild bard whom he may assist into the limelight by helping to create the new Edinburgh edition of the poems.

Blair, though, was much more successful in shaping Macpherson's work than he was in moulding that of Burns, who opined in his Second Commonplace Book that

Dr Blair is merely an astonishing proof of what industry and application can do. Natural parts like his are frequently to be met with; his vanity is proverbially known among his acquaintants; but he is justly at the head of what may be called fine writing; and a Critic of the first, the very first rank in Prose; even in Poesy a good Bard of Nature's making can only take the pass of him. (*Letters*, ii. 440.)

If Burns paid lip-service to Blair's suggestions, this grudgingly admiring passage spells out the fact that the poet's true allegiances lay elsewhere. Burns's main act of literary homage in Edinburgh was not to the superfine literati who 'spin their thread so fine, that it is neither fit for weft or woof', but to Robert Fergusson (1750–74), who, though educated at St Andrews University, had chosen to write in Scots and whose work went unmentioned in Blair's lectures.[79] Burns's request to the Bailies of the Canongate to be allowed to place a simple stone on Fergusson's grave reads like an indictment of metropolitan Edinburgh for its treatment of contemporary native literary distinction:

[77] Low, *Robert Burns*, 82.   [78] See Blair's letter quoted in *Letters*, ii. 398.
[79] Burns, *Poems and Songs*, iii. 1540.

Gentlemen,

I am sorry to be told that the remains of Robert Ferguson [*sic*] the so justly celebrated Poet, a man whose talents for ages to come will do honor, to our Caledonian name, lie in your church yard among the ignoble Dead unnoticed and unknown. (*Letters*, i. 90.)

Burns's tributes to the 'Heaven-taught Fergusson' (*PS*, 258), with whom he clearly identifies, 'my elder brother in Misfortune | By far my elder Brother in the muse', signal clearly that he was aware of the limitations of Edinburgh's academic literary world. For him, the ultimate talent 'Confounds rule and law, reconciles contradiction', and 'if these mortals, the Critics, should bustle, | I care not, not I, let the Critics go whistle!' (*PS*, 371). Burns explained to John Moore: 'For my part, my first ambition was, and still my strongest wish is, to please my Compeers, the rustic Inmates of the Hamlet, while ever-changing language and manners will allow me to be relished and understood.' (*Letters*, i. 88.) This may be as sly and truthful as calling the university-trained Fergusson 'Heaven-taught', but it does show Burns's significant wish to assert the values of his small local community above those of the metropolitan and cultured. His subsequent career frequently bears out such desires, producing not only a constant stream of locally grounded poems, but also such explicit statements as his 'Prologue' (*PS*, 396), written in 1789 for the Dumfries theatre company, the opening lines of which both jokily acknowledge the pleasures of metropolitan taste and counter them with the pleasure of relying on one's own resources:

> No song nor dance I bring from yon great city,
> That queens it o'er our taste—the more's the pity:
> Tho' by the bye, abroad why will you roam?
> Good sense and taste are natives here at home . . .

These sentiments look back to Allan Ramsay. The conventional (and, up to a point, sensible) view of Scottish literary history defines Burns as a distinctively Scots poet who is the poetic descendant of Ramsay and Fergusson as a vernacular writer. However, the Burns who so cunningly juxtaposed small and great in unsettling unions, and who so subtly engaged with the literary teachers, can be seen as representing (though only occasionally consciously advocating) a model of Britishness precisely because of the language which he deployed.

In terms of language, the most densely Scots piece Burns wrote was not a poem but a letter, written in Carlisle on 1 June 1787 to his Edinburgh friend William Nicol (*Letters*, i. 120–1). The letter dates from a tour of the Borders which Burns made after the several months' stay in Edinburgh which first brought him into sustained contact with the literati. Like his coining of the term 'bardship', this Scots letter surely represents a balancing manœuvre, helping him, while on the other side of the English border, to shake the linguistic dust of Belles Lettres off his heels. The exceptional language of this letter reminds us unintentionally not just that Burns habitually corresponded in English, but that he could also speak it correctly. Dugald Stewart, who met the poet in Ayrshire in 1786 before his first Edinburgh visit, noted that 'Nothing perhaps was more remarkable among his various attainments, than the fluency, and precision, and originality of his language, when he spoke in company; more particularly as he aimed at purity in his turn of expression, and avoided more successfully than most Scotchmen, the peculiarities of Scottish phraseology.'[80]

No doubt the 'purity' with which Burns engaged in English conversation with Professor Stewart was more studied than that of the language which he used with the Scots-speaking companions of his own class with whom he drank in Ayrshire pubs. From earliest infancy Burns had been educated in two cultures—that of the *Best English Authors* featured in Masson's anthology, and that of his family background, where his mother sang him Scots songs and he enjoyed listening with terror to the vast repertoire of supernatural tales which his mother's maid rejoiced in telling him. If these two cultures could be detected in his conversation and his correspondence, then a few early observers noticed that his verse contained a much wider spectrum of language than might have been expected of this supposed child of nature from a Scots-speaking background who once called himself 'Robert Ruisseaux' (*PS*, 257), punning, one might suspect, not only on his own surname but also on that of Rousseau. Particularly acute in this respect was Robert Anderson, the Edinburgh doctor who was responsible for helping to publicize the Kilmarnock *Poems* in the *Edinburgh Magazine* and who, between 1792 and 1807, became editor of the fourteen volumes of *Poets of Great Britain*.

It was, I know, a part of the machinery, as he called it, of his poetical character to pass for an illiterate ploughman who wrote from pure inspiration.

[80] Ibid. 1534.

When I pointed out some evident traces of poetical imitation in his verses, privately, he readily acknowledged his obligations, and even admitted the advantages he enjoyed in poetical composition from the *copia verborum*, the command of phraseology, which the knowledge and use of the English and Scottish dialects afforded him . . .[81]

This matter of Burns's 'use of the English and Scottish dialects' has been of considerable interest to modern scholars. Thomas Crawford pointed out in 1978 that different forms of Scots were spoken in eighteenth-century Scotland, and that a speaker able to do so would 'derive special effects by moving from one level of discourse to another'. Crawford makes general distinctions between four kinds of language used in Burns's verse: 'English English', 'Scots English' or 'Anglo-Scots', a 'General Scots', which would be understood across the whole of Scotland, and his own 'regional dialect' of Ayrshire and the south-west.[82] More recently, Carol McGuirk has looked at Burns's 'blending of English with Scots in his best vernacular poetry', and has concluded that 'Burns's diction, like his poetic world, seems "natural" but is designed and invented: a mixture of local dialect, archaic Middle Scots, dialect words of regions other than his own, sentimental idioms, and "high" English rhetoric.'[83] In this conclusion McGuirk relies on her own observation as well as on various earlier analyses of Burns's language, including that of Raymond Bentman, which she quotes: 'Burns wrote some poems in pure English . . . but he wrote no poems in pure vernacular Scottish. The "Scottish" poems are written in a literary language, which was mostly, although not entirely English, in grammar and syntax, and, in varying proportions, both Scottish and English in vocabulary.'[84] Bentman's 1987 study of *Robert Burns* reinforces this view, arguing that Burns's 'new idiom' was 'created out of his ability to merge the English with the Scottish literary traditions'.[85]

Bentman and McGuirk point out that such a synthesis had been under way for some time in eighteenth-century Scottish Literature, with Bentman showing that, while the poets of the vernacular revival led by Ramsay presented themselves as writing in the language of the older Scottish poets, these older poets had written by and large in an

---

[81] Ibid. 1537–8.

[82] Thomas Crawford, *Burns: A Study of the Poems and Songs*, 2nd edn. (Edinburgh: James Thin, The Mercat Press, 1978), pp. viii–x.

[83] McGuirk, *Burns and the Sentimental Era*, p. xxii.   [84] Ibid.

[85] Raymond Bentman, *Robert Burns* (Boston: Twayne Publishers, 1987), p. i.

elegant and courtly language rather than in the vernacular. The poets of the revival, more interested in asserting a symbolic continuity with Scottish poetic ancestors than in continuing accurately the language of those ancestors, were reconstructing Scottish literary identity. They wrote in what Bentman describes as 'a literary language, combining elements of older Scottish diction, contemporary colloquial Scottish of various dialects, spoken sophisticated Scottish, spoken English, and literary English'.[86] This view allows both Bentman and McGuirk to demonstrate the ease with which Burns assimilated Augustan modes into his verse, and to see him as continuing a process of synthesising of Scottish and English language and modes which had been initiated by earlier writers of the vernacular revival.

Ramsay had been just such a synthesizer, and his defence of his linguistic strategy offers us a way of describing Burns's work which provocatively realigns it. Though Ramsay usually presented his original poems and anthologies in terms of simple Scottish patriotism, he indicated that his language could be seen as 'British'. In the Preface to his 1721 *Poems* he writes of 'the *Scots* and *English* tongue'—not 'tongues'—as if describing one language. His reference to 'beautiful Thoughts dress'd in *British*' reinforces the idea that, for Ramsay, 'British' seems to describe the mixed speech which is the language of Scotland.[87] Though speculating that some languages may be deficient for poetry, suffering from rough sounds and scantiness of vocabulary, he emphasizes that

These are no Defects in our's, the Pronunciation is liquid and sonorous, and much fuller than the *English*, of which we are Masters, by being taught it in our Schools, and daily reading it; which being added to all our native Words, of eminent Significancy, makes our Tongue by far the completest: for Instance, I can say, *an empty House, a toom Barrel, a boss Head*, and *a Hollow Heart.*—Many such Examples may be given, but let this one suffice.[88]

Ramsay goes on to write that, in some of the poems in his book, 'tho the Words be pure *English*, the Idiom or Phraseology is still *Scots*'.[89] Again he is describing a union of the two tongues, for which the best term is surely his own word 'British'. According to his lights, Adam Smith was right to claim that Ramsay did not write like a gentleman.[90] Where the Scottish teachers of Rhetoric and Belles Lettres sought to

[86] Ibid. 14–15.
[87] Allan Ramsay, *Poems* (2 vols.; Edinburgh: Thomas Ruddiman, 1721), i, p. vi.
[88] Ibid. p. vii.   [89] Ibid.
[90] Anon., 'Anecdotes . . . of the Late Adam Smith' (1791), repr. in Smith, *Lectures*.

purify their language of Scotticisms and so achieve a language fit for the conduct of British affairs, Ramsay, too, was edging towards a British tongue, but one which, following the political Union of 1707, united Scots and English. It is specifically because he had a knowledge of Scots that Ramsay could write British rather than merely English language. The same can be said of Burns, whose linguistic spectrum extends more widely than either Scots or English, taking in both. Burns often writes as if the political Union of 1707 has affected a linguistic union, giving him an enlarged territory in which to operate. Linguistically, he is the most brilliantly distinguished eighteenth-century example of a British poet.

Perhaps the 'Scotch bard' would have disliked this description, though, like Smollett, he could sign himself feelingly 'A BRITON' (*Letters*, i. 335), and, like James Thomson, he could extol the liberty of the 'independant British mind' (*Letters*, ii. 209). More characteristically, he might question the fairness of a political union between Scotland and England 'which should ever after make them one people' (*Letters*, i. 373), but has resulted in inequalities. In this, Burns followed earlier British Scottish writers such as Thomson and Smollett. He also lamented the dominating effect of purely English taste in literature.

I had often read & admired the Spectator, Adventurer, Rambler, & World, but still with a certain regret that they were so thoroughly & entirely English. —Alas! have I often said to myself, what are all the boasted advantages which my Country reaps from a certain Union, that can counterbalance the annihilation of her Independance, & even her very Name! (*Letters*, ii. 23–4.)

Yet if we say that Burns is a British poet, in the sense that he fully utilized the spectrum of British language, then it is clear that he did this precisely because he was a Scottish writer. His deployment of a mixture of Scots and English is fully consonant with the characteristic delight of his imagination in combining high and low, little and large. It represents a mingling of the low, dominated Scots language with the high, dominant language of 'proper English'. Burns saw the world in terms of 'intermingledoms of the good & the bad, the pure & impure' (*Letters*, ii. 143). Taken as a whole, his language is an 'intermingledom' of the language considered pure and that considered impure. It is a *Variorumsprach* whose strength is in its magnificent mixedness. Child of that Scottish eclectic age which produced the *Encyclopaedia Britannica* (whose founders he knew) and the *Statistical*

Account of Scotland (to which he contributed), Burns presents a different eclectic achievement in his verse, a *Lingua Britannica* which, in its constantly shifting mix, illustrates how a truly 'British' language might have operated.

Burns's eclectic use of language complements his work as a collector of songs, acting as a cultural broker between the oral folk-culture and the cultivated reading public of Edinburgh, where his songs appeared in James Johnson's *Scots Musical Museum* and the more 'elegant', but less accurate, *Select Collection of Original Scotish Airs* edited by George Thomson, with whom Burns pleaded: 'if you are for *English* verses, there is, on my part, an end of the matter.—Whether in the simplicity of *the Ballad*, or the pathos of *the Song*, I can only hope to please myself in being allowed at least a sprinkling of our native tongue.' (*Letters*, ii. 149.) This suggests that the poet sees the songs as a blend of English and Scots. His 'sprinkling' of Scots in his songs is small but crucial, preventing the material from becoming 'pure English'. It is, if you like, the Britishness of the songs' language which makes them Scottish.

While individual poems range from dense Scots to 'English English', the case for Burns's linguistic Britishness rests on the language spectrum of his *œuvre*. Yet, unlike Fergusson, Burns gained some of his characteristic effects by mixing English and Scots in the one poem, nowhere more energetically than in the splendidly British achievement of 'Tam o' Shanter', whose roots go deep into the union of English and Scottish cultures which formed Burns's childhood. On the one hand, 'Tam o' Shanter', with its dangerous crossing of a bridge between hellish excitement and domestic safety, recalls the central image of Addison's 'Vision of Mirzah', which sees human life as a bridge.[91] Burns recalled this as 'The earliest thing of Composition that I recollect taking pleasure in', and juxtaposed this 'Composition' from the world in which he 'made an excellent English scholar' with the world of Scots folklore which so clearly fuels the energetic listings of 'Tam o' Shanter':

In my infant and boyish days too, I owed much to an old Maid of my Mother's, remarkable for her ignorance, credulity and superstition.—She

---

[91] Addison's 'Vision of Mirzah' comprises issue 159 of the *Spectator* (1 Sept. 1711); it is tempting to believe that Burns was also familiar with the versified version of this piece which exaggerates the supernatural and dance motifs as well as the moralizing on vanity. This anonymous poem, *The Vision of Mirzah: Versified from the Spectator* (London: J. Payne, 1753), has been attributed to the (presumably Scottish) poet David McDuff in the copy in St Andrews University Library.

had, I suppose, the largest collection in the county of tales and songs concerning devils, ghosts, fairies, brownies, witches, warlocks, spunkies, kelpies, elf-candles, dead-lights, wraiths, apparitions, cantraips, giants, inchanted towers, dragons and other trumpery.—This cultivated the latent seeds of Poesy; but had so strong an effect on my imagination, that to this hour, in my nocturnal rambles, I sometimes keep a sharp look-out in suspicious places . . . (*Letters*, i. 135).

If 'Tam o' Shanter' (*PS*, 443) looks back to Burns's bicultural British education, the specific publication for which it was written also represented a demonstration of Britishness in its full sense. Captain Grose, for whose work Burns wrote the poem, had already published the *Antiquities of England and Wales* (1773–87), and was now at work collecting material for his *Antiquities of Scotland* before going on to begin an Irish volume. Burns found Grose (a collector like himself) personally likeable; moreover, Burns was contributing to the extension of a work which initially had covered only England and Wales into a work which would take account of Britain as a whole.[92] Both in its background and in itself, 'Tam o' Shanter' is both a Scottish and a British production.

When he sent the poem to Grose, Burns described it as being 'in Scots verse' (*Letters*, ii. 62). When we read the poem, it becomes clear that the linguistic spectrum is wider than this phrase suggests. For, if the bulk of the poem is grounded in the Scots language and culture suggested by the epigraph from Gavin Douglas, then, as has often been pointed out, the poem also contains lines such as 'Gathering her brows like gathering storm, | Nursing her wrath to keep it warm' (ll. 11–12), which exhibit no indication of being other than correct English. The longest and most obvious of these passages is the section beginning 'But pleasures are like poppies spread' (l. 59). Burns's embedding of such sections in the poem represents not only an extension of the linguistic palette to create verse which operates through a British union of Scottish and English diction, but also an enlivening of the poem's architecture by this broadening. For 'high' and 'low' forms of diction are allowed to play against one another, letting the poem decelerate into English and accelerate into Scots.

The poem is structured around a number of counterpoised elements. The comfort of the pub plays against the violence of the storm and the witches; the threatening hour when '*Tam* maun ride', 'o' night's black arch the key-stane' (ll. 69–70), is counterbalanced by the promised safety of the 'key-stane' (l. 206) of the bridge which Tam must reach to save himself; the amusingly threatening Kate is set against the threateningly funny wearer of the 'cutty sark' (l. 171) and her hellish companions. The setting of English against Scots is simply another of the poem's architectural juxtapositions. 'Tam o' Shanter' is a triumph of emotional and linguistic range, moving from drowsy contentment to the most violent excitement, and from the most formal English to the raciest Scots. Deploying this range of language, Burns's work highlights the linguistic diversity of Britain as it was present in the bifocal culture of Lowland Scotland.

No contemporary English poet achieved anything similar, nor would he or she have wished to do so. Throughout his work Burns does something very different from Thomson, Boswell, and Smollett. If they can often be seen as allies of the teachers of Rhetoric and Belles Lettres, Burns, because of his fierce deployment of an energetically impure language, may be seen as opposing them. It is startling to realize that, in his very deployment of that language, Burns, just as much as Thomson, Boswell, and Smollett, demonstrates that it is because he is a Scottish writer that he exemplifies the development of a fully British Literature. It is for that reason, paradoxical as it may seem, that Burns must be considered as one of the major figures in the Scottish invention of British Literature.

Yet at the same time it must be acknowledged that Burns was profoundly uneasy with Britishness. Though Ramsay's linguistic term 'British' may be used to describe Burns's poetic language, that language was manifestly not the language which the teachers of Rhetoric and Belles Lettres wished for North Britain, and 'British' is not the term Burns uses to describe it. Burns may sign himself 'A BRITON', but he could also denounce the Scottish signatories to the Act of Union as 'a parcel of rogues in a nation' (*PS*, 511). In 'The Author's Earnest Cry and Prayer, to . . . the Scotch Representatives in the House of Commons' Burns might express hostility towards Boswell's 'gab' (*PS*, 151), but in both prose and verse Burns, like Boswell, liked to play a wide variety of the roles made available to Scots by Britishness. There is a British aspect to Burns's work, but it is a potentially explosive Britishness, because the poet is often hovering near the touch-paper of Scottish republican nationalism. Burns is popularly regarded as a nationalist poet, yet, whatever sentiments he may have entertained privately at times, he never quite made that touch-paper blaze. The great Scottish writer who followed

him, who was, like Burns, a skilful negotiator with Scottish Enlightenment culture, developed and reconstructed the Scottish invention of
British Literature in a safer (though still manifestly Scottish) political
image, an image which successfully established Walter Scott as the
greatest novelist of Britishness.

3
Anthropology and Dialect

SCOTT'S 'ANGLIFIED ERSE'

HERE REST
THE MORTAL REMAINS OF
ADAM FERGUSON, L.L.D.
PROFESSOR OF MORAL PHILOSOPHY IN THE UNIVERSITY OF EDINBURGH.
HE WAS BORN AT LOGIERAIT, IN THE COUNTY OF PERTH, ON THE
20TH JUNE, 1723,
AND DIED IN THIS CITY OF SAINT ANDREWS, ON THE 22D DAY OF
FEBRUARY, 1816.
UNSEDUCED BY THE TEMPTATIONS OF PLEASURE, POWER, OR AMBITION,
HE EMPLOYED THE INTERVAL BETWIXT HIS CHILDHOOD AND HIS GRAVE
WITH
UNOSTENTATIOUS AND STEADY PERSEVERANCE IN ACQUIRING AND IN
DIFFUSING KNOWLEDGE,
AND IN THE PRACTICE OF PUBLIC AND OF DOMESTIC VIRTUE.
TO HIS VENERATED MEMORY
THIS MONUMENT IS ERECTED BY HIS CHILDREN,
THAT THEY MAY RECORD HIS PIETY TO GOD AND BENEVOLENCE TO MAN,
AND COMMEMORATE THE
ELOQUENCE AND ENERGY, WITH WHICH HE INCULCATED THE PRECEPTS
OF MORALITY,
AND PREPARED THE YOUTHFUL MIND FOR VIRTUOUS ACTIONS.
BUT A MORE IMPERISHABLE MEMORIAL OF HIS GENIUS EXISTS IN HIS
PHILOSOPHICAL
AND HISTORICAL WORKS,
WHERE CLASSIC ELEGANCE, STRENGTH OF REASONING, AND CLEARNESS
OF DETAIL,
SECURED THE APPLAUSE OF THE AGE IN WHICH HE LIVED,
AND WILL LONG CONTINUE TO DESERVE THE GRATITUDE AND COMMAND
THE
ADMIRATION OF POSTERITY.

Few short texts are as redolent of the Scottish Enlightenment as this memorial tablet set in the perimeter wall of St Andrews Cathedral. In its emphases on industry, education, eloquence, and social virtue, the tribute to the author of the *Essay on the History of Civil Society* is at one with the spirit of eighteenth-century Scottish neo-classicism. The author of the inscription was the writer who, two years before, had published his first novel, *Waverley*.[1]

To realize that the Ferguson inscription and *Waverley* are almost contemporary is graphically to perceive how far Scott's activities are grounded in the Scottish eighteenth century, particularly in its philosophical and historical considerations of the progress of society, the best-known of which is Ferguson's *Essay*. Duncan Forbes has traced Scott's 'overriding interest . . . in the study of social man' to a paper on the origins of the feudal system which Scott produced when he was 19, and points out that, though Scott seldom, if ever, wrote straight 'philosophical history', he never questioned its presuppositions. Adam Ferguson's son was Scott's close friend, and Scott, the venerator of Ferguson senior, 'whom I have known and looked up to for thirty years and upward', was to continue the Scottish Enlightenment's great gathering of information about primitive and developed societies for the purposes of cultural comparison.[2] In 1818 he wrote to Southey that

the history of colonies has in it some points of peculiar interest as illustrating human nature. On such occasions the extremes of civilized and savage life are suddenly and strongly brought into contact with each other and the results are as interesting to the moral observer as those which take place on the mixture of chemical substances are to the physical investigator.[3]

Though Duncan Forbes does not use the word, his scholarship is crucial because it alerts us to the *anthropological* dimension in Scott's thought. Forbes points out, for instance, that

The Antiquary, in whom there is something of Scott himself, says: 'To trace the connections of nations by their usages and the similarity of the implements which they employ has long been my favourite study', and he tries to interest his nephew in a fisherman's funeral and in 'the resemblances which I will point out betwixt popular customs on such occasions and those of the ancients'.[4]

[1] R. G. Cant, *The University of St Andrews*, rev. edn. (Edinburgh: Scottish Academic Press, 1970), 90.
[2] Duncan Forbes, 'The Rationalism of Sir Walter Scott', *Cambridge Journal*, Oct. 1953, 23; on Scott's relations with Ferguson, 'my learned and venerated friend', see ibid. 23–4.   [3] Quoted ibid. 33.   [4] Ibid. 29.

While the Antiquary represents a degree of self-mockery on the part of his creator, he not only recalls the eclectically minded investigators of the Scottish Enlightenment, but also, in his comparative method and manic eclecticism, looks forward to that line of Scottish nineteenth-century anthropologists which culminates in J. G. Frazer, and for which (as the present chapter will demonstrate) Scott's example was so important.

In beginning to examine this, one can suggestively align Scott with Burns. Far removed from 'the Scotch bard' in terms of class and native tongue, Scott, too, was bicultural. Though born into a genteel household, son of a mother who had learned 'correctness of speech and writing and something of history and belles-lettres', Scott's earliest literary delight was in the ballad material dismissed by Adam Smith as almost entirely 'rubbish'.[5] Inside his copy of Ramsay's *Tea-Table Miscellany* Scott wrote: 'This book belonged to my grandfather, Robert Scott, and out of it I was taught Hardiknute by heart before I could read the ballad myself. It was the first poem I ever learnt—the last I shall ever forget.'[6] Ramsay's anthology may have been Scots verse adapted to British tea-tables; none the less, it pointed Scott crucially towards vernacular Scots culture. In collecting his own *Minstrelsy of the Scottish Border*, he was furthering Ramsay's enterprise but also, like Burns, collecting the material of a folk tradition which was to prove richly seminal for his own work, and, in collecting it, constantly remaking it. Again, though he came from the 'high' side of the cultural fence, Scott the collector, like Burns, was crossing a boundary betwen the world of the vernacular and the world of the dominant Anglicized culture. He was moving between societies and kinds of language. What distinguishes the Scottish eclecticism of Ramsay, Burns, and Scott from the collecting work of English anthologists like Ritson is that, first, eclecticism was particularly intense in Enlightenment Scotland, and, secondly—and more importantly—in Scotland it was intimately bound up with the output of major creative writers. Ramsay, Burns, and Scott were as much major collectors as major creative artists. Writing in a culture under pressure, each sought to bind that culture together, to preserve it and celebrate it through anthology, which was closely bound up with creative endeavour. Sometimes, as in the case of the Ossianic works, collection

---

[5] Edgar Johnson, *Sir Walter Scott, The Great Unknown* (2 vols.; London: Hamish Hamilton, 1970), i. 5; Anon., 'Anecdotes . . . of the late Adam Smith', repr. in Adam Smith, *Lectures on Rhetoric and Belles Lettres*, ed. J. C. Bryce (Oxford: Clarendon Press, 1983), 230.   [6] Johnson, *Sir Walter Scott*, 11.

and creation become so confused as to be virtually inextricable, but the mixed urge to preserve and, at the same time, to build or develop a tradition is shared by Macpherson and Scott, along with other major Scottish eclectic writers, most notably Burns.

As a 5-year-old child, Scott had returned to Edinburgh with an English accent after a year in the south (where he met John Home of *Douglas* fame). He soon lost that accent and spoke 'in a tone and accent broadly Scotch'.[7] Transitions in language and accent would be of great importance to him as a novelist. So would transitions between the civilized and the uncivilized. One of the childhood books he remembered with particular enjoyment was *The Voyages, Dangerous Adventures, and Imminent Escapes of Captain Richard Falconer*. His aunt read him the story of the white man who had lived among the North American Indians, married an Indian wife, and then escaped.[8] In outline, this seems like a straightforward version of Lismahago's North American adventures in *Humphry Clinker*, and the teenage Scott was an admirer of Smollett. Later, writing about that other Scottish novelist and collector, he would call *The Expedition of Humphry Clinker* a 'delightful work' and 'the most pleasing of his compositions', noting in particular how the book (despite malicious criticism) was not compromised by its justified defence of Scottish mores against English opprobrium.[9] Scott's own first novel may be seen as a reworking of much that is most important in Smollett's last.

For *Waverley*, too, is a novel which takes the visitor from the south through different Scottish cultural territories, encouraging the reader to make comparisons and evaluations. It is, like *Humphry Clinker*, very much a novel about prejudice, and, like Smollett the admirer of Adam Smith's *Theory of Moral Sentiments*, Scott seeks to develop and use sympathy in order to break down the opprobrium and prejudice of both his better characters and of some of his readers. *Waverley*, like *Humphry Clinker*, is a novel of cultural investigation; it has a marked and seminal anthropological focus; it is also, as Andrew Hook and others have indicated, a book about the healing of internal divisions within Britain.[10] Its concern with Britishness is very much about

<hr>

[7] Ibid. 23, 27.   [8] Ibid. 24.

[9] Ibid. 54; Sir Walter Scott, *Biographical Memoirs of Eminent Novelists* (2 vols.; Edinburgh: Robert Cadell, 1834), i. 160–2.

[10] *W*, 26; because of the acuteness of its editorial material and its wide availability, references to the novel in this present text are keyed to Hook's edition, using the prefix *W*; for ease of reference, each page reference is preceded by the number of the appropriate chapter from which the quotation comes.

Britain as a full cultural amalgam, rather than about Britain as a synonym for England. Scott is not only the inheritor of the eighteenth-century Scottish tradition of philosophical history, as Duncan Forbes has argued; he is also the great focal point of the British concerns in eighteenth-century Scottish Literature.

Familiar from childhood with Ramsay's anthologizing, Scott was quickened by Bishop Percy's *Reliques of Ancient English Poetry*, which gave further literary sanctification to a love of ballads and ballad-collecting. The young Scott met and was much impressed by Burns; his own earliest poetry reads like an imitation of a Burns song:

> Lassie gin ye'll love me weel,
> Weel I'll love ye in return,
> While the salmon fills the creel
> While the flower grows by the burn.[11]

Like the work of Ramsay and Burns, this is in a 'British' diction—an English laced with, and made different by, a Scots infusion. Scott's roots in eighteenth-century Scottish letters went deep. An early mentor was Alexander Fraser Tytler, who had helped to see the first Edinburgh edition of Burns's poems through the press. In 1789–90, while attending Dugald Stewart's philosophy lectures at Edinburgh University, Scott also attended Tytler's lectures on universal history. Tytler's idea of history was a very broad one, corresponding to the widest modern sense of the word 'culture'. When, in 1801, he published a slightly expanded version of his 'Course of Lectures on General History, delivered for many years in the University of Edinburgh', he emphasized in the Preface that 'As the progress of the Human Mind forms a capital object in the Study of History, the State of the Arts and Science, the Religion, Laws, Government, and Manners of Nations, are material parts, even in an elementary work of this nature. The History of Literature is a most important article in this study . . .'.[12]

Tytler's wide, anthropological view of history, and his bonding of literature to it, anticipates Scott's work. Like other Scottish Enlightenment historiographers, he was eclectic in his comparative materials, referring to modern arguments that compare Spartans to 'rude' Red Indians, while cautioning the reader against leaping to inaccurate

---

[11] Johnson, *Sir Walter Scott*, 54, 60–2, 67.
[12] Alexander Fraser Tytler, *Elements of General History, Ancient and Modern*, 2nd edn. (2 vols.; Edinburgh: William Creech, 1803), i, pp. v–vi.

conclusions without knowledge of detailed facts. Tytler sought to avoid bigotry, and desired an anthropologically broad knowledge of history which 'extended to that of the whole species in every age and climate'. He referred both to the *Essay on the History of Civil Society* by Adam Ferguson and to the work of Montesquieu. He was interested in universal features of human nature as well as in 'such particulars as mark the genius and national character', and it was surely this which led him to take a particular interest in the culture of his own country.[13] Tytler would write books on Petrarch, and India, and France, yet he was also an editor and critic of Allan Ramsay as well as an encourager of Burns. His lectures on universal history point out that 'William Wallace [was] one of the greatest heroes whom history records', and celebrate his restoration of 'the fallen honours of his country'. Tytler's *General History* paid particular attention to Scotland, and it may well have been his remark that, during the fourteenth and fifteenth centuries, 'in no country in Europe had the feudal aristocracy attained to a greater height than in Scotland' which helped to motivate the young Scott to write the paper 'On the Origins of the Feudal System', delivered to a university literary society in 1789 and submitted as an essay to Dugald Stewart.[14]

Certainly, it was from his Enlightenment teachers that Scott inherited the desire for comparative scope in the investigation of manners. Where one of Burns's most admired correspondents, the novelist Dr John Moore, a friend of Smollett, had preceded his career in fiction by writing a *View of Society and Manners in France, Switzerland and Germany* (1779) and *A View of Manners and Society in Italy* (1781), so Scott was a cultural investigator before he was a novelist. His student paper 'On the Manners and Customs of the Northern Nations' drew on comparative folklore, while another student paper (sceptical in tone) 'On the Authenticity of Ossian's Poems' demonstrated that he was also eager to pursue cultural investigation closer to home.[15]

Tytler's version of Scottish history made clear the destabilizing impact of the 'secret faction' which the English sovereigns maintained within the Scottish nobility, and showed how, even in the time of James VI and I, the King's ideal of uniting his kingdoms north and south of the Border was one which 'the mutual prejudices of the two nations were as yet too violent to bear'.[16] Scott the lover of Border

---

[13] Ibid. 3, 43, 86–7, 174.

[14] Ibid. ii. 117–18, 152; Johnson, *Sir Walter Scott*, 77.

[15] Johnson, *Sir Walter Scott*, 77.

[16] Tytler, *Elements of General History*, ii. 162, 288.

lore was enthralled by that prejudicial violence, and, in writing on Ossian, he was interested in investigating the other 'primitive' native culture which his country had to offer him. In his own researches, as in Tytler's teaching, he found examples of the breadth and distinctiveness of the different cultural groups to be found within Scotland. Yet Tytler and the other Scottish Enlightenment teachers also taught him the British ideal. Outlining the constitutional basis of the United Kingdom, Tytler concluded with a conservative flourish: 'Such are briefly the outlines of the admirable fabric of the British constitution. *Esto perpetua!*'[17] Attached to the distinctiveness of Scottish culture yet fully in favour of the British identity, Tytler had adopted the common eighteenth-century position of the Scottish literati. It was a position which, with his own unique complications, Scott too would occupy.

Yet, in beginning a legal career, Scott entered an area of Scottish life most distinctly different from English culture; like his collecting of ballads and folklore material, the law gave Scott direct access to Scotland's cultural heritage. He complained of difficulties in balancing his literary and legal careers, but there were ways in which they were mutually reinforcing as he amassed materials for the three volume *Minstrelsy of the Scottish Border* (1802–3). This great editorial homage to Scottish tradition forms one of the bases of Scott's art. His own poem *The Lay of the Last Minstrel* (1805), set in the ballads' Border territory, deals with the mysterious potency of an ancient Scottish text. Anticipating Keats's device at the end of 'The Eve of St Agnes', Scott forcefully details past actions, and then abruptly jolts us back across the centuries into the present, making the past appear both vivid and remote. Scott's poem is carefully structured so that the potent, often angry voices of the characters are contained within the aged, lamenting voice of the Last Minstrel, whose words are themselves framed by an outer narrative voice which, like that of an editor, provides us with an Introduction, further historicizing the Minstrel's song and events within it. Deriving from his own work as a collector and editor, this technique would be redeployed in Scott's novels, where he would also make use of the immediacy of adventurous narrative at the same time as introducing a historicizing note which frequently takes the form of editorial and antiquarian footnotes. Such paratextual material lends an elegiac, distancing tinge to the adventure, a scholarly wordiness which at once fully explains and consciously distances the action. Scott's wish for both immediacy and

17 Ibid. 321.

distancing is present in the very subtitle of *Waverley; or, 'Tis Sixty Years Since*, the latter phrase being repeated throughout the course of the novel. An elegiac note often sounded in the novels is also found in the very title of *The Lay of the Last Minstrel*, a poem which constantly stresses the 'lastness' of the figure described Ossianically in its seventh line as 'last of all the Bards'. Scott looks to his ancestral Borders for many of the features which the creative collector Macpherson sought in his ancestral Highlands—past wildness, tragic feuding, a distinctive yet faded tradition capable of yielding a literary experience which would enrich Scottish culture. As Macpherson collaborated with the eighteenth-century literati to translate the Highlands through elegiac tones into a model that could be accepted by the British state whose imperial ends he would later serve, so Scott the devoted advocate of full Britishness slants *The Lay* (like *Waverley*) towards the celebration of a marriage between previously hostile northerner and southerner, converting Scott's source material about Border feuds into a poem about the ending of such intersocietal prejudice, a poem about the possibility of Britishness.

Scott's early work as a collector had an anthropological as well as a creative dimension. His *Minstrelsy* was much encouraged by the 'extraordinary' Scottish poet and polymath John Leyden, of whom the English book-collector Francis Douce wrote: 'Had he lived he would have been another Crichton.'[18] Editing the anti-English *Complaynt of Scotland* while he collected ballads for Scott's *Minstrelsy*, Leyden had already published translations from Greek, Norse, Arabic, Icelandic, and Persian, and had compiled his remarkable eclectic anthropological and historical work, *A Historical and Philosophical Sketch of the Discoveries and Settlements of the Europeans in Northern and Western Africa at the Close of the Eighteenth Century* (1799). Later, after emigrating to the Far East, he would write his *Comparative Vocabulary of the Barma, Malayu and T'hai Languages* (1810), though he is remembered as exclaiming: 'Learn English! . . . Never! It was trying to learn that language that spoiled my Scots.'[19] Patriot, and inspiration of the greatest Scottish word-collector, Sir James Murray of the *New English Dictionary*, Leyden was also, like Scott, a poet and recorder of Scottish manners. He edited *Scottish Descriptive Poems* in 1802, and, as cultural explorer, travelled in the Highlands at the beginning of the

---

[18] From an initialled manuscript note by Francis Douce, written inside the front cover of his copy of Leyden's 1801 edition of *The Complaynt of Scotland*, now in the Bodleian Library, Oxford.   [19] Quoted in Johnson, *Sir Walter Scott*, 167.

nineteenth century, investigating Ossianic material. Like Macpherson, Leyden fashioned his own creative work partly out of found materials and lore which he recast. Scott recalled, for instance, in a note to *The Lord of the Isles*, how 'The ballad entitled "Macphail of Colonsay, and the Mermaid of Corrievrekin" was composed by John Leyden, from a tradition which he found while making a tour through the Hebrides about 1801.'[20] *The Lord of the Isles* (1815) followed in Leyden's footsteps, paying tribute to his 'varied lore'—lore which, like Scott's own work, bridged poetry and anthropology.[21]

The *Minstrelsy*, for whose scope Leyden was partly responsible, is, then, an anthropological as well as a literary achievement. Its apparatus of detailed footnotes and commentaries, those potent *seuils* and *paratextes* whose importance Gerard Genette has emphasized, link literature with history in a way which would have delighted Tytler.[22] The effect of this apparatus is to present the ballads themselves as part of an investigation of Scottish primitive culture. Fascinating as primitive bards supplying information about a particular society, Scott's minstrels, like Macpherson's Ossian, were subjects for anthropological as well as literary research, as the general introduction makes clear.

Here, therefore, we have the history of early poetry in all nations. But it is evident that, though poetry seems a plant proper to almost all soils, yet not only is it of various kinds, according to the climate and country in which it has its origin, but the poetry of different nations differs still more widely in the degree of excellence which it attains. This must depend in some measure, no doubt, on the temper and manners of the people, or their proximity to those spirit-stirring events which are naturally selected as the subject of poetry, and on the more comprehensive or energetic character of the language spoken by the tribe. But the progress of the art is far more dependent upon the rise of some highly-gifted individual, possessing in a preeminent and uncommon degree the powers demanded, whose talents influence the taste of a whole nation, and entail on their posterity and language a character almost indelibly sacred. In this respect Homer stands alone and unrivalled, as a light from whose lamp the genius of successive ages, and of distant nations, has caught fire and illumination; and who, though the early poet of a rude age, has purchased for the era he has celebrated, so much reverence, that, not daring to bestow on it the term of barbarous, we distinguish it as the heroic period.[23]

---

[20] Sir Walter Scott, *Poetical Works* (12 vols.; Edinburgh: Robert Cadell, 1833), x. 151. [21] Ibid.

[22] Gerard Genette, *Seuils* (Paris: Seuil, Coll. Poétique, 1987).

[23] Sir Walter Scott, 'Introductory Remarks on Popular Poetry', prefaced to the *Minstrelsy of the Scottish Border*, in *Poetical Works*, i. 11–12.

One of the aims of the *Minstrelsy*, like the activities of Macpherson, was to show that the apparently barbarous could be viewed as the heroic, if seen through the proper lens. Poetry can function as a subject for anthropological investigation; the footnotes of the *Minstrelsy*, detailing strange lore, are one manifestation of this.

Apparently primitive literary material might also stimulate modern literature. Scott's activities as a collector (like those of Burns and Macpherson) shade imperceptibly from recording to altering to re-creating the found material. If eighteenth-century Scottish manners, historiography, philosophy, and science could rival those of the supposedly more cultivated English culture, there remained a wish for a more powerful imaginative literature. In 1794, in Dugald Stewart's drawing-room, Scott had been inspired by a reading of Bürger's German ballad 'Lenore' to translate Bürger for himself, investigating recent German interest in the ballad.[24] In the *Minstrelsy* he quotes an observation by Henry Mackenzie which spelt out why Scottish writers might be attracted to the model of German Literature, where the revival of a supposedly primitive form—the ballad—could help invigorate a modern literature.

'Germany,' he observed, 'in her literary aspect, presents herself to observation in a singular point of view; that of a country arrived at maturity, along with the neighbouring nations, in the arts and sciences, in the pleasures and refinements of manners, and yet only in its infancy with regard to writings of taste and imagination. This last path, however, from these very circumstances, she pursues with an enthusiasm which no other situation could perhaps have produced, the enthusiasm which novelty inspires, and which the servility incident to a more cultivated and critical state of literature does not restrain.'[25]

Scott's son-in-law, disciple, and biographer Lockhart (translator of Schlegel), and Thomas Carlyle would follow the gaze of Mackenzie and Scott. Wordsworth and Coleridge, at odds with metropolitan taste in *Lyrical Ballads*, would do likewise, but Scott recalled of his turn-of-the-century *Minstrelsy* that, at the time of its publication, 'The curiosity of the English was not much awakened by poems in the rude garb of antiquity, accompanied with notes referring to the obscure feuds of barbarous clans, of whose very name civilized history was

24 Johnson, *Sir Walter Scott*, 128.
25 Sir Walter Scott, 'Essay on Imitations of the Ancient Ballad', in *Poetical Works*, iv. 40.

ignorant.'[26] If English taste was not yet quite adapted to the *Minstrelsy*, Scott, educated in the Scottish eclectic tradition, where creative and anthological energies were usually closely bonded, was able to build quickly upon the *Minstrelsy* project and create his *Lay of the Last Minstrel*.

At this stage he acknowledged no similarities between his own anthologizing in the 'rude' Borders and the Highland project of the Macpherson whose authenticity he had questioned as a student. But Scott did make an imaginative link between the Borders and Highlands, those two supposedly primitive regions of Scottish feuding, for he wrote in November 1805 that he had no more plans for poetic projects, 'unless I should by some strange accident reside so long in the Highlands, and make myself master of their ancient manners, as to paint them with some degree of accuracy in a kind of *companion* to the *Minstrel Lay*'.[27] By this time, though, Scott had already begun, and then abandoned for the time being, his novel *Waverley*, where *The Lay*'s chronological and editorial devices and its elegiac tone would flourish. *Waverley* is a work which enacts a cultural exploration. Its genesis lies in the explorations of the *Minstrelsy* and *Lay*, as well as in a more immediate Ossianic stimulus.

If Leyden, after guiding two German boys round the north and west of Scotland in 1800, had returned claiming that the Ossianic poems were 'far from being a forgery of Macpherson's', his friend Scott remained much more sceptical, though he continued to be aware of the Ossianic legacy, and not least of the exploration of Highland culture which it stimulated.[28] Scott's admired Henry Mackenzie, dedicatee of *Waverley* and one of the founders of the Highland Society of Scotland, chaired a committee in 1804 which promoted wide-ranging investigation into 'the Nature and Authenticity of the Poems of Ossian'. In July 1805, at the very time Scott was beginning his first attempt at *Waverley*, the *Edinburgh Review* published a piece of his on the report of Mackenzie's committee and on Malcolm Laing's new edition, pointedly entitled *The Poems of Ossian, &c., containing the Poetical Works of James Macpherson, Esquire*. Written at a time when he was also discussing the Ossianic poems in his correspondence, Scott's long, considered essay reveals interest in, and sympathy with, Gaelic matters, as its author ventures into dangerous cultural territory between the 'Highland claymore' of the defender

---

[26] Quoted in Johnson, *Sir Walter Scott*, 211.
[27] Ibid. 242.   [28] Ibid. 177.

of Ossian and the 'Orcadian battle-axe' of the editor Laing (an Orkneyman).[29]

Scott regrets that the authenticity of the poem 'has been unnecessarily and improperly made a *shibboleth*, to distinguish the true Celt from his Saxon or Pictish neighbours'; he uses the word 'unnecessarily' because, for him, Celtic (Highland) culture is clearly distinguished from Saxon (English) or Pictish (Lowland Scottish) culture.[30] Yet he is fascinated by the passionate prejudices involved, quoting a minister called Gallie who assumes that, simply because he is an Orcadian, Laing will be anti-Highland, and finding even the normally 'unprejudiced . . . philosophical Dr Fergusson [*sic*]' liable to errors of judgement in a matter which relates to his own Highland upbringing. Modern scholars point out how frequently Scott refashioned the originals of his ballads, and it is just this that interests him about Macpherson, 'a scholar and a poet'.[31] Comparing an ancient original with Macpherson's version, Scott notes that 'In the original ballad there is no splendid scenery, no sentimental exclamation, no romantic effusions of tenderness or sensibility; it is a matter-of-fact statement . . .'[32] Precisely such 'splendid scenery' and romantic effusions would oscillate with a realistic presentation of the Highlands in *The Lady of the Lake* (1810). Scott greatly admires Macpherson's ability to draw on fact and lore to produce 'a web in which truth and falsehood should be warped and blended together in inseparable union', and so produce just the proper sort of recipe for 'a cake of the right leaven for the sentimental and refined critics, whom it was his object to fascinate' using material which 'few were displeased to recognize in a garb so different from its native and rude dress, as to interest the admirers of poetry through all Europe'.[33]

Describing Macpherson's achievement, Scott also sketches out a direction for his own work. Amusement mingles with admiration for the Fingal who is no barbarian or savage but 'has all the strength and bravery of Achilles, with the courtesy, sentiment, and high-breeding of Sir Charles Grandison'. Though Macpherson may be a fabricator, Scott is impressed with his manufacture of a compelling myth, proud that 'a remote, and almost barbarous corner of Scotland, produced in the 18th century, a bard, capable not only of making an enthusiastic

---

[29] Sir Walter Scott, Review of Henry Mackenzie, *Report of the Committee of the Highland Society*, and Malcolm Laing, *The Poems of Ossian, &c.*, Edinburgh Review, 6/2 (July 1805), 429.   [30] Ibid. 436.   [31] Ibid. 437.
[32] Ibid. 441.   [33] Ibid. 445.

impression on every mind susceptible of poetic beauty, but of giving a new tone to poetry throughout Europe.'[34] He wishes for more detailed investigation of Highland culture and its poetry, an investigation which will both be beautiful and reveal 'many curious circumstances of manners, and perhaps even of history'. Scott marvels that 'Many great chieftains retained their bards till within half a century' (*Waverley*'s original subtitle was *'Tis Fifty Years Since*).[35] Throughout this piece he broods on differences between the civilized and the savage, as well as on the 'warm feelings of enthusiasm, provoked by any investigation of Highland culture'.[36] An aside links Ossian to the recent saga of Celtic generations, *Castle Rackrent* by Maria Edgeworth, to whose Irish novels Scott would pay tribute in the 1829 General Preface to *Waverley* as not only celebrating the nature of the Irish, but also supporting an idea of Britain which took full account of Ireland as a part of that cultural amalgam. Edgeworth's work encouraged Scott with *Waverley*.[37] So did his consideration of Ossian.

Most striking is the way in which the concerns which prompted this review—concerns with investigations of Highland culture and their attendant prejudices, and with the successful marketing of an attractive Highland identity—still inform at the deepest level the novel which Scott began in the summer of 1805. That novel drew on, and developed, Macpherson's fabrication of Highland identity. It also made fiction out of the business of cultural exploration, taking to itself an anthropological dimension. This, surely the paradigmatic Scott novel, is arresting both in its own right and in its deployment of the techniques and pleasures of much modern writing. *Waverley*, like *Humphry Clinker* and Boswell's *Life of Johnson*, is a book of cultural investigation, a book about prejudice and the crossing of boundaries. The borders crossed are both cultural and linguistic. Edward Waverley, the young and naïve chivalric English hero, passes from his native country into Scotland, which is revealed as itself comprising several languages and cultures: English-speaking Scotland, Scots-speaking Scotland, and the Gaelic-speaking Highlands. Waverley picks up some of the traits of each of these societies, and both he and the reader are encouraged to compare them.

If *The Lay of the Last Minstrel* grew out of Scott's collecting work on *The Minstrelsy*, so in a sense did *Waverley*. For, particularly in its earlier chapters, the novel frequently appears to be an anthology,

---

[34] Ibid. 446, 462.   [35] Ibid. 460.   [36] Ibid. 453.
[37] Ibid. 457; *W*, 523.

presenting the reader in a short space of time with a poem by Edward Waverley (*W*, V. 60–1), parts of Spenser (*W*, IX. 80), old Lowland Scottish songs and ballads (*W*, IX. 82–3), French songs (*W*, XI. 96), and modern Scottish songs and poems (*W*, XIII. 112–13), not to mention a fresh 'Highland Minstrelsy' (*W*, XXII. 171–80). Within this anthological format, we have homages to the Scott family (*W*, XXIV. 191) as well as to the wider Scottish traditions which Scott, the admirer of Burns (*W*, XXVIII. 218), would invoke. No mere anthology of songs, *Waverley* also displays a delight in verbal eclecticism, in the simple collecting of words and documents. Whereas in *The Lay of the Last Minstrel* Scott had used the figure of the Minstrel periodically to distance his vivid account of the past and to heighten its vividness, so in *Waverley* it is the author himself who comes forward, apparently in his own voice, to remind us of the construction of the tale, and to recall that all this vivid action took place 'sixty years since' (*W*, XXXVI. 268; LIV. 377; LXVI. 453; LXVII. 461). A good example of this comes just after Waverley has first become infatuated with Flora Mac-Ivor, and has lain down to dream of her. Chapter XXIV opens with authorial deliberations which remind us that we are dealing with an eclectically structured text, one pieced together out of various documents by an author who is himself an ardent collector of words. An extended quotation best proves this point.

It is true that the annals and documents in my hands say but little of this Highland chase; but then I can find copious materials for description elsewhere. There is old Lindsay of Pitscottie ready at my elbow, with his Athole hunting, and his 'lofted and joisted palace of green timber; with all kind of drink to be had in burgh and land, as ale, beer, wine, muscadel, malvaise, hippocras, and aquavitae; with wheat-bread, main-bread, ginge-bread, beef, mutton, lamb, veal, venison, goose, grice, capon, coney, crane, swan, partridge, plover, duck, drake, brissel-cock, pawnies, black-cock, muir-fowl, and capercailzies;' not forgetting the 'costly bedding, vaiselle, and napry,' and least of all, the 'excelling stewards, cunning baxters, excellent cooks and pottingars, with confections and drugs for the desserts.' Besides the particulars which may be thence gleaned for this Highland feast (the splendour of which induced the Pope's legate to dissent from an opinion which he had hitherto held, that Scotland, namely, was the—the—latter end of the world) —besides these, might I not illuminate my pages with Taylor the Water Poet's hunting in the braes of Mar, where,

> Through heather, mosse, 'mong frogs, and bogs, and fogs,
> 'Mongst craggy cliffs and thunder-battered hills,

Hares, hinds, bucks, roes, are chased by men and dogs,
 Where two hours' hunting fourscore fat deer kills.
Lowland, your sports are low as is your seat;
 The Highland games and minds are high and great.

But without further tyranny over my readers, or display of the extent of my own reading, I shall content myself with borrowing a single incident from the memorable hunting at Lude, commemorated in the ingenious Mr Gunn's Essay on the Caledonian Harp, and so proceed in my story with all the brevity that my natural style of composition, partaking of what scholars call the periphrastic and ambagitory, and the vulgar the circumbendibus, will permit me. (*W*, XXIV. 186–7.)

Novelists had presented themselves as editors before, but the intrusive, self-mocking, and elaborately self-conscious tone of Scott's editorial voice is a new one. Here is an author who delights in the synthetic nature of his text, and wishes its eclecticism to be part of the reading experience. This author, the collector of the *Minstrelsy*, of Abbotsford's artefacts, and old Scots locutions, provides versions of himself not only in the Antiquary, but also in the Waverley whose early life is filled with 'much curious, though ill-arranged and miscellaneous information' (*W*, III. 48) and in the Bradwardine who is 'a scholar, according to the scholarship of Scotchmen, that is, his learning was more diffuse than accurate' (*W*, VI. 65–6). But while Scott gently mocks Bradwardine, whose 'language and habits were as heterogeneous as his external appearance' (*W*, X. 87), he is ultimately interested in winning the reader's sympathy for Bradwardine and the Lowland Scottish culture he represents. It is undeniable that, just as medieval Scottish authors like Dunbar, or the writer of *The Complaynt of Scotland*, or the translator of Rabelais, Sir Thomas Urquhart, loved a swollen, heterogeneous language which gave energy to their texts, so Scott revels in the Baron's heterogeneous, eclectic speech. Bradwardine is, like his creator, a delighter not only in ancient Scottish tradition, but also in the mixing of his language, the piecing together of quotations, pieces of information, and odd locutions in order to generate a form of expression remarkably synthetic and eclectic:

'Upon the honour of a gentleman,' he said, 'but it makes me young again to see you here, Mr Waverley! A worthy scion of the old stock of Waverley-Honour—*spes altera*, as Maro hath it—and you have the look of the old line, Captain Waverley, not so portly yet as my old friend Sir Everard—*mais cela viendra avec le tems*, as my Dutch acquaintance, Baron Kikkitbroeck, said of the *sagesse* of Madame son *épouse*.—And so ye have mounted the cockade? Right,

right; though I could have wished the colour different, and so I would ha'
deemed might Sir Everard. But no more of that; I am old, and times are
changed.—And how does the worthy knight baronet, and the fair Mrs
Rachel?—Ah, ye laugh, young man! In troth she was the fair Mrs Rachel in
the year of grace seventeen hundred and sixteen; but time passes—*et singula
praedantur anni*—that is most certain. But once again, ye are most heartily
welcome to my poor house of Tully-Veolan!—Hie to the house, Rose, and
see that Alexander Saunderson looks out the old Chateau Margaux, which
I sent from Bourdeaux to Dundee in the year 1713.' (*W*, x. 88.)

This speech brings the reader through English, Latin, French, and a
note of Scots; through references to England, Holland, Scotland, and
France; tabs and tags of information pierce the text, for the Baron,
like the author, is a lover of footnotes. Scott's deployment of footnotes
about language, lore, and the history of the societies with which he is
dealing gives his text an anthropological feeling, so that we read in a
suspension between creative writing and historiography. It is as if this
is a development from Hugh Blair's idea that the novel was a kind of
'fictitious history'.[38] Scott's texts came out of a culture where
information and its dissemination were particularly treasured. His
novels, full of their lists, catalogues, and investigations of social
mores, presuppose a readership with a taste for factual textures as
well as lyrical ones. They also presuppose a readership interested in
the crossing of cultural borders.

Those cultural borders are highlighted in *Waverley*, where we are
made aware that, to some Englishmen, Scotland is a state of 'utter
darkness' (*W*, VI. 68). Waverley is an innocent abroad, for whom
moving from England into Scotland is a move into 'a new world' (*W*,
VII. 72). But Scotland turns out to be not one new world but several.
Waverley's destination, Tully-Veolan, is peopled by Scots-speaking
inhabitants of a culture which (in a movement of thought reminiscent
of eighteenth-century Scotland) appears so alien and 'primitive' as to
resemble the Red Indian world of North America. The inhabitants
are 'almost in a primitive state of nakedness', with 'children . . .
whose skins were burnt black' beside 'a miserable wigwam' (*W*, VIII.
74, 76). Yet the apparently primitive nature of this culture is offset by
Bradwardine's eclectic learning, so that what gradually emerges to
the reader (whose likely prejudices are invoked before his powers of

---

[38] Hugh Blair, *Lectures on Rhetoric and Belles Lettres* (2 vols.; London: W. Strahan and
T. Cadell, 1783); Lecture 37, which dealt with novels, was entitled 'Fictitious
History'.

sympathy are worked on) is a culture unlike, but not unattractive in comparison with, the one which Waverley has left. Indeed, in its curiousness and promise of love and adventure, it may come to be more attractive than Waverley Honour.

The oddly heterogeneous Lowland diction epitomized by Bradwardine gives way to a different kind of language experience as Waverley crosses his next cultural border, going over the Highland Line into a territory whose Gaelic speech is often represented in Scott's text by an Ossianic sort of 'translatorese'. On his way north, Waverley has already heard not only Bradwardine's heterogeneous learned Scottish language, but also the ordinary Scots diction represented by an English energized with Scots:

'Out, ay. Did not ye hear him speak o' the Perth bailie? It cost that body five hundred merks ere he got to the south of Bally-Brough. And ance Donald played a pretty sport. There was to be a blythe bridal between the Lady Cramfeezer, in the howe o' the Mearns (she was the auld laird's widow, and no sae young as she had been hersell), and young Gilliewhackit, who had spent his heirship and moveables, like a gentleman, at cock-matches, bull-baiting, horse-races, and the like . . .' (*W*, XVIII. 151).

Now, as he moves towards the Highlands, Waverley hears this representation of Scots in collision with the differently represented diction of the Gael.

'That,' quoth Evan, 'is beyond all belief; and, indeed, to tell you the truth there durst not a Lowlander in all Scotland follow the fray a gun-shot beyond Bally-Brough, unless he had the help of the *Sidier Dhu*.'
'Whom do you call so?'
'The *Sidier Dhu*? the black soldier; that is what they call the independent companies that were raised to keep peace and law in the Highlands. Vich Ian Vohr commanded one of them for five years, and I was sergeant myself, I shall warrant ye. They call them *Sidier Dhu*, because they wear the tartans— as they call your men, King George's men, *Sidier Roy*, or red soldiers.' (*W*, XVIII. 150.)

As Waverley moves into the Highlands, this new sort of speech becomes more frequent. Reproductions of it can cover a range which Andrew Hook has characterized aptly as 'a kind of pidgin English, realistically trying to catch the struggle of the Gaelic speaker to use and pronounce a foreign tongue. At other times his Highlanders speak with a grave and splendid rhetoric, remarkably like that which

James Fenimore Cooper attributes to his American Indians.'[39] To
this we might add that, in the chapter 'Highland Minstrelsy', where
we actually enter an anthology of another culture, the Gaelic characters also speak not only in sometimes Ossianic tones of quasi-Ossianic
matters (about which their strong feelings are apparent), but also in
the accents of the anthropological guide, as if the folkloric footnotes
have become part of the direct speech of the text.

'The recitation' she said, 'of poems, recording the feats of heroes, the
complaints of lovers, and the wars of contending tribes, forms the chief
amusement of a winter fireside in the Highlands. Some of these are said to be
very ancient, and if they are ever translated into any of the languages of
civilized Europe, cannot fail to produce a deep and general sensation. Others
are more modern, the composition of those family bards whom the chieftains
of more distinguished name and power retain as the poets and historians of
their tribes. These, of course, possess various degrees of merit; but much of it
must evaporate in translation, or be lost on those who do not sympathise with
the feelings of the poet.' (*W*, XXII. 173.)

This speech, made to Waverley by the sister of a Highland chief, is
part of a larger dialogue about the difficulties of translating Gaelic
which is intended to help 'shoehorn' or transport both the hero and
the reader into an alien cultural milieu. For Waverley, ever quick on
the uptake as far as Gaelic goes, is not only the Englishman whom
Scott physically translates through the various zones of Scotland,
Scottish societies, and Scottish politics; he is also assigned the role of
both linguistic and cultural translator, becoming a cultural broker
who can interpret between the various communities of the novel.
This is seen linguistically when Waverley is introduced to his first
Highland chieftain:

'And did you ever see this Mr Mac-Ivor, if that be his name, Miss
Bradwardine?'
'No, that is not his name; and he would consider *master* as a sort of affront,
only that you are an Englishman, and know no better. But the Lowlanders
call him, like other gentlemen, by the name of his estate, Glennaquoich; and
the Highlanders call him Vich Ian Vohr, that is, the son of John the Great;
and we upon the braes here call him by both names indifferently.'
'I am afraid I shall never bring my English tongue to call him by either one
or other.' (*W*, XV. 128.)

Waverley initially expresses his alien, staunch English prejudice
and apparently determined inability to come to terms with the various

[39] *W*, 25.

foreign cultural locutions with which he is confronted. Soon, though, as his name suggests, he wavers, becoming a creative waverer between traditions and languages. He grows close to the wild Highlanders, learns Gaelic, and comes to unite three language cultures—English, Scots, and Gaelic—not to mention French and Italian. Waverley visits cave-dwellings, is sartorially translated into the kilt, sees peculiar Scottish wildlife and romantic scenery, and learns much of Highland lore. The reader learns with him. Like *Humphry Clinker*, *Waverley* is an educational book, particularly as regards the matter of Scotland and her stages of society, which the novel sets out with the detail of a Scottish Enlightenment philosophical historian.

So Waverley hears his own last minstrel—'one of the last harpers of the Western Highlands'—and enjoys the 'poetical language' of this land of the 'solitary' hill, with its various 'armed tribes' (*W*, XXII. 176–7; XXX. 236). He seems in a violent, yet not unattractive, zone beyond that England where 'everything was done according to an equal law' (*W*, XXVIII. 220). Circumstances involving the taking of prisoners and interrogation prompt Scott to talk of 'North American Indians . . . at the stake of torture', or 'crawling on all-fours with the dexterity of an Indian', or moving 'Indian file' (*W*, XXXI. 250; XXXVIII. 281). Waverley has to choose between the romantic Flora Mac-Ivor and the Lowland Rose, as he continually finds himself moving between speakers of Scots, of 'the English dialect', and that 'uncouth and unknown language' which Scott (a non-Gaelic speaker) often conveniently translates for us into Ossianic English (*W*, XLVI. 333).

Language differences are continually highlighted. The reader is encouraged to enjoy the text as a linguistic as well as a cultural amalgam. *In extremis*, this leads to the atmosphere of a babel when the mixture of speakers of Scots, English, Gaelic, and French makes for total confusion in Prince Charlie's Jacobite army. Yet even that gives the text a linguistically eclectic energy, counterpointing Bradwardine's conversation. The upset of the battlefield is a linguistic as well as a military one.

A hundred tongues were in motion at once. The Baron lectured, the Chieftain stormed, the Highlanders screamed in Gaelic, the horsemen cursed and swore in Lowland Scotch.

'Messieurs les sauvage Ecossois—dat is—gentilmans savages, have the goodnes d'arranger vous.' (*W*, LVIII. 398, 400.)

The linguistic mix here is comic as well as grim. As readers, we may smile, but we also regret the carnage visited upon cultures that we have come to know. Although the plurality of tongues is seen here as laughably disastrous, it normally makes for the entertainment of the reader, who comes to see the value of bringing together the different cultural traditions (ultimately in marriage or political union), while wishing to see them maintained. So the reader is eventually invited to laugh at, rather than share, the cultural imperialism of the English Colonel Talbot who exclaims of the wild Highlanders:

'Let them stay in their own barren mountains, and puff and swell, and hang their bonnets on the horns of the moon, if they have a mind: but what business have they to come where people wear breeches, and speak an intelligible language? I mean intelligible in comparison with their gibberish, for even the Lowlanders talk a kind of English little better than the negroes in Jamaica. I could pity the Pr——, I mean the Chevalier himself, for having so many desperadoes about him.' (*W*, LVI. 387.)

Scott's text has set out to make intelligible to the sympathetic reader what to Talbot is only a provincial babel. The un-English parts of the novel are strikingly assertive, giving it much of its distinctive flavour; *Waverley* is a Scottish and a British book. Partly about the need to be able to cross the boundaries of prejudice which divide the societies within Scotland, it is not advocating the erasure of these distinctive societies, but celebrating their diversity. *Waverley* matters most as a multicultural novel. If it has a 'target', then that target is mainly the oppressive prejudice represented by Colonel Talbot. His speech condemning those whose barbaric speech he considers little better than that of 'the negroes in Jamaica' hints at the way in which the cultural and linguistic issues raised by the book are universal ones.

More specifically, this speech also shows the importance of name-changing. Not only do characters disguise themselves and adopt different languages; their names, too, alter according to circumstances. To the Hanoverians, Charlie is the Pretender, to the Jacobites, he is the Prince; neutrally, he is the Chevalier (*W*, LVI. 386). Scott's novel (his novels, in fact) is about the construction of a new, culturally eclectic unity—Great Britain—but it is also about the need to preserve the cultures within that unity; where Scott sees these threatened, his imagination quickens, as it does to 'the last Vich Ian Vohr' (*W*, LXIX. 472). The wild Highlanders seem to exist outside the British legal framework, and must be brought inside it, but Scott the professional lawyer sees all the tragedy involved. As Waverley learns from Vich Ian Vohr, who is about to be executed for high treason,

'This same law of high treason,' he continued, with astonishing firmness and composure, 'is one of the blessings, Edward, with which your free country has accommodated poor old Scotland: her own jurisprudence, as I have heard, was much milder. But I suppose one day or other—when there are no longer any wild Highlanders to benefit by its tender mercies—they will blot it from their records, as levelling them with a nation of cannibals. (*W*, LXIV. 474.)

The pathetic point here is that, though they may appear primitive in comparison with English or Scottish Lowland commercial society, the Highlanders, with their own distinctive culture, are far from being cannibals. Scott may often see their way of life as doomed to extinction: 'the last Vich Ian Vohr', like Ossian and the Last Minstrel, seems a cultural remnant. But the evolution which may help them to develop is to come through Britishness, through the building of a multicultural, fully British society, just as it is in such a future that Waverley, himself seen early in the book as 'the last of that race', may find hope (*W*, VI. 64).

The conclusion of *Waverley*, like the endings of so many Scott novels, celebrates union, not only a marital union but also a social and political one between Scotland and England. It might be argued that Scotland, in the person of Rose Bradwardine, has the subservient, woman's part in this eighteenth-century marriage, but in many ways she has already demonstrated herself to be stronger and more resourceful than her English husband. Not the least of Scott's novelistic gifts is his ability to show women of strong character, such as Jeanie Deans, whose journey southwards from Scotland in *The Heart of Midlothian* is in some ways the reverse of Waverley's. His first novel has moments both of resentment and powerful pathos, yet throughout it runs a gently mocking note. If the Highlanders are presented as intensely romantic, the wish for romantic charm is also mocked. Scott's use of humour, like Smollett's, is designed to engage the reader's sympathy, easing the crossing into an unfamiliar culture. Aimed in part at an audience to whom Scotland is alien and England the norm, *Waverley* also mocks that audience's potential inflexibility through the person of Colonel Talbot. Written with the anthropological-cum-conservationist urge to preserve 'some idea of the ancient manners of which I have witnessed the almost total extinction', the book is no mere lament, but a plea for recognition of the validity of the cultural differences within a Scotland which must now be seen as a loyal and important part of Great Britain (*W*, LXXII. 492–3). Anthropological and eclectic, designed, like *Humphry Clinker*, to defuse anti-Scottish prejudice, *Waverley* is the consummate British novel.

The Scott who turned down Edinburgh's Chair of Rhetoric and Belles Lettres brought to the fore in his text just the sort of Scottish quirks which the teachers of that subject had sought to purge.[40] But English predominates in Scott's text, and it may be that, just because English was becoming the dominant medium in Scotland, there was greater toleration for Scots and Gaelic, which now stood no chance of dominance. Scott continually celebrates the persistence of these native cultures. In *Redgauntlet*, another novel of cultural cross-over, borders, and name-changing, it is the Scots folk-culture shared by blind bard Wandering Willie and the hero, Darsie Latimer, which (in Chapter IX, itself another of Scott's anthologies) makes possible the secret communication between the two which is unintelligible to Darsie's captors. Ending with another celebration of union and Britishness, *Redgauntlet*, too, manifests a continuing loyalty to the variety of Scottish cultural traditions.

Repeatedly, Scott's novels celebrate a cultural amalgam, taking protagonists—whether Major Neville (under the name change 'Lovel') in *The Antiquary*, or Ivanhoe as the 'Disinherited Knight' in *Ivanhoe*—across cultural borders. With its Saxons and Normans, Christians and Jews, *Ivanhoe* is as much about a multicultural society as is *Old Mortality*, where the focus is on conflicting religious sects for whom a measure of tolerance is the only answer—other than disaster. Like the interest in cultural assemblage, the delight in linguistic eclecticism runs through Scott's *œuvre*. So, for instance, in *Ivanhoe* the Jewess Rebecca speaks a language so heavily biblical that its 'translatorese' parallels the Ossianic English of *Waverley*'s Highlanders.

'God,' said Rebecca, 'is the disposer of all. He can turn back the captivity of Judah, even by the weakest instrument. To execute his message the snail is as sure a messenger as the falcon. Seek out Isaac of York—here is that will pay for horse and man—let him have this scroll. I know not if it be of Heaven the spirit which inspires me, but most truly do I judge that I am not to die this

---

[40] See H. W. Meikle, 'The Chair of Rhetoric and Belles-Lettres in the University of Edinburgh', *University of Edinburgh Journal*, Autumn 1945, 95.

death, and that a champion will be raised up for me. Farewell! Life and death are in thy haste.'[41]

Scott's Introduction to *Ivanhoe* makes it clear, with anthropological care, that he wants the reader to be aware of linguistic differences, and that, as in *Waverley*, the prototype of his fiction, he is interested in the coming-together of distinct cultures.

Early reviewers' complaints that *Guy Mannering*, the cultural eclecticism of which ranges through Scotland, England, Holland, and even India, was 'too often written in a language unintelligible to all except the Scotch' were very much to the point.[42] For Scott was giving linguistic (and so cultural) difference a prominence which challenged the prejudices of (particularly English) reviewers who were used to a monodialectal and monolingual standard English text, rather than one which moved from standard English to what one reviewer of *The Antiquary* complained of as the 'dark dialect of Anglified Erse'.[43] Scott's compositional strategies challenged an audience used to thinking of itself as monocultural and monolingual. Though he also played his part in the northward movement of the Romantic imagination away from metropolitan London, this anthropological, linguistically daring multiculturalism was his greatest achievement. He sought a devolution of sensibility.

With his linguistic and cultural eclecticism went Scott's delight in self-conscious anthologizing, in manifestly piecing together a work from various documents and kinds of language. Moving among these, as much as following any particular adventure which the novel is 'about', is one of the great pleasures of his texts. Scott the magpie novelist, the obsessive anthropological collector who is at once being a fully British writer and, in so doing, developing his native Scottish, un-English tradition, produces work which is not only Romantic according to the conventional use of that term, but also proto-Modernist. The proto-Modernist aspect of his writing would be his great gift to American Literature. It also nourished Scottish Literature, most immediately in the work of Carlyle. Scott's impact on the currents of American writing which flowed into Modernism will be examined in the next chapter of this book. The rest of this chapter looks at his linguistic and other legacies to Carlyle and to a powerful line of Scottish nineteenth-century anthropological writing which also

[41] Sir Walter Scott, *Ivanhoe* (1819; repr. London: Collins, 1953), 392 (Ch. XXXVIII); see also 54–5 (Ch. I). [42] E. Johnson, *Sir Walter Scott*, 468.
[43] Ibid. 519.

fuelled the work of Modernist writers in developing their preoccupation with the crossing of cultural boundaries.

CARLYLE'S 'ALLGEMEINE-MID-LOTHIANISH'

While recent criticism has, rightly, seen Carlyle's work both as anticipating Modernism (particularly through Carlyle's American impact) and, at its most adventurous, as prefiguring the structuralist project of Barthes and others, this has only added to the awkwardness of Carlyle's place in Scottish writing.[44] It is easy to ignore his Scottishness. His most obvious debts are to Germany—to the universalism of Goethe and the stylistic eccentricities of Jean Paul Richter. His debts to English humanists such as Swift and Sterne have been stressed.[45] Looking at Carlyle in a Scottish context does not erase these moulding pressures, but it does supplement them, and helps to rewrite nineteenth-century Scottish literary history so as to make clear why the Scottish example is of use in an examination of Modernism.

Though it might have surprised Carlyle and his contemporaries, his place is with Scott. Where Scott was a novelist so strongly drawn to fact that he became, on various occasions, a historian and biographer, Carlyle was a historian and biographer utterly fascinated by the devices of fiction. Scott produced the historical novel; in the *History of the French Revolution* Carlyle produced the novelistic history. If one of the characteristic pleasures of *Waverley* is crossing linguistic and cultural barriers, so that an act of translation is performed on the reader, this is what Carlyle strives for in *Sartor Resartus*. Both writers looked obsessively to the previous century, a great age of historiography which had also produced an upheaval in society that demanded a cure through some new sort of synthesis. Scott looked to the British civil war of the 1745 Rebellion; Carlyle to the decisive modern European conflict of the French Revolution, which signalled dissatisfaction with an old order and the danger of so oppressing the poor as to

[44] See in particular the 'Carlyle' section of Steven Helmling, *The Esoteric Comedies of Carlyle, Newman, and Yeats* (Cambridge: Cambridge University Press, 1988); also Roderick Watson, 'Carlyle: The World as Text and the Text as Voice', in Douglas Gifford (ed.), *The History of Scottish Literature*, iii. *Nineteenth Century* (Aberdeen: Aberdeen University Press, 1988), 153–68. The ongoing research of Mr Ralph Jessop, Churchill College, Cambridge, is likely to help reveal the full extent of Carlyle's debt to the Scottish philosophical tradition.

[45] See e.g. G. B. Tennyson, *'Sartor' Called 'Resartus'* (Princeton, NJ: Princeton University Press, 1965), 51, 53, 279–81.

produce a fissure between the two nations of rich and poor which required a reconciling reunion. Different politically, both Carlyle and Scott developed an interest in medieval society and in modern autocracy in the person of Napoleon. Both writers were Scottish Borderers enriched as authors by family tradition and by contact with Goethe and German Romanticism. Each fuelled his own work by German translation, and became (like J. G. Lockhart) a premier interpreter of German Literature to the British public. For different reasons, each published his first major prose work anonymously, and was closely involved with the business of periodical publishing and reviewing. Carlyle's elaborate use of a spurious editor in *Sartor Resartus* appears a natural development of one of Scott's favourite devices. Each caricatured himself as an eccentric eclectic with a passion for history—an Antiquary, a Teufelsdröckh. Each took as a major theme the workings of history on society, viewing his subject with an anthropological eye that delighted in comparing different cultures. Each became the most famous Scottish writer of his generation. Given all these similarities, it is remarkable how seldom Scott and Carlyle are seen as linked—not least, perhaps, by an anxiety of influence.

For the young Carlyle, Scott was not a novelist but a poet, for whose work he showed less enthusiasm than he showed for that of Thomas Campbell.[46] Yet when Carlyle read *Waverley* in 1814, his response was perceptive and notably enthusiastic. He thought it 'the best novel that has been published these thirty years', admiring the anonymous book's '*Scottean* colouring', and singling out for praise the 'Cervantic vein of humour' in the Scots-speaking 'characters of Ebenezer Cru[i]kshank[s] mine host of the garter, the Reverend Mr. Gowkthrapple and Squire Bradwardian [*sic*]', all of whom are particularly marked out by their diction.[47] Scholarship has revealed that, even in early 1814 in the *Dumfries Courier*, Carlyle was cultivating an eccentric style, mingling neologism, allusion, and foreign phrasing with contorted syntax; he was ripe to respond to Bradwardine.[48] Scattered admiring references to *Guy Mannering* in 1815, *Rob Roy* in 1818, the 'Letters of Malachi Malagrowther' in 1826, *The Life of Napoleon Buonaparte* in 1827, Scott on German Literature in 1828, *Tales of a Grandfather* in 1829, *Old Mortality* in 1816, and *Kenilworth* in

---

[46] *The Collected Letters of Thomas and Jane Welsh Carlyle*, ed. C. R. Sanders, K. J. Fielding *et al.* (Durham: Duke University Press, 1970–), i. 11.
[47] Ibid. 29.   [48] Ibid. 8–9.

1821 demonstrate that, like his contemporaries, Carlyle was deeply impressed by Scott, the most obvious model for a young Scottish writer of his generation.[49] In 1818 Carlyle's friend Robert Mitchell advised him to follow Scott's example, combining a literary with a legal career.[50] Carlyle declined the law, but an early fable written to encourage his own struggles holds up as fit for emulation the 'confused' schoolboy who became 'Sir Walter Scott of the Universe'.[51]

In 1822 Carlyle followed Scott further. His first prose fiction, the short story 'Cruthers and Jonson', set in the Scottish Borders, hinges round a love-rivalry during the 1745 Jacobite Rebellion.[52] In its subject-matter, its use of an introductory narrator, rich landscape description, and humorous commentary on local customs, the story looks awkwardly towards Scott; but in the area of German translation, with his first major publications—versions of Goethe's *Wilhelm Meisters Lehrjahr* (1824) and *Wilhelm Meisters Wanderjahre* (1827)—Carlyle nimbly overtook his master. The Carlyle who found a 'new Heaven and new Earth' in German Literature was well aware that Scott had been there before him.[53] As he wrote in 1827:

Sir Walter Scott's first literary enterprise was a translation of *Götz von Berlichingen*: and if genius could be communicated like instruction, we might call this work of Goethe's the prime cause of *Marmion* and the *Lady of the Lake*, with all that has followed from the same creative hand. Truly, a grain of seed that has lighted in the right soil![54]

When Carlyle wrote this, Scott was still actively involved with German Literature. If the scarcely known Carlyle was writing on E. T. W. Hoffmann in 1827, then so, in 1828 (as Carlyle pointed out to Goethe), was the famous Sir Walter Scott. While Carlyle struggled as a literary apprentice, he observed Scott 'a hundred times' in Edinburgh yet never managed to meet him.[55] Planned assignations fell through, most notably when, in connection with some medals sent by Goethe, Carlyle wrote to the Scott to 'whom I in common with

---

[49] Ibid. 47, 121; iv. 51, 365; v. 17; Thomas Carlyle, *Reminiscences*, ed. James Anthony Froude (2 vols.; London: Longmans, Green, and Co., 1881), i. 177, 252.

[50] Fred Kaplan, *Thomas Carlyle: A Biography* (Cambridge: Cambridge University Press, 1983), 56.

[51] Thomas Carlyle, *Works* (30 vols.; London: Chapman and Hall, 1897-9), xxvi. 472.

[52] See Tennyson, *'Sartor' Called 'Resartus'*, 48-9; Carlyle, *Works*, xxx. 175.

[53] Carlyle, *Letters*, i. 268.   [54] Carlyle, *Works*, xxiii. 15.

[55] Carlyle, *Letters*, iv. 365, 375.

other millions owe so much', received no reply, and was clearly piqued.[56] In June 1828 Carlyle wrote to his brother: 'Walter Scott I did *not* see, because he was in London; nor hear of, perhaps because he was a busy or uncourteous man.'[57]

Privately, Carlyle had begun to question the eminence of Scott, a 'sufficient "hodman"', less concerned with metaphysical speculation than with public entertainment (Carlyle's 1827 attempt at a novel, *Wotton Reinfred*, was ruined by metaphysical speculation swamping the plot).[58] On his first (1832) visit to Craigenputtock, Emerson noted with surprise that Carlyle classed Scott with figures who were 'no heroes of his'.[59] This defiant dethroning of Scott was part of Carlyle's bid for literary power, and reached its peak in the 1838 essay 'Sir Walter Scott', where the young Carlyle's enthusiastic response to *Waverley* is replaced by the retrospective judgement that Scott's first novel was an event more important in 'British Bookselling' than in 'British Literature'.[60] No simple hatchet job, this essay sees Scott as the heir of John Knox ('little as he dreamed of debt in that quarter!'), emphasizing that 'No Scotchman of his time was more entirely Scottish that Walter Scott.' The effect is to distance Scott by reminding the reader that 'his time' is no longer the present.[61] But the odd angle of approach also has the effect of siting Scott as a precursor of Carlyle, for it is surely easier to see Carlyle as deeply indebted to Knox's Scottish Presbyterianism than it is to see the Anglican Scott in this light. Accenting Scott's datedness, the essay also establishes covert links with Carlyle: a Border childhood, the importance of translating Goethe, ambition. This essay is Carlyle's attempt to inscribe himself into Scottish tradition without being overshadowed by its most impressive recent representative. Carlyle's discussion of Scott's attitude to history, and of what we might call Scott's comparative method, is crucial here:

In fact, much of the interest of these Novels results from what may be called contrasts of costume. The phraseology, fashion of arms, of dress and life, belonging to one age, is brought suddenly with singular vividness before the eyes of another. A great effect this; yet by the very nature of it, an altogether temporary one. Consider, brethren, shall not we too one day be antiques, and

---

[56] Ibid. 354.   [57] Ibid. 382–3.

[58] Ibid. 271; for *Wotton Reinfred*, see *Last Words of Thomas Carlyle* (1892; repr. Farnborough: Gregg International, 1971), 1–148.

[59] Carlyle, *Letters*, vii. 262.   [60] Carlyle, *Works*, xxix. 60.

[61] Ibid. 44.

grow to have as quaint a costume as the rest? The stuffed Dandy, only give him *time*, will become one of the wonderfulest mummies. In antiquarian museums, only two centuries hence, the steeple-hat will hang on the peg next to Franks and Company's patent, antiquarians deciding which is uglier: and the Stulz swallowtail, one may hope, will seem as incredible as any garment that ever made ridiculous the respectable back of man. Not by slashed breeches, steeple-hats, buff-belts, or antiquated speech, can romance-heroes continue to interest us; but simply and solely, in the long-run, by being men. Buff-belts and all manner of jerkins and costumes are transitory; man alone is perennial.[62]

Deliberately or not, this passage clearly suggests that *Sartor Resartus* begins where the Waverley Novels leave off. If Scott is singularly alert to the 'contrasts of costume'—linguistic, sartorial, and behavioural—which differentiate societies, then so is the author of '*Die Kleider, ihr Werden und Wirken* (Clothes: Their Origin and Influence)'. Teufels-dröckh develops Scott's anthropological eye.[63] Carlyle's portrait of Scott as a provider of curios for 'antiquarians' and 'antiquarian museums' draws on Scott's own partial self-portrait as Jonathan Oldbuck in *The Antiquary* (1818), who lives surrounded by

book-shelves, greatly too limited in space for the number of volumes placed upon them, which were, therefore, drawn up in ranks of two or three files deep, while numberless others littered the floor and the tables, amid a chaos of maps, engravings, scraps of parchment, bundles of papers, pieces of old armour, swords, dirks, helmets, and Highland targets.[64]

This apartment parallels the inner sanctum of the Teufelsdröckh, given over to the territory of 'the Antiquary and Student of Modes' (*SR*, 36). His room is

full of books and tattered papers, and miscellaneous shreds of all conceivable substances, 'united in a common element of dust'. Books lay on tables, and below tables; here fluttered a sheet of manuscript, there a torn handkerchief, or nightcap hastily thrown aside; ink-bottles alternated with bread-crusts, coffee-pots, tobacco-boxes, Periodical Literature, and Blücher Boots. (*SR*, 18.)

The similarities between the worlds created by Scott and by Carlyle extend beyond this one instance. *Sartor Resartus*, like so many Scott novels, displays a revelling in arcane knowledge which is both in-dulged and mocked. Scott drew attention to the constructed nature of his text in *Waverley*. He continually relished editorial *seuils*, lists, and

---

[62] Ibid. 76–7.   [63] *SR*, 5.
[64] Sir Walter Scott, *The Antiquary* (1816; repr. London: Dent, 1923), 31 (Ch. III).

parodies of scholarship. *Rob Roy*, for instance, is buttressed by an advertisement, an Introduction with six appendices, and a postscript. Typical of Scott's games with diction is Blattergowl in *The Antiquary*, whose harangue mingling 'Latin forms of feudal grants . . . with the jargon of blazonry, and the yet more barbarous phraseology of the Teind Court of Scotland' is but one of 'three *strands* of the conversation, to speak the language of a ropework . . . again twined together into one indistinguishable string of confusion'.[65] Beside this can be set *Waverley*'s Culloden babel or Bradwardine, whose 'language and habits were as heterogeneous as his external appearance' (*W*, X. 87). Scott's play with diction strongly prefigures the un-English Teufelsdröckh's.

rich, idiomatic diction, picturesque allusions, fiery poetic emphasis, or quaint tricksy turns; all the graces and terrors of a wild Imagination, wedded to the clearest Intellect, alternate in beautiful vicissitude. Were it not that sheer sleeping and soporific passages; circumlocutions, repetitions, touches even of pure doting jargon, so often intervene! On the whole, Professor Teufelsdröckh is not a cultivated writer. (*SR*, 24.)

Like Bradwardine in his village, Teufelsdröckh from the village of Entephul ('Duckpond') is an obscure, barbarous provincial who excites laughter, yet also (eventually) respect for his 'reading and literature in most known tongues, from *Sanchoniathon* to *Dr Lingard*, from your Oriental *Shasters*, and *Talmuds*, and *Korans*, with Cassini's *Siamese Tables*, and Laplace's *Mécanique céleste*, down to *Robinson Crusoe* and the *Belfast Town and Country Almanack*' (*SR*, 23). German rather than Scottish, Teufelsdröckh, like Scott's characters, is used to question the language and mind-set of the dominant metropolitan monoculture. When Carlyle's work was attacked by his friend Sterling for its 'positively barbarous' style, Carlyle looked towards Scott as the first of a number of linguistic precursors of his own inventive, eclectic, non-Standard English diction, relishing an age 'with whole ragged battalions of Scott's-Novel Scotch, with Irish, German, French, and even Newspaper Cockney (when 'Literature' is little other than a Newspaper) storming in on us, and the whole structure of our Johnsonian English breaking up from its foundations,—revolution *there* as visible as anywhere else!'[66] Prefiguring Browning's diction and the committed eclecticism of modernist texts which draw freely on extremities of the linguistic spectrum, those sentiments also show

---

[65] Ibid. 174–5 (Ch. VII).   [66] Carlyle, *Letters*, viii. 135.

how Carlyle found Scott's language useful in escaping from the restriction of that correct English favoured by the teachers of Rhetoric and Belles Lettres. Though mooted as a candidate for the Edinburgh Chair of Rhetoric in 1834, Carlyle showed little interest.[67] That year he linked *Sartor* to

> my view . . . that now at last we have lived to see all manner of Poetics and Rhetorics and Sermonics, and one may say generally all manner of *Pulpits* for addressing mankind from, as good as broken and abolished: alas, yes; if you have any earnest meaning, which demands to be not only listened to, but *believed* and *done*, you cannot (at least I cannot) utter it *there*, but the sound sticks in my throat, as when a Solemnity were *felt* to have become a Mummery; and so one leaves the pasteboard coulisses, and three Unities, and Blair[']s Lectures, quite behind; and feels only that there is *nothing sacred*, then, but the *Speech of Man* to believing Men![68]

This pronouncement looks to Wordsworth's 'selection of language really used by men', but also to the Scottish efforts of Burns and Scott, which had escaped Blair's restrictions.[69] After the 'ragged battalions of Scott's-Novel Scotch' comes the odd diction of Carlyle the translator of German into what William Maginn satirically called 'the Allgemeine-Mid-Lothianish of Auld Reekie'.[70]

German Literature was uniquely important to Carlyle, offering him its attraction to universal history and its splendidly eccentric Jean Paul Richter, author of the 1788 *Teufels Papieren* or *Selection from the Papers of the Devil*, self-appointed 'Professor of his own History', questioner of the respect paid to clothes, and master of an endlessly digressive style, in whose 'boundless uproar' Carlyle delighted. Richter's language was so odd that 'a dictionary of his works has actually been in part published for the use of German readers'.[71] C. F. Harrold, Hill Shine, G. B. Tennyson, and other able Carlyle commentators have underlined the importance of Carlyle's German studies. Carlyle himself linked Richter with Sterne.[72] Yet it is signi-

---

[67] Ibid. vii. 124.   [68] Ibid. 265.

[69] William Wordsworth, 'Preface' to 1805 edn. of William Wordsworth and Samuel Taylor Coleridge, *Lyrical Ballads*, ed. Derek Roper, 2nd edn. (London: MacDonald and Evans, 1976), 21.

[70] 'William Maginn's Portrait of Carlyle, June 1833', repr. in Kerry McSweeney and Peter Sabor (eds.), *Sartor Resartus* (Oxford: Oxford University Press, 1987), 229.

[71] Carlyle, *Works*, xxvii. 120, 128, 103, 144; xxvi. 4.

[72] C. F. Harrold, *Carlyle and German Thought, 1819–1834* (New Haven, Conn.: Yale University Press, 1934); Hill Shine, *Carlyle's Early Reading to 1834* (Lexington, Ky.: University of Kentucky Press, 1953); Tennyson, *'Sartor' Called 'Resartus'*, *passim*, esp. 69–85; see also Carlyle, *Works*, xxvii. 151.

ficant that Carlyle came to this German material after reading Scott, whose continual relishing of 'the circumbendibus' and 'heterogeneous' relies, as in *The Antiquary*, on the old Scottish ideal of the virtuoso skilled in various areas of learning and educated in the tradition of generalism hymned by George Davie.[73]

Carlyle, who described himself in 1827 as an intellectual 'Jack of all Trades', was a product of this Scottish generalist tradition, a native eclectic who relished 'the contemplation of things in general' in his early correspondence, long before appointing Teufelsdröckh to a Chair in that subject.[74] His 1827 essay on 'The State of German Literature' sees Dugald Stewart as preparing the way for Kant, pointing out that 'the Saxon School corresponded with what might be called the Scotch'.[75] Carlyle described his own metaphysics as 'the oddest mixture of Scotch and German'.[76] Critics overemphasize Carlyle's Germany at the expense of Carlyle's Scotland. His German borrowing fitted well into a Scottish milieu. *Sartor*'s formal jokes acknowledge this. Not only is Teufelsdröckh (a name from Richter) a modification of the German for asafoetida, but G. B. Tennyson notes that 'Wilson has pointed out that Scots for asafoetida is *Deil's dirt*, giving thus a Scoto-German appropriateness to the word similar to Weissnichtwo, which reflects Sir Walter Scott's Kennaquhair in *The Monastery*.'[77] Subtly, Carlyle relishes such Scoto-German parallels in his text. The narrator uses the German word '*Orte*, or as the Scotch say, and we ought to say, *Airts*', and is presented as an Englishman, but the book is sprinkled with Scots puns (such as the title of the 'Duke of Windlestraw') and references—to the 'Presbyterian Witchfinder' of *Satan's Invisible World Displayed*, to the 'Scottish Brassmith' James Watt, to John Baliol, whose wife was buried in Sweetheart Abbey close to the Craigenputtock where Carlyle was writing *Sartor* (*SR*, 15, 48–9, 35, 96, 191). Carlyle not only has the editor describe Teufelsdröckh's work as 'like a Scottish haggis', he even has Teufelsdröckh himself suggest a Scottish origin for his writings, when he describes how it was on seeing a leather-worker's sign 'in the Scottish Town of Edinburgh' that he first 'conceived this work on Clothes' (*SR*, 233, 232).

---

[73] George Davie, *The Democratic Intellect* (Edinburgh: Edinburgh University Press, 1961). [74] Carlyle, *Letters*, iv. 225; ii. 34.

[75] Carlyle, *Works*, xxvi. 79, 68. [76] Carlyle, *Letters*, iv. 225.

[77] Tennyson, '*Sartor*' Called '*Resartus*', 220–1.

Carlyle's early translations had been censured both for Germanisms and for Scotticisms. In *Sartor Resartus* he turned his disadvantages as a provincial Scot into a stylistic coup, producing a text deliberately and purposefully barbarous. In giving the appearance of translating from a non-existent original, Carlyle was following a pattern already laid down by 'Ossian' Macpherson, by Scott the producer of 'translatorese' Gaelic, and by Fenimore Cooper (whose work Jane Carlyle also liked) the producer of 'translated' Indian dialects.[78] Writing covertly as a Scot, overtly as a German, Carlyle's language as 'Teufelsdröckh' is so deliberately and aggressively impure that it threatens the diction of the English 'editor'. 'Thus has not the Editor himself, working over Teufelsdröckh's German, lost much of his own English purity?' (*SR*, 234.) The book is in part an attack on standard English language and attitudes. Written by a Scottish writer living in the depths of Dumfries-shire (the Oxonian Froude would describe Craigenputtock as 'the dreariest spot in the British dominions'), yet one who, from there, is corresponding with Goethe and pondering much of the most advanced thought of his day, *Sartor Resartus* unites barbarism and sophistication to celebrate an invented philosopher who is, like his creator, both obscure provincial and man of international vision.[79] In its combination of the provincial and the international, as in other ways, *Sartor Resartus* heralds the Modernist enterprise.

Like the Carlyle who waited at Craigenputtock for letters from Goethe, *Sartor*'s editor waits for news of Teufelsdröckh. The text, which delights in exposing its own eclectic construction, depends for its existence on modern technologies of information—long-distance postal services, encyclopaedias, reviews, and journals. Presenting his book as a heavily footnoted assembly, Carlyle is following not just Scott but also Boswell. Teufelsdröckh's biographer hangs 'on the Professor with the fondness of a Boswell for his Johnson' (*SR*, 20). Just after the original short draft of 'Teufelsdreck' in 1830 and the composition of the book-length *Sartor Resartus*, Carlyle discussed Boswell's *Life of Johnson* at length, quoting from the Teufelsdröckhian writings of a supposed 'Professor Gottfried Sauerteig', who writes in a discussion of 'Fictitious Biographies' about the 'Invention of

[78] Carlyle, *Letters*, vi. 342.
[79] James Anthony Froude, *Thomas Carlyle: A History of the First Forty Years of His Life* (1882; 2 vols., repr. London: Longmans, Green, and Co., 1908), ii. 24.

Reality'—of the symbols which seem most real to an age.[80] This was very much the theme of the fictitious biography of Teufelsdröckh. In one way Carlyle was following in the footsteps of the Boswell who 'unconsciously works together for us a whole *Johnsoniad* . . . a kind of Heroic Poem. The fit *Odyssey* of our unheroic age was to be written, not sung; of a Thinker, not of a Fighter; and (for want of a Homer) by the first open soul that might offer,—looked such even through the organs of a Boswell.'[81] In *Sartor* Carlyle, too, wished to write his age's epic through biography. His writing on Boswell's *Life of Johnson* alerts us to links between that work and his own, not the least of which is the sense of an author who has to cross boundaries of prejudice and cultural difference in order to deal with his subject. 'Consider what an inward impulse there must have been, how many mountains of impediment hurled aside, before the Scottish Laird could, as humble servant, embrace the knees (the bosom was not permitted him) of the English Dominie!'[82] The editor of *Sartor* is also a figure who has to cross or remove 'mountains of impediment' to become attuned to his subject. To achieve and dramatize this crossing, Carlyle redeploys the Scottean device of linguistic and cultural translation. Aware of Scott and Boswell, while Carlyle ostensibly connects himself to the German intellectual tradition, he also inscribes himself deeply in the Scottish.

The French Revolution takes its place in Scottish literary tradition not simply because its writing was a determined slog that only Carlyle's Calvinist determination brought to completion ('Pray that I may quit me like a true Scottish man, in this matter'), nor because it is a Boswellian 'Epic' whose events constitute 'the Grand Poem of our Time', but because it is the successor to Scott's work.[83] Carlyle's 'Sir Walter Scott' suggests this when, after censuring Scott's antiquarian 'mummeries', Carlyle about-turns to associate Scott with his own recent endeavours:

Secondly, however, we may say, these Historical Novels have taught all men this truth, which looks like a truism, and yet was as good as unknown to writers of history and others, till so taught: that the bygone ages of the world were actually filled by living men, not by protocols, state-papers, controversies and abstractions of men. Not abstractions were they, not diagrams and theorems; but men, in buff or other coats and breeches, with colour in their cheeks, with passions in their stomach, and the idioms, features and

---

[80] Carlyle, *Works*, xxviii. 49, 54.   [81] Ibid. 75.   [82] Ibid. 70.
[83] Carlyle, *Letters*, vii. 255.

vitalities of very men. It is a little word this; inclusive of great meaning!
History will henceforth have to take thought of it. Her faint hearsays of
'philosophy teaching by experience' will have to exchange themselves everywhere for direct inspection and embodiment: this, and this only, will be
counted experience; and till once experience have got in, philosophy will
reconcile herself to wait at the door. It is a great service, fertile in consequences, this that Scott has done . . .[84]

The most obvious literary consequences of Scott's imbuing of history
with the vividness of actual events was *The French Revolution*, published
by Carlyle one year before he wrote these words. In a footnote
Jerome Rosenberg writes of Scott: 'That Carlyle credits a novelist
with a major innovation in historiography suggests how fluid are the
boundaries separating history from fiction.' For the young Carlyle,
heir to the Scottish tradition of 'fictitious history' and interested in
the process of symbolism and interpretation leading to the 'Invention
of Reality', this was particularly so. Rosenberg also suggests that
Scott had drawn on accounts of Parisian revolutionary mobs for his
depiction of the Porteous Riots in *The Heart of Midlothian*, and that, as
'A keen student of Scott', Carlyle had 'in turn found hints in Scott's
novels for his account of the storming of the Bastille in *The French
Revolution*'. In failing to see Carlyle as Scott's successor, most critics
have problematized the reading of nineteenth-century Scottish
Literature.[85]

In one other significant area Carlyle inherited Scott's mantle—
periodical journalism. In Carlyle, as in Scott and other Scottish
predecessors, the creative and editorial functions were bonded. His
first literary employer was the polymath Sir David Brewster, editor of
the *Edinburgh Encyclopaedia*, for which Carlyle wrote several articles,
such as that on Montesquieu, which fuelled his own work.[86] 'As
Montesquieu wrote a *Spirit of Laws*', observes our Professor [Teufels-
dröckh], 'so could I write a *Spirit of Clothes*.' (*SR*, 27.) One of the
flowerings of Scottish eclecticism, the development of such encyclopaedias as the *Britannica*, the *Edinburgh*, and the reference works of
W. and R. Chambers, was nourished by most major nineteenth-
century Scottish writers, and, in turn, fed their imaginations. *Sartor*'s
editor mocks the first part of Teufelsdröckh's volume on 'Clothes
Philosophy' as 'much more likely to interest the Compilers of some
Library of General, Entertaining, Useful, or even Useless Knowledge

[84] Carlyle, *Works*, xxix. 76–7.
[85] John D. Rosenberg, *Carlyle and the Burden of History* (Oxford: Clarendon Press,
1985), 34.   [86] See Tennyson, *'Sartor' Called 'Resartus'*, 28, 333, 340–2.

than the miscellaneous readers of these pages' (*SR*, 28). *Sartor* itself
is a parody of such a '*Library*', authored by the encyclopaedically
minded Carlyle who may have mocked the 'MacVey-Napier Truth'
of mere fact but who admired the Supplement to the *Encyclopaedia
Britannica*, of which MacVey Napier (one of Carlyle's correspondents)
became the editor.[87] Writing to Goethe in 1830, the year *Sartor* was
begun, Carlyle remarked particularly on the growth of organs for
popularizing knowledge.

> To say nothing of our *Societies for the Diffusion of useful Knowledge*, with their
> sixpenny treatises, really very meritorious, we have I know not how many
> *Miscellanies, Family Libraries, Cabinet Cyclopedias* and so forth; and these not
> managed by any literary Gibeonites [menials] but sometimes by the best men
> we have: Sir Walter Scott, for instance, is publishing a History of Scotland by
> one of these vehicles . . .[88]

Carlyle himself had just been approached by G. R. Gleig, editor of
the Library of General Knowledge, and he was involved in Edinburgh encyclopaedism as well as becoming more closely immersed in
the world of the reviews and periodicals. He was an avid reader of
the *Edinburgh Review* and *Blackwood's Magazine*. Co-authored by the
German-translator Lockhart, the 1817 satire of the 'Chaldee Manuscript', an attack on the *Edinburgh Review* which was published in
Blackwood's and took the form of a pseudo-translation written in
archaic Old Testament 'translatorese', may be seen as another Scottish
ancestor of *Sartor*. Certainly, it excited Carlyle.[89]

   Carlyle the encyclopaedist discussed becoming a journal editor
several times (once with Brewster and Lockhart), and produced
a proposal for an 'Annual Register' which would have provided
'a compressed view of the actual progress of *mind* in its various
manifestations during the past year', with 'biographical portraits of
distinguished persons lately deceased' as well as 'essays, sketches,
miscellanies of various sorts, illustrating the existing state of literature, morals, manners'.[90] Though Carlyle never actually produced
such a register of things in general, he acquired a deep familiarity
with editorial demands and procedures, becoming not just a contributor but also an obsessive devourer of periodicals in this great age of
Scottish literary journalism. Written while he felt at the mercy of 'the

[87] Carlyle, *Letters*, v. 13; iii. 123.   [88] Ibid. v. 85.
[89] Ibid. 81–2; i. 114; Tennyson, '*Sartor*' Called '*Resartus*', 135.
[90] Carlyle, *Letters*, iv. 22, 167; Froude, *Thomas Carlyle*, i. 391.

uncertainty and unintelligible whimsicality of Review Editors',
Sartor's mockery of editor–author relationships surely draws on its
author's experience.[91] About Craigenputtock, Carlyle wrote to
Goethe in 1828, 'as to its solitude, a Mail Coach will any day
transport us to Edinburgh, which is our British Weimar[.] Nay even
at this time, I have a whole horse-load of French, German, American,
English Reviews and Journals, were they of any worth, encumbering
the tables of my little Library.'[92] Where the magazines juxtaposed
science with philosophy, poetry with criticism and comment, *Sartor*
follows. Teufelsdröckh produces an ' "extensive Volume" of boundless, almost formless contents' (*SR*, 6). In his 'paper bags' is material
as varied as that of a nineteenth-century journal or Annual Register:
'Then again, amidst what seems to be a Metaphysico-theological
Disquisition, "Detached Thoughts on the Steam-engine", or, "The
continued Possibility of Prophecy", we shall meet with some quite
private, not unimportant Biographical fact.' (*SR*, 61.) Such a pot-
pourri mixes Carlyle's enthusiasm for Richter with his Scottish
experience of encyclopaedias and journals, his rootedness in a native
eclecticism.

   Steven Helmling has pointed to the Carlyle whose Teufelsdröckh
discovers that 'ours is a fictile world; and man is the most fingent of
creatures'. Helmling sees that, though for Teufelsdröckh 'the very
Rocks and Rivers' of material reality are '*made*' by our senses, none
the less, 'Carlyle could not follow Teufelsdröckh so far as to dissolve
the explanatory category of "nature" entirely into "culture" '.[93]
Helmling does not consider how powerfully Carlyle's dealings with
his own culture and ancestry have a bearing on this. In 1834, writing
about his late father's 'inspiring example' in the *Reminiscences*, Carlyle
hopes the dead man will 'still live even here in me', and presents him
as having a head 'strikingly like that of the poet Goethe'. The un-
educated James Carlyle appears in important ways as both father and
literary father-figure, his speech 'full of metaphors' like that of the
hero of the recent *Sartor*. Hoping to 'act worthily of him' in the world
of literature and learning, Carlyle seems to wish to link his literary
work and his own development to his father, yet he continually senses
the gap between son and an orthodoxly religious parent who appears
to him '*Ultimus Romanorum*'. A sense of separation from, and of
longing for communion with, the spiritual and cultural milieu into

91 Carlyle, *Reminiscences*, ii. 161.   92 Carlyle, *Letters*, iv. 408.
93 Helmling, *Esoteric Comedies*, 27, 71.

which he was born pervades Carlyle's memories of his 'peasant father', whose example taught him all about 'Gold and the guinea stamp—the Man and the clothes of the man'. Carlyle's allusion here is to Burns's song 'For a' that and a' that', with its Teufelsdröckhian emphasis that clothes are irrelevant to human worth and that 'The rank is but the guinea's stamp, | The Man's the gowd for a' that'.[94] Carlyle connects his own father with the Burns of whom he wrote elsewhere: 'Poetry, as Burns could have followed it, is but another form of Wisdom, of Religion; is itself Wisdom and Religion.'[95]

This is a powerful misreading which sites Burns as the literary ancestor of the Carlyle who seeks, in *Sartor* and elsewhere, to see through old forms and vestures, and so to preserve an essential spirit. The search for truth in *Sartor* is an attempt to enact Carlyle's loyalty to his upbringing, to save the seemingly lost world of his father, who

> was among the last of the true men which Scotland on the old system produced or can produce . . . He was never visited with doubt . . . So quick is the motion of transition becoming, the new generation almost to a man must make their belly their God, and alas, find even that an empty one. Thus, curiously enough and blessedly, *he* stood a true man on the verge of the old, while his son stands here lovingly surveying him on the verge of the new, and sees the possibility of also being true there. God make the possibility, blessed possibility, into a reality.[96]

Carlyle's use of the word 'true' here balances precariously between the sense of 'according to the truth' and the sense of being 'loyal' to an earlier system of belief. Teufelsdröckh seeks truth, only to find that it relies on metaphors and, like all human ideas, is a kind of constructed fiction. Yet *Sartor*'s author, longing to be true to the religion of his father, instead of emphasizing the constructed, relative nature of all belief, writes in the language of the miraculous, re-channelling religion into a natural supernaturalism which resites and recites the splendours of God. The elect of his father's Calvinist theology are replaced by the 'Happy few' who are able to cross the strange bridge offered by *Sartor* and so be translated into the new metaphysical kingdom (*SR*, 214). Nietzsche was angry because that kingdom, however obliquely presented, was still put forward in terms of a kingdom of God.[97] But the use of such terminology, like so much of Carlyle's work, is the result of a wish to locate the truth for a

[94] Carlyle, *Reminiscences*, i. 6, 27, 8, 58, 15; *PS*, 602.
[95] Carlyle, *Works*, xxvi. 314. [96] Carlyle, *Reminiscences*, i. 8–9.
[97] See McSweeney and Sabor, *Sartor Resartus*, pp. xxviii, xxvii.

modern age while remaining true to the faith of his father which came out of the Scottish Reformation. The Carlyle whose thought appears to us so modern and international in its resonances wishes also to be located in a particular Scottish tradition.

For Carlyle, Mahomet, upsetter of 'Idols and Formulas', can be seen as a predecessor of 'the most Scottish of Scots', John Knox.[98] Seeing through vestures idolized by most men, Teufelsdröckh is a successor to Knox, and Carlyle, too, with his friend Edward Irving 'among the last products' of his father's old Seceder clergy, is a preacher. In 1815, at Edinburgh's Divinity Hall, Carlyle had delivered Latin 'exegesis' on the question '*Num detur religio naturalis*'.[99] Though he abandoned his training for the Christian ministry, the theme of that 'exegesis' became one of his mainstays and he remained a preacher. Considering the way in which 'Our pious fathers' revered churches and pulpits, Carlyle argues: 'The Writer of a Book, is not he a Preacher preaching not to this parish or that, on this day or that, but to all men in all times and places? . . . I many a time say, the writers of Newspapers, Pamphlets, Poems, Books, these *are* the real working effective Church of a modern country.' (*Heroes*, 159, 162.) In this way Carlyle reforms the Church of his upbringing into the literary world of his adulthood. Preaching to the Victorian audience, he is, and is not, true to his father.

Carlyle is the preacher of a reformation which continues to see religion as crucial, but this reformation is a translation of his father's religion into the new notion of 'hero-worship' which stands as a further factor linking the comparative and eclectic perspective of Scott to the later development of nineteenth-century Scottish writing on comparative religion and anthropology. Carlyle's recollection in the *Reminiscences* of 'the old excessive Edinburgh hero-worship' grounds this concept in his own background.[100] With its half-ironic reference to Boswell's *Life*, *Sartor* can be viewed as a slant piece of hero-worship. In the 1832 essay 'Boswell's Life of Johnson', written between the completion of *Sartor* and its preparation for serial publication, Carlyle saw Boswell as 'what the Scotch name *flunky*', but also a unique example of, and even a '*martyr*' to, 'that Loyalty, Discipleship, all that was ever meant by *Hero-worship*'.[101] Carlyle's Irvingite rhapsody for Boswell sites him as a Scottish precedent for Carlyle's later (1840) lectures *On Heroes and Hero-Worship*, where again, when

---

[98] *Heroes*, 56; Carlyle, *Works*, xxx. 319.   [99] Carlyle, *Reminiscences*, i. 83, 92.
[100] Ibid. ii. 41.   [101] Carlyle, *Works*, xxviii. 70, 74.

he wishes to demonstrate hero-worship's ineradicable spirit, he turns first to Boswell.

These lectures have an anthropological scope, aiming to link explicitly 'all religion hitherto known' (*Heroes*, 11). Like Frazer, Carlyle implies the inclusion of Christianity in his survey without explicitly discussing it: 'The greatest of all Heroes is One—whom we do not name here!' (*Heroes*, 11.) The effect is a tactful questioning of the conventional interpretation of Christianity by implicitly comparing it with a wide variety of other beliefs. Ruth apRoberts has perceptively remarked that, in these lectures, 'Carlyle has laid the groundwork for James Frazer', and she has demonstrated how an implicit parallel between Christ and Odin can be seen in the first lecture on 'The Hero as Divinity'.[102] In 'The Hero as Prophet' Carlyle subtly invites us to consider Christ, when he writes: 'We have chosen Mahomet not as the most eminent Prophet; but as the one we are freest to speak of.' (*Heroes*, 43.) In a strategy reminiscent of Carlyle's Scottish Presbyterian heritage, 'zeal against Idolatry' is used to link Mahomet with Luther, Knox, and all prophets (*Heroes*, 120); another unexpected contact with Scotland is made when Carlyle writes of 'that vein on . . . [Mahomet's] brow, which swelled-up black when he was in anger: like the *'horse-shoe* vein' in Scott's *Redgauntlet'* (*Heroes*, 53). His Scottish inheritance is still active in Carlyle's mind.

Strongly influenced by German thought, these lectures, with their comparative, anthropological dimension, also play their part in a Scottish cultural genealogy, recalling the ambitions of eighteenth-century Scottish thinkers not just in a stray reference to Adam Smith's *Essay on Language*, but in their vast ranging across the epochs—from primitive 'child-man' to the modern world of Napoleon (*Heroes*, 24, 7). *On Heroes and Hero-Worship* juxtaposes religions and societies—Scandinavian, Islamic, Christian, German, Scottish, even Confucian —in a text of remarkable comparative eclecticism, as it strives to find analogues between cultures while stressing the centrality of reformation, just as the Reformer Knox is seen as crucial to Scottish culture. Along with other aspects of the book, this Scottish-accented vast hunting for analogues looks toward Frazer's comparative method. The lectures' attention to Knox and Burns heightens that Scottish accent. When Carlyle admires Luther's 'rugged honesty,

---

[102] Ruth apRoberts, *The Ancient Dialect: Thomas Carlyle and Comparative Religion* (Berkeley, Calif.: University of California Press, 1988), 101, 73–86.

homeliness, simplicity', or the silent bravery of Burns's father, he admires the virtues he prized in his own father (*Heroes*, 139). Delivering these lectures, Carlyle simultaneously continues traditions of preaching, reformation, and valorizing of religion which he knew in his youth, yet he also employs a comparative perspective geared to the supplanting of his father's religion. This loyalty to, and erosion of, a father's Scottish religious values again anticipates Frazer. In these lectures Carlyle mapped out his own future career and one of the most important directions to be taken by nineteenth-century Scottish writing.

Though *Past and Present* (1843) shows he could follow Scott to the Middle Ages as well as employ a comparative perspective to juxtapose two societies, the lectures *On Heroes and Hero-Worship* mark a shift of Carlyle's energies towards biography. His career developed in the direction of fact, not fiction, despite a continuing rich interplay between the two. This interplay recalls the career of Scott, a novelist, biographer, essayist, and historian who liked to blur those generic distinctions. Tensions between the claims of the factual and the demands of art would be prominent in later anthropologically minded Scottish writers such as Andrew Lang, Stevenson, and Frazer, who, like Carlyle, treasured a Scottish cultural background while often describing themselves publicly as English. Underneath the surface of the Sage of Chelsea, confused chauvinisms and Scottish concerns bubbled. As editor of *Oliver Cromwell's Letters and Speeches* (1845), Carlyle was determined to attest that 'John Knox was the author, as it were, of Oliver Cromwell', and Carlyle could be as vehement as any eighteenth-century Scot in claiming Scotland's part in a full Britishness, as the near-racist patriotism of his rectorial address at Edinburgh University in 1866 demonstrates when he proclaims: 'I believe that the British nation—including in that the Scottish nation,—produced a finer set of men than you will find it possible to get anywhere else in the world [*Applause*].'[103]

Carlyle's Scottishness haunted him, but it had, as that metaphor suggests, a ghostly quality. 'My thoughts go often to Scotland, go daily thither; but all is becoming more and more *spectral* for me there. Ah me!', he wrote in a letter in 1840.[104] In the same year, delivering one of his lectures *On Heroes*, he spoke (apparently naturally) of 'we English' (*Heroes*, 224). In many ways Carlyle has come to be accepted

---

[103] Carlyle, *Works*, xxix. 457–8.   [104] Carlyle, *Letters*, xii. 75.

as part of the slippery entity 'English Literature', and he was himself capable of using national terms with a curious slipperiness. As early as 1828, in a formal correspondence with his German mentor Goethe, he was enthusing about Scottish Literature yet writing that 'We English, especially we Scotch, love Burns.'[105] Yet one only has to read the painful posthumous *Reminiscences* (1881) to see Carlyle's concentration on a private Scottish group that was enormously important to him. As part of the devolution of 'English Literature', we need to be more alert to the complex but insistent Scottish strains in Carlyle's writing that look back to Scott and forward to the later nineteenth century, particularly to those strains seen most clearly at different stages of his work and focused in *Sartor Resartus*: dialect and anthropology. If dialect and non-standard language were to receive a potent new life in the work of the Modernists, then so would anthropological thought, particularly the thought and writing of one of the main figures of late nineteenth-century Scottish culture, James George Frazer.

SCOTT'S ANTHROPOLOGISTS

Before considering Frazer's work in detail, it is worth indicating that the lines that lead eighteenth- and early nineteenth-century Scottish culture towards later nineteenth-century Scottish anthropology are manifold. A typical figure in whom such connections are clear is Carlyle's 'greatest favourite', the Ossian-admiring poet Allan Cunningham, Burns's acquaintance, devotee, and editor who, in 1808, had walked all the way from Dalswinton (near Dumfries) to Edinburgh, hoping to catch a glimpse of his other hero Scott.[106] Cunningham supplied Cromek, editor of the *Reliques of Burns*, with what Cromek later published as his own *Remains of Nithsdale and Galloway Song: Historical and Traditional Notices Relative to the Manners and Customs of the Peasantry* (1810), which, rich in folklore and anthropological materials, followed Scott's *Minstrelsy* in mixing editing with creative writing and in relating the poems to the manners of the society which produced them, detailing many Border customs. Scott's own *Letters on Demonology and Witchcraft* (1830) have been described as

[105] Ibid. iv. 407.
[106] Kaplan, *Carlyle: A Biography*, 94; David Hogg, *Life of Allan Cunningham* (Dumfries: John Anderson and Co., 1875), 238, 41, 31.

'the first full-length treatise in English on what would be called folk-lore', and Richard Dorson shows how Scott's work also inspired the folklore-collecting of Robert Chambers, encyclopaedist and editor of Burns, friend of Cunningham and Scott, who dedicated the second volume of his *Traditions of Edinburgh* (1824) to Scott. Again, the eclectic Hugh Miller, whose works included *Scenes and Legends of the North of Scotland* (1835), was stimulated by Scott, as were other Scottish folklorists detailed by Dorson.[107] Scott and his eighteenth-century teachers continued to remain a powerful presence. Dugald Stewart, for example, the teacher of James Mill, Jeffrey, and Scott, lived until 1835, and the standard edition of his works was being produced by Carlyle's friend Sir William Hamilton in the 1850s.

It is with a bow towards Dugald Stewart that J. F. McLennan ends his seminal anthropological work of 1865, *Primitive Marriage*.[108] Born in Inverness in 1827 and educated at Aberdeen and Cambridge, McLennan shared with Carlyle a 'moral feeling, which attached itself most of all to the necessity of truth and hard work'.[109] Like so many of the other Scottish anthropological writers, he had a strong taste for creative writing. Like Scott, he entered the Edinburgh legal profession, where awareness of Scottish cultural tradition was particularly strong, but he had earlier tried his hand at literary work, writing a collection of Pre-Raphaelite verse in the 1850s and being associated in London with W. M. Rossetti, G. H. Lewes, George Eliot, and the *Leader* at a time when that periodical declared 'Scottish metaphysics are not and never have been dead'.[110] McLennan's satirical piece 'Concerning Easy-Writing' (1866) contains a send-up of literary eclecticism, but its rather jealous tone suggests that its author might like to have been one of the Scottish literary 'men of genius like Alexander Smith, or George MacDonald'.[111] When he returned to Scotland in 1855, McLennan became a friend and collaborator of the poet Smith, but, despite another literary attempt in 1860, 'Some Recollections of an Old Street', which colourfully paints the 'poor

[107] Richard Dorson, *The British Folklorists: A History* (London: Routledge and Kegan Paul, 1968), 115, 123, 107–59.

[108] John F. McLennan, *Primitive Marriage* (Edinburgh: Adam and Charles Black, 1865), 291.

[109] Anon., 'J. F. M'Lennan', Obituary, the *Scotsman*, 20 June 1881, 4, repr. in J. F. McLennan, *Primitive Marriage*, ed. Peter Rivière (Chicago: University of Chicago Press, 1970), p. xiii. Unless otherwise indicated, other biographical details about McLennan are drawn from the Introduction to this edition, hereafter abbreviated as *PM*, ed. Rivière.   [110] *PM*, ed. Rivière, p. xxviii.

[111] J. F. McLennan, 'Concerning Easy-Writing', *Argosy*, 1, Apr. 1866, 328.

and ill-conditioned people' of the lower-class, small-town Scotland of McLennan's boyhood, it was not in Belles Lettres but in anthropology that McLennan was to acquire fame.[112]

His article on 'Law' for the 1857 *Encyclopaedia Britannica* builds on the eighteenth-century Scottish tradition in its echoing of Ferguson's title in the phrase 'the origins of civil society', and in its references to Adam Smith's *Theory of Moral Sentiments*. McLennan's observation that 'in some localities the old names for law and marriage are interchangeable' looks to his later anthropological work in *Primitive Marriage*.[113] Literary aspirations remain apparent in his 1861 'Marriage and Divorce: The Law of England and Scotland', which opens with mention of novels, including *Heart of Midlothian* and *Adam Bede*. Written from a comparative perspective, this piece is designed to counter English prejudices against apparently 'barbarous' Scottish customs.[114] In *Primitive Marriage*, hailed by an eminent modern anthropologist as 'the first really systematic attempt to make a comparative study of primitive societies', McLennan, who has been called the father of modern anthropology, opens his first chapter with a sentence about the 'history of civil society', and closes his last chapter with the bow to Dugald Stewart.[115] Citing Kames and drawing on the author's knowledge of Highland marriage customs around Inverness and Dingwall, *Primitive Marriage* obviously grows from Scottish roots as clearly as McLennan's work nurtures the work of his Scottish friend and Frazer's mentor, William Robertson Smith.[116]

Brought up discussing Carlyle, among many other subjects, the polymath Robertson Smith was, like Carlyle's brother John, David Masson, and Thomas Stevenson (father of the novelist), a member of the Edinburgh Evening Club, which Smith's biographers describe as 'the last embodiment of the corporate intellect of Edinburgh'. Smith

[112] 'Richard Futiloe, sen.', 'Some Recollections of an Old Street', *Macmillan's Magazine*, Apr. 1860, 433. *PM*, ed. Rivière, p. xlix, identifies McLennan as the author of this piece.

[113] 'J. F. M'L.', 'Law', in *Encyclopaedia Britannica*, 8th edn. (Edinburgh: Adam and Charles Black, 1857), xiii. 256, 264, 267.

[114] Anon., 'Marriage and Divorce: The Law of England and Scotland', *North British Review*, Aug. 1861, 187, 198. *PM*, ed. Rivière, p. xlix, identifies McLennan as the author of this piece.

[115] E. E. Evans-Pritchard, *The Comparative Method in Social Anthropology*, L. T. Hobhouse Memorial Lecture 33 (London: Athlone Press, 1963), 4; George W. Stocking, jun., *Victorian Anthropology* (London: Collier Macmillan, 1987), 297–8; McLennan, *Primitive Marriage*, 5, 291.

[116] McLennan, *Primitive Marriage*, 36–9, 106–7; *Victorian Anthropology*, 260, 287.

met McLennan in 1869, when the Evening Club was being set up and when McLennan's articles on 'The Worship of Animals and Plants' were appearing in the *Fortnightly Review*.[117] These essays further explored the theme of the 'Totem', about which McLennan had written in *Chambers's Encyclopaedia* in 1868.[118] By 1871, when McLennan had moved to London, he and Robertson Smith were corresponding about totemism. By 1874 Smith was writing a variety of articles for the ninth edition of *Encyclopaedia Britannica*, at the invitation of Spencer Baynes, Professor of Logic, Rhetoric, and Metaphysics at St Andrews, with whom he became the *Britannica*'s co-editor in 1881.[119]

A dedicated encyclopaedist, Robertson Smith also motivated an international encyclopaedia of Islam and the *Encyclopaedia Biblia*, though it was his own writings on biblical criticism for the *Britannica* which led to his expulsion from Scottish academia and his move to Trinity College, Cambridge, in 1883.[120] South of the Border, he continued to keep in touch with Scotland through visits, which he said 'always did for him what the earth did for Antaeus', and through correspondence. Smith's contact with McLennan had stimulated his own work on 'Animal Worship and Animal Tribes among the Arabs and in the Old Testament' (1880), and he championed his protégé J. G. Frazer as the person to write an article on 'Totemism' for the *Britannica*. Robertson Smith's own great work, *Kinship and Marriage in Early Arabia* (1885), built on McLennan's ideas, the 'general principles' of which Smith thought 'not likely to be shaken', and he continued to be loyal to his Scottish friend and mentor, whose posthumous *Studies in Ancient History* he reviewed for *Nature* in 1886.[121] Notorious in his own day for his reformist views on biblical criticism, and celebrated as a pioneering anthropological thinker, Smith should be seen not only as the Scottish mentor of Frazer (a role carefully detailed by Robert Ackerman), but also alongside McLennan, Andrew Lang, and others as important in nineteenth-century Scottish writing.[122]

[117] John Sutherland Black and George Chrystal, *The Life of William Robertson Smith* (London: Adam and Charles Black, 1912), 29, 67, 139–40, 116.

[118] See *PM*, ed. Rivière, 1.

[119] Black and Chrystal, *William Robertson Smith*, 143, 157, 453.

[120] Ibid. 628, 543; for a clear, short account of Smith's expulsion, see the entry on him in *DNB*.

[121] Black and Chrystal, *William Robertson Smith*, 530, 369, 494, 623.

[122] See Robert Ackerman, *J. G. Frazer: The Life and Work* (Cambridge: Cambridge University Press, 1987), 58–63; see also Robert Fraser, *The Making of 'The Golden Bough'* (Basingstoke: Macmillan, 1990).

Scott and McLennan come together in motivating the work of Andrew Lang, a Borderer who admired the novelist from childhood, began to study folktales as a St Andrews undergraduate, and was spurred to interest in 'savages' at the start of the 1870s by the essays on totemism by McLennan, whose works he described as 'revelations'. In 1880 Lang hoped to edit McLennan's notes, and throughout the 1880s he developed his own comparative anthropological studies in works such as *Myth, Ritual and Religion* (1887).[123] Like McLennan, Lang had a great love for creative writing, his anthropological work going hand in hand with poetry, fiction, editorial projects, and folktale studies. Something of a virtuoso, Lang contributed articles on a variety of subjects to the *Encyclopaedia Britannica*, under Robertson Smith's co-editorship.[124] Like Carlyle's linking of 'all religion hitherto known', Lang's project is hugely eclectic in its aspirations. His early examination of myth and folk-tale contended that 'there has existed an order whose primitive form of human life has been changeless, a class which has put on a mere semblance of new faiths, while half consciously retaining the remains of immemorial cults'.[125] His celebrated 1873 essay, 'Mythology and Fairy Tales', was later held to be 'the first full statement of the anthropological method applied to the comparative study of myths', and made it clear that Lang saw the origins of his own study in Scott: 'More than sixty years have passed since Scott, in a note to "The Lady of the Lake", first called attention in England to the scientific importance of these fictions . . .'.[126]

Homages to Scott resound throughout Lang's career and he continued to produce comparative anthropological works such as *Custom and Myth* (1884), *Modern Mythology* (1897), *The Making of Religion* (1898), *The Origins of Religion* (1908), and *Method in the Study of Totemism* (1911). Alongside these, he was working on Scott's texts, producing his compendiously annotated Border edition of the Waverley Novels (1892–4), as well as a biography of Scott in 1906 and a study of *Sir Walter Scott and the Border Minstrelsy* in 1910. Lang

---

[123] Roger Lancelyn Green, *Andrew Lang: A Critical Biography* (Leicester: Edmund Ward, 1946), 9, 23; Andrew Lang, 'Edward Burnett Tylor', in W. H. R. Rivers, R. R. Marett, and Northcote W. Thomas (eds.), *Anthropological Essays Presented to Edward Burnett Tylor in Honour of his 75th Birthday* (Oxford: Clarendon Press, 1907), 1; Andrew Lang, *Adventures among Books* (London: Longmans, Green, and Co., 1905), 36; *PM*, ed. Rivière, p. xxxv.   [124] Green, *Andrew Lang*, 65.

[125] Andrew Lang, '*Kalevala*; or, The Finnish National Epic', *Fraser's Magazine*, June 1872, 676.

[126] Salomon Reinach, in the *Quarterly Review*, Apr. 1913, quoted in Green, *Andrew Lang*, 70; Andrew Lang, 'Mythology and Fairy Tales', the *Fortnightly Review*, May 1873, 619.

shared his admiration of Scott with his friend R. L. Stevenson, whom he met in 1874 and whose work he championed, supplying Stevenson with information when the latter was in Samoa.[127] Lang would write that 'he [Stevenson] and I had a common forbear with Sir Walter Scott, and were hundredth cousins of each other'.[128] His friendship with Stevenson is one of the strongest indications of the way in which associated literary and anthropological enterprises emanated from Scott and flowered in Scottish culture later in the nineteenth century. But the figure who demonstrates this most strongly, and whose work connects this flowering with the development of literary Modernism, is not Lang, but his correspondent and fellow admirer of McLennan, James George Frazer, whose work cries out for a reading which places it in the context of nineteenth-century Scottish Literature.

The scholarship of Robert Ackerman, Robert Fraser, and Robert Alun Jones has done much to reinstate the importance of Scottish intellectual tradition in our view of Frazer.[129] Ackerman and Fraser draw particular attention to the part played by Scottish Enlightenment philosophy in Frazer's intellectual make-up, but both these commentators pass too easily over other formative experiences of Frazer's first twenty-one years in Scotland.

Like Carlyle, Frazer was a work-driven Scot who travelled south and then felt compelled to memorialize his Scottish parents and the upbringing which moulded him. Describing his father (whose library he inherited), Frazer stresses both religion and literature, which came together on Sundays when 'in the evening our father read to us a good or edifying book'. Most interesting of these from the point of view of Frazer's later career is *The Land of the Book* by William McLure Thomson. Theologically acceptable to the 'staunch Presbyterian' Daniel Frazer, Thomson was an American who had spent thirty years as a missionary in Syria and Palestine.[130] His 1859 book, subtitled *Biblical Illustrations Drawn from the Manners and Customs, the*

---

127 Green, *Andrew Lang*, 38–9, 145–6.   128 Ibid. 1.

129 Ackerman, *J. G. Frazer*; Fraser, *The Making of 'The Golden Bough'*; Robert Alun Jones, 'Robertson Smith and James Frazer on Religion', in George W. Stocking (ed.), *Functionalism Historicized: Essays on British Social Anthropology* (Madison, Wis.: University of Wisconsin Press, 1984), 31–58.

130 J. G. Frazer, 'Memories of My Parents', in *Creation and Evolution in Primitive Cosmogonies and Other Pieces* (London: Macmillan, 1935), 132–3; entry for 'Thomson, William M.', in S. A. Allibone, *A Critical Dictionary of English Literature* (3 vols.; Philadelphia: J. B. Lippincott, 1858–71). For directing me to the latter work, I am grateful to Anthony Esposito of the Biographies Research Unit, University of Glasgow.

Scenes and Scenery of The Holy Land, presents its subject geographically, as the author vividly transports his audience across the terrain of the Middle East, detailing numerous local customs and antiquities. Many, but not all of these, are related to scriptural passages, and Thomson likes to work into his text evidence drawn from a wide variety of earlier writers and travellers, ranging from Herodotus to Lane. The reader is often addressed directly, as if he were right beside Thomson moving through the landscape past old temples and traces of local curiosities, which sometimes involves considerable narrative digressions. The treatment of the subject-matter, and the guidebook technique may seem odd to the modern reader, but they look less peculiar when set beside that text which Frazer's mother clutched on her death-bed: her son's edition of Pausanias's *Guide to Greece*.

All through his career, Frazer delighted in presenting landscape set pieces and relating these to local practices and beliefs. Some of the most vivid scene-painting in the third edition of *The Golden Bough* comes in the two volumes of *Adonis Attis Osiris*, where Frazer evokes Middle Eastern scenes which he had never visited—though, thanks to the Bible and McLure Thomson, these landscapes had been familiar from his Scottish boyhood. The famous opening of *The Golden Bough*, with its description of the sacred grove at Nemi, is but the best known among hundreds of landscape passages. Some of these were even anthologized separately in collections like Frazer's *Studies in Greek Scenery, Legend and History* (1917). From his earliest childhood, Frazer was familiar with writing of this kind. It comes as no surprise to find that, fifty years after Frazer's father had first read it to him, William McLure Thomson's work was listed in the bibliography to the third edition of *The Golden Bough*.

Frazer knew also that his own land was invested with sacred sites, and that topography, lore, and landscape were bound together. Among his father's books, he recalled, '*Tales of the Convenanters* and Wilson's *Tales of the Borders* served to kindle or nurse the fires of piety and patriotism in my Scottish heart.'[131] Robert Pollok's Covenanting stories, such as *Heather of the Glen* and *The Banks of the Irvine*, presented tales of Claverhouse, religious persecutions, and violent, sectarian pursuits over a Scottish territory that had already been traversed by Scott in *Old Mortality*. John Wilson's *Tales of the Borders* also followed in Scott's footsteps in its anthologizing of traditional materials drawn

---

[131] Frazer, *Creation and Evolution*, 134.

from the geographical area covered by the *Minstrelsy of the Scottish Border*. Wilson, a Border printer, knew Scott. His *Tales*, collected in the 1830s, are not simply fictions; they are also folklore, bonded to a particular landscape.

Frazer, then, even as a very young child, was familiar with a number of works in the Scottish topographical/folklore tradition, but, more than anything else, he was familiar with the works of Scott. Frazer's 'Memories of My Parents' points out that Scott was a favourite author not just with Daniel Frazer, but also with his son, who later wrote about Daniel's library.

> Among his books was, I think, a complete set of Sir Walter Scott's works, including what is called the author's favourite edition of the Waverley Novels. When the first popular edition of the Waverley Novels was published in the sixties of last century my father presented the volumes to me as they appeared, and I read them from first to last with keen enjoyment.[132]

The edition of Scott which Frazer's father possessed (and which seems to have passed into Frazer's own library) was the Magnum Opus edition, with its full, antiquarian footnotes.[133]

To have read right through the Waverley Novels in one's late teens is an achievement no doubt much more common in mid-nineteenth-century Scotland than it is today, and Frazer was a bookish boy. For him, the particular attractions of Scott must have been several. First, Scott was a writer dealing with Frazer's own country, with a topography which Frazer loved, and with places which he knew. When Frazer first matriculated as a student at Glasgow University, it was still on its original site, he remembered, next to the 'recreation ground, which Scott has immortalized by making it the scene of the famous duel in *Rob Roy*'.[134] *Rob Roy* seems to have impressed Frazer, who once described Samson as 'a doughty highlander and borderer, a sort of Hebrew Rob Roy, whose choleric temper, dauntless courage, and prodigious bodily strength marked him out as the champion of Israel in many a wild foray across the border into the rich lowlands of Philistia'.[135] Thinking about *Rob Roy* suggests another, important

---

[132] Ibid.

[133] Information about Frazer's library comes from the catalogues among his papers in the Library of Trinity College, Cambridge. I am very grateful to Dr Alice Crawford of Dundee University for providing this information.

[134] Frazer, *Creation and Evolution*, 134.

[135] J. G. Frazer, *Folk-Lore in the Old Testament: Studies in Comparative Religion, Legend and Law* (3 vols.; London: Macmillan, 1918), ii. 481.

reason for Frazer's fondness for Scott: the novelist's juxtaposition of primitive and civilized societies, and his examination of tensions between them. Like *Waverley*, *Rob Roy* takes its young 'standard English' hero through a series of different societies and landscapes. The Highland clans and their chiefs are regarded by most of the Englishmen in the book as 'artful savages' (Chapter XXXII), with their 'savage, uncouth, yet martial figures' (Chapter XXXI). The Scottish Lowlanders see them as having violent customs 'that nae civilised body kens or cares onything about' (Chapter XXVI). At the same time, *Rob Roy* typifies Scott's presentation of Highland mountains, and landscape set pieces in general. As Francis Osbaldistone and Bailie Nicol Jarvie journey from Glasgow into the Highlands,

On the right, amid a profusion of thickets, knolls, and crags, lay the bed of a broad mountain lake, lightly curled into tiny waves by the breath of the morning breeze, each glittering in its course under the influence of the sunbeams. High hills, rocks, and banks, waving with natural forests of birch and oak, formed the borders of this enchanting sheet of water; and, as their leaves rustled to the wind and twinkled in the sun, gave to the depth of solitude a sort of life and vivacity. (Chapter XXX.)

This is the sort of landscape, and landscape description, that Frazer loved for his set pieces.

The lake of Nemi is still as of old embowered in woods, where in spring the wild flowers blow as fresh as no doubt they did two thousand springs ago. It lies so deep down in the old crater that the calm surface of its clear water is seldom ruffled by the wind. On all sides but one the banks, thickly mantled with luxuriant vegetation, descend steeply to the water's edge.[136]

When Frazer wrote his famous description of the lake of Nemi for the first edition of *The Golden Bough*, he had not yet visited the site. He was not to do so until 1901, after the second edition. The details which he added in consequence are, for the most part, minor and technical. They scarcely alter the nature of the landscape presented.[137] The craggy mountains, the solitary lakes, the waterfalls, and the sunsets over imposing hills, as well as the desolate expanses that fill Frazer's texts, are all developments of the Scottscape he knew from childhood both from his reading and from the historically rich landscape around Helensburgh—sites such as Dumbarton Castle, which

---

[136] J. G. Frazer, *The Golden Bough*, 3rd edn. (12 vols.; London: Macmillan, 1906–15), i. 2.   [137] I am grateful to Robert Fraser for this information.

Frazer remembered for its 'natural strength and picturesque grandeur, rising as it does in sheer precipices from a dead flat'.[138]

We are told that one of Frazer's favourite passages from his own work was the account of the Witch of Endor (which appears in *Folk-Lore in the Old Testament*).[139] It seems appropriate that, as a reader of Scott's *Letters on Demonology and Witchcraft*, Frazer should have been attracted by this account of a journey over a desolate mountain track to visit the cottage of a mysterious woman in the north. Frazer himself loved mountains, whether writing about them and the 'glens' of Greece or walking in them, as he did when he returned to Scotland, most notably in 1884, ten years after he had graduated from Glasgow, when he made a walking-tour with his closest friend, Robertson Smith, who was himself deeply attracted to the landscape sanctified by Scott. As Frazer enthusiastically recalled in 1897, writing of this walking-tour with his fellow Scot,

He loved the mountains, and one of my most vivid recollections of him is his sitting on a hillside looking over the mountains and chanting or rather crooning some of the Hebrew psalms in a sort of rapt ecstatic way. I did not understand them, but I suppose they were some of the verses in which the psalmist speaks of lifting up his eyes to the hills. He liked the absolutely bare mountains, with nothing on them but the grass and the heather, better than wooded mountains, which I was then inclined to prefer. We made an expedition in a boat down the loch and spent a night in a shepherd's cottage. He remarked what a noble life a shepherd's is. I think he meant that the shepherd lives so much with nature, away from the squalor and vice of cities, and has to endure much hardship in caring for his flock. After returning from our long rambles on the hills we used to have tea (and an exceedingly comfortable tea) at the little inn and then we read light literature (I read French novels, I forget what he read), stretched at ease one of us on the sofa, the other in an easy chair. These were amongst the happiest days I ever spent . . .[140]

This side of Robertson Smith reinforced in Frazer the combination of elements which had impressed him so strongly in John Veitch, his Professor of Moral Philosophy and Rhetoric at Glasgow University. Though Veitch's philosophy may have been outdated, that did not worry Frazer, who proudly linked him to Hutcheson, Dugald Stewart, and Sir William Hamilton. Frazer's library contained

---

138 Frazer, *Creation and Evolution*, 137.

139 Frazer, *Folk-Lore*, ii. 520; Sarah Campion, 'Autumn of an Anthropologist', *New Statesman and Nation*, 13 Jan. 1951, 36.

140 Quoted in Ackerman, *J. G. Frazer*, 61.

Veitch's 1869 *Memoir of Sir William Hamilton*. But what most impressed Frazer was that Veitch

> was a man of true poetical feeling, and I well remember the quiet but deep enthusiasm with which he recited verses of Wordsworth and of the fine old Scottish ballad 'Sir Patrick Spens'. His teaching made on my mind a profound impression which has never been worn out. It opened up an intellectual vista of which I had never dreamed before, and which has never since been wholly closed or obscured by later and very different studies.[141]

Frazer's love of the Scottish ballads must have been encouraged by Veitch, and is clear, for instance, from the notes to the anthropologist's *Passages of the Bible*, where sections from Judges are compared with a stanza from 'Sir Patrick Spens' and with Scott's ballad 'Lord Soulis'. Veitch's great love was expressed in the title of his 1878 book, *The History and Poetry of the Scottish Border*. Frazer's father presented his son with a copy in 1878 which is now in Trinity College Library. Veitch's book blends the literary and the antiquarian. A friend, writing of Veitch's love of 'solitary wanderings amongst the hills', wrote of how Veitch 'certainly did much to re-create, and re-vivify the interest which Sir Walter [Scott] started'. The academic year in which Veitch's course so impressed Frazer was 1871–2—this was the period during which Veitch's following in Scott's footsteps was at its most intense.[142] Like Andrew Lang, both Frazer and Veitch were minor poets. Veitch's *Hillside Rhymes* were published in 1872, and Frazer owned Veitch's book of poems *The Tweed* (1875). However, Veitch's main work, like Frazer's, was written in a prose that was Romantically tinged.

Frazer took his topographical Romanticism with him as part of his Scottish upbringing. He had grown up, after all, in what he called 'the beautiful natural surroundings of Helensburgh, situated at the mouth of the lovely Gareloch and looking across its calm water to the wooded peninsulas of Roseneath'.[143] After childhood reading of McLure Thomson, Wilson, Pollok, and Scott, he had been inspired by the literary personality of John Veitch, as he was to be by the similarly oriented Robertson Smith. One of Frazer's poems, 'June in Cambridge', suggests that his massive and laborious studies took the place of the Scottish environment he had known.

[141] Frazer, *Creation and Evolution*, 123.
[142] William Knight, *Some Nineteenth Century Scotsmen* (Edinburgh: Oliphant, Anderson and Ferrier, 1908), 255, 259.
[143] Frazer, *Creation and Evolution*, 119.

I shall not feel the breezes,
    I may not smell the sea
That breaks to-day in Scotland
    On shores how dear to me!

.   .   .   .   .   .

Still, still I con old pages
    And through great volumes wade,
While life's brief summer passes,
    And youth's brief roses fade.

Ah yes! Through these dull pages
    A glimmering vista opes,
Where fairer flowers are blowing
    Than bloom on earthly slopes.

The dreamland world of fancy!
    There is my own true home
There are the purple mountains
    And blue seas fringed with foam.[144]

The poem is conventional enough, but valuable, since it draws attention to the way in which study and scholarship were imaginative activities for Frazer, and deeply Romantic ones at that. The paper landscape of Frazer's imaginative prose grows out of the Scottish Romantic topographical tradition rooted in Scott's work, which he had known and loved since boyhood.

Frazer's anthropological prose, though, is not uniform. It tends to oscillate between purplish landscape descriptions and factual card-indexing. Here again the connection with Scott is strong, if we remember that Scott of the lengthy antiquarian and folkloric foot-notes who not only mythologized landscape but also juxtaposed cultures, and who was fascinated by (though he could also smile at) obsessive scholarly collecting, as we can see from *The Antiquary*. Frazer's own immediate environment—'untold reams of paper'— and his antiquarian mania make him resemble a latter-day version of Scott's *Antiquary* self-portrait.[145] Certainly, Scott's elaborate introductions and his passion for recording fact and folklore as well as fiction also have their counterpart in Frazer.

[144] J. G. Frazer, *The Gorgon's Head and Other Literary Pieces* (London: Macmillan, 1927), 439–40.

[145] Sir James Frazer and Lady Frazer, *Pasha the Pom: The Story of a Little Dog* (London: Blackie and Son, 1937), 11.

Furthermore, if there are biographical and stylistic reasons for viewing Frazer in relation to Scott, there are also reasons in terms of cultural genealogy. If Carlyle may be seen as Scott's successor, then no later Scottish writer did more than Frazer to bring to fruition the Sage of Chelsea's ambition to connect 'all religion hitherto known'. His enormous, rewritten, and encyclopaedic blend of literary craftsmanship and eclectic historical research, *The Golden Bough*, produced a modern prose epic as suitable for his age as *The French Revolution* had been for a previous generation. Like Carlyle, the eclectic Frazer served his apprenticeship as an encyclopaedist, another aspect of his work which demonstrates the propriety of seeing him in the context of a Scottish cultural genealogy. For, as indicated above, it was Robertson Smith who commissioned Frazer to write on 'Totemism' for the *Encyclopaedia Britannica*. The strong, extensive connections between the work of Robertson Smith and Frazer have been set out most extensively by Robert Ackerman, but we should take note, too, of the extent to which the Frazer who thought McLennan's work 'epoch-making' was part of a longer and wider Scottish line.[146] Many aspects of Frazer's work and career make this manifest: McLennan and Robertson Smith were near-lifelong friends who, with Frazer, shared the experience of going to Trinity College, Cambridge, after studying in Scottish universities; Frazer points out that the influence of McLennan on Robertson Smith 'was deep and lasting'; like Scott and Frazer, McLennan had taken a law degree; Scott, Robertson Smith, Lang, Frazer, and McLennan all wrote for the *Encyclopaedia Britannica* during the period when it remained a Scottish-dominated production.[147]

What all this means is that, when Robertson Smith introduced the young Glaswegian J. G. Frazer to Smith's Scottish friend William Wright, Professor of Arabic at Cambridge, as 'one of the Scotch contingent', this phrase should alert us to a cultural and intellectual genealogy whose roots were grounded in eighteenth-century Scotland and in Scott, a genealogy in which Frazer plays a crucial part.[148] Frazer's 'Scotch contingent' at Cambridge may invite recollection of the sort of Scottish support networks in England which were highlighted in *Roderick Random* and which had assisted Scottish writers in the south since at least the days of James Thomson.

[146] Ackerman, *J. G. Frazer*, 58–63, 322 n. 24.

[147] Frazer, *The Gorgon's Head*, 284; *DNB* entries for each man.

[148] Ackerman, *J. G. Frazer*, 60.

Robert Ackerman contends, rightly, that the Preface of the second edition of *The Golden Bough* is the piece of writing which, perhaps more than any other, displays Frazer the man, but particular attention might be drawn to the concluding words of this Preface.[149] Frazer has revealed himself as divided between the beauty of religious traditions and the violence that he feels he must do to them in order to further his scientific work. The revelation of the truth in this anti-holy war, with its great field guns, is all that matters: 'Whatever comes of it, wherever it leads us, we must follow truth alone. It is our only guiding star: *hoc signo vinces*.'[150] This passage is indeed a daring use of a Christian text against Christianity, another example of Frazer's familiar technique of tactfully undermining Christian foundations. Like Carlyle, Frazer manages to use language so as to be both true and untrue—loyal and disloyal—to the religion of his pious father. But after this resounding declaration about the pursuit of truth, the paragraph that follows is a remarkable one:

> To a passage in my book it has been objected by a distinguished scholar that the church-bells of Rome cannot be heard, even in the stillest weather, on the shores of the Lake of Nemi. In acknowledging my blunder and leaving it uncorrected, may I plead in extenuation of my obduracy the example of an illustrious writer? In *Old Mortality* we read how a hunted Covenanter, fleeing before Claverhouse's dragoons, hears the sullen boom of the kettledrums of the pursuing cavalry borne to him on the night wind. When Scott was taken to task for this description, because the drums are not beaten by cavalry at night, he replied in effect that he liked to hear the drums sounding there, and that he would let them sound on so long as his book might last. In the same spirit I make bold to say that by the Lake of Nemi I love to hear, if it be only in imagination, the distant chiming of the bells of Rome, and I would fain believe that their airy music may ring in the ears of my readers after it has ceased to vibrate in my own.[151]

After declaring that 'we must follow truth alone', Frazer seems to be making a volte-face, or at least to be suggesting that there is a 'poetic' truth that should be allowed to prevail over prosaic, literal truth. He leaves us with his own work aligned beside that of Scott. He leaves us less with *The Golden Bough* as science, than *The Golden Bough* as literature. This impulse becomes clearer in the Preface to the third edition, when Frazer hopes that, while he will not be 'sacrificing the solid substances of a scientific treatise' by expanding his book, he has

---

[149] Ibid. 164; Frazer, *The Golden Bough*, i, pp. xxvi–xxvii.
[150] Frazer, *The Golden Bough*, i, p. xxvi.   [151] Ibid. pp. xxvi–xxvii.

been able to cast his 'materials into a more artistic mould and so perhaps attract readers, who might have been repelled by a more strictly logical and systematic arrangement of the facts'. Frazer presents the 'mysterious' as happening among 'picturesque natural surroundings'.[152] The stress on facts remains, as it does in the elaborate apparatus that accompanies Scott's fictions and anthological work, but more and more attention is being paid to the *literary* presentation of the data.

Frazer's reference to Scott in the conclusion of his Preface to *The Golden Bough*'s second edition seems to me to be a crucial point of reference. The novel mentioned, *Old Mortality*, takes us back to Frazer's childhood reading of Pollok's *Tales of the Covenanters*, and then to Scott himself. *Old Mortality* is relevant, too, because it is a book about religious bigotry, superstition, and its alleviation through the adoption of more enlightened views. It opens with a landscape set piece description of an old burial ground among the hills, and its theme of 'old mortality' is very much Frazer's theme also. The particular passage which Frazer mentions occurs early in Chapter VI of Scott's novel. It clearly impressed Frazer, who mentions it again in the notes to his *Passages of the Bible* as conveying an unusually 'vivid impression to ear and eye . . . of a cavalry regiment on the march by moonlight'.[153] Yet surely the modern reader is not struck by this as one of the most arresting passages in Scott. It describes how Henry Morton

did not think it necessary to take a light, being perfectly acquainted with every turn of the road; and it was lucky he did not do so, for he had hardly stepped beyond the threshold ere a heavy trampling of horses announced, that the body of cavalry whose kettle-drums they had before heard, were in the act of passing along the high-road which winds round the foot of the bank on which the house of Milnwood was placed. He heard the commanding officer distinctly give the word *halt*. A pause of silence followed, interrupted only by the occasional neighing or pawing of an impatient charger. (Chapter VI.)

Aesthetically, this passage pleased Frazer, yet its real significance for him becomes clear only when we relate it to Scott's footnote and to the passage where Frazer ties it to the bells at Nemi. Scott's footnote reads:

---

[152] Ibid. p. viii.
[153] J. G. Frazer, *Passages of the Bible Chosen for Their Literary Beauty and Interest* (London: A. and C. Black, 1895), 429.

¹ Regimental music is never played at night. But who can assure us that such was not the custom in Charles the Second's time? Till I am well informed on this point, the kettle-drums shall clash on, as adding something to the picturesque effect of the night march.

In summarizing Scott, Frazer subtly shifts the emphasis of the note. Scott implies, at least, that if he were to receive precise factual information, and is eventually 'well informed on this point', he might alter the passage. The struggle here is between Scott the novelist and Scott the antiquarian collector of precise and curious facts. The novelist is allowed to win, but the hint is that the fact-lover *may* eventually prevail. Frazer presents Scott as saying that because 'he liked to hear the drums sounding there . . . he would let them sound on so long as his book might last'—a rather more confident and determinedly novelistic statement than the one Scott made. Frazer focuses on this obscure point not because it is obscure, but because it puts in a nutshell the constant and contrary pulls in his own work—between factual recording and tabulation on the one hand, and creative writing on the other. What is particularly interesting here is that, where Scott the novelist teasingly suggests that the antiquarian may eventually prevail, Frazer the scholar comes down on the side of true fiction against mere fact.

Stanley Edgar Hyman, whose book *The Tangled Bank* (1962) is the only study to consider Frazer extensively as an imaginative writer, thinks that the basic form of *The Golden Bough* is the travelogue, with a strong presence also of the tragic genre. Hyman makes no connection with Scott, but the travelogue is a favourite narrative device of Scott's, and there is often a gloom hanging over his text which comes from the realization that the innumerable picturesque customs and traditions of an earlier world are being lost, tragically destroyed by the modern ethos. This emotion is complicated by the fact that, however much he may lament the destruction of the fascinating old ways, and however much he may memorialize them, Scott, like Frazer, is part of the newer world which is hastening that destruction. Travelogue and tragedy are certainly important to both Frazer and Scott, and Mary Beard has argued that it was as a 'voyage of exploration' into the Other, a voyage which both excited readers and returned them safely home, that *The Golden Bough* sold so splendidly to the late Victorians and to T. S. Eliot's generation (between 1922 and 1933 the new abridged edition sold 33,000 copies).[154] Such pleasures of a 'voyage of exploration' were just what *Waverley* and Scott's ensuing fictions had offered readers in the early nineteenth-century, and what some of Lang's and Stevenson's tales continued to offer. Yet the experience of reading *The Golden Bough* calls also for another form of description, one that does not put the text in the genre of fiction, but does place it in a meaningful relation to fiction, and again highlights the connections between the author of *The Golden Bough* and the author of the Waverley Novels.

The most suitable term of description is Frazer's own, and it comes from his essays on the Australian anthropologists whom he had befriended and whose work had contributed a good deal to his own endeavours. Frazer's essay on 'Fison and Howitt' is interesting for various reasons. First, Frazer points out how childhood imaginative reading had an impact on Fison's future development. Secondly, Frazer stresses the importance of the use of the imagination in considering anthropological material—particularly for 'all who view savages through a telescope, whether from a club or a college window'. Frazer contends that 'If we are really intent on knowing the truth', then

by long and patient effort we may come to see in the magic mirror of the mind a true reflection of a life which differs immeasurably from our own. Yet this reflection or picture must itself be pieced together by the imagination; for imagination, the power of inward vision, is as necessary to science as to poetry, whether our aim is to understand our fellowmen, to unravel the tangled skein of matter, or to explore the starry depths of space.[155]

Though he warns that imagination has its hazards for the anthropologist, Frazer clearly states the importance of the imaginative, quasi-poetic vision for the anthropological writer. Again, Frazer is obviously attracted to the conventionally 'poetical' landscape passages in the writing of Howitt and Fison—set pieces very similar to those in his own work.

It is such landscape descriptions which fire Frazer's enthusiasm for Baldwin Spencer's *Wandering in Wild Australia*. That book *is* a kind of travelogue, but it is also, for Frazer, something more:

[154] Mary Beard, 'With the Wind in Our Shrouds', *London Review of Books*, 26 July 1990, 7–8.   [155] Frazer, *The Gorgon's Head*, 301–2.

In the directness and simplicity of its style, in the impression which it leaves of truth to nature, in the fascination of its description of strange folk and ever shifting scenery, *Wanderings in Wild Australia* may be compared to the *Odyssey*. If the writer did not tread enchanted ground, at least he moved among people who firmly believed in the power of enchantment and constantly resorted to it for the satisfaction of their wants and the confusion of their foes; if he did not encounter monsters like the Cyclopes or Scylla and Charybdis, at least he beheld with his own eyes the rocky pool in which the dreadful dragon, the Wollunqua, was believed to lurk, ready to dart out and devour its human victims. All this serves to invest the story of Spencer's wanderings in Australia with an atmosphere of romance, and to lend it the character of an anthropological epic.[156]

'Anthropological epic' seems a term particularly fitted to describe *The Golden Bough* as well as Frazer's conception of the Bible. Perhaps the travelogue is strongest in the 1890 edition, but, as the text develops, we move toward something else. *The Golden Bough* seems to become much less one work than a series of related volumes, panoramic in their scope, crammed with cultural data—an 'anthropological epic'. All these features are common to both Frazer's work and to the Waverley Novels. For if Scott set out in *Waverley* to write romance, then, in the related but independent series of novels taken as a whole, he ended up writing an epic much wider in its scope than Ossian's, an eclectic prose epic which, like Boswell's, prefigures Carlyle's desire to write the prose epic fitted to his own era. It is highly improbable that Frazer always wrote in conscious imitation of Scott. Scott's impact was more subtle than that; consciously, Frazer tended to align himself with English eighteenth-century writers, and critics have followed him in this, linking him especially with Gibbon. In retrospect, though, it makes best sense to align the full *Golden Bough* with the Waverley Novels, particularly in view of Frazer's almost lifelong familiarity with them, not to mention his cultural and intellectual descent from Scott's milieu. Scott is a strong part of that 'romanticism' in Frazer which, Ackerman writes, 'went underground in the work of the mature man, but never totally disappeared'.[157] In those landscape set pieces, particularly, that Romanticism moulds the surface of Frazer's prose. In other ways, too, the workaholic Scot and the more eclectically fact-loving of the Romantic writers should be placed beside Frazer, and their works aligned to show that it was not

156 Frazer, *Creation and Evolution*, 68–9.   157 Ackerman, *J. G. Frazer*, 9.

only Carlyle and Andrew Lang, but also J. G. Frazer who was one of the descendants of Scott in the tradition of Scottish writing.

Seeing Frazer as a descendant of Scott completes a restructuring of the nineteenth-century literary map, and prompts the positioning of Frazer beside another exiled Scottish writer, one who is much more conventionally seen as inheriting many of Scott's concerns: Stevenson. This is not to suggest that Stevenson influenced Frazer, but one may indicate points of contact between them that help to show why the two should be considered as complementary members of a Scottish cultural tradition. Stevenson was born four years before Frazer. Both men left Scotland at nearly the same time. Stevenson's adventurous life was very different from that of the Fellow of Trinity, but, if we take his work as a whole, connections may become clearer. For Stevenson wrote not only romances of the Scottish mountains such as *Kidnapped* and *Catriona*, he also wrote about primitive life in the South Seas in novellas like 'The Beach of Falesá' and *The Ebb-Tide* which reveal a concern with the coming-together of Western and Pacific man—of so-called 'civilized' and 'savage'. Often the connections between the two are ironically related. A concern with the animalistic primitive and the civilized modern is what underlies *Dr Jekyll and Mr Hyde*. Stevenson was also a folklore collector, an aspect of his work which has only recently received much study, and he was interested in cultural juxtapositions.[158] Frazer and Stevenson never met; in anthropology, Stevenson's great friend was Lang. But Frazer, like Robertson Smith, was a keen reader of Stevenson, particularly in the late 1880s, when he was working on the first edition of *The Golden Bough*. The 1907 catalogue of Frazer's library notes nine volumes by Stevenson, including *Kidnapped*, *The Master of Ballantrae*, and *Memories and Portraits* in the first editions. Two more books are present in the 1935–6 catalogue. Though Frazer and Stevenson never corresponded, there are other strong links between them. Stevenson's *Vailima Letters* are addressed to one of his closest friends, Sidney Colvin, a friend of Lang, and later Stevenson's editor. Colvin was a Fellow of Trinity, and advised Frazer in 1884 about his work on Pausanias.[159] A much closer connection exists in the doctoral thesis which Frazer's step-daughter Lilly M. Grove wrote at the University of Paris and pub-

---

[158] Robert I. Hillier, 'Folklore and Oral Tradition in Stevenson's South Seas Narrative Poems and Short Stories', *Scottish Literary Journal*, Nov. 1987, 32–47.

[159] Ackerman, *J. G. Frazer*, 56; *DNB* entry for Sidney Colvin.

lished in 1908 as *Robert Louis Stevenson: Sa vie et son œuvre, étudiée surtout dans les romans écossais*. In discussing Stevenson, Lilly Grove discusses a man whose family background and early literary orientations are remarkably similar to those of her own stepfather. Stevenson's father, she writes, was a reserved man, but much loved and respected by his family. He was a staunch Calvinist who brought up his son on the Bible and stories of the Covenanters. Grove discusses Stevenson's great debt to Scott, and the way in which he was able to convey such a powerful sense of landscape. But most striking is her quotation of a long passage from Stevenson's 'Genesis of the Master of Ballantrae', in which he describes the workings of his imagination as the idea for that novel came to him while he was living in the Adirondack mountains. Grove urges us to compare the passage with the text of a letter Stevenson sent to Sidney Colvin on Christmas Eve 1887; here is part of the passage from 'Genesis of the Master of Ballantrae', as quoted in Grove's published thesis:

Come, said I to my engine, let us make a tale, a story of many years and countries, of the sea and the land, savagery and civilisation . . . There cropped up in my memory a singular case of a buried and resuscitated fakir, which I had often been told by an uncle of mine . . . On such a fine frosty night, with no wind and the thermometer below zero, the brain works with much vivacity; and the next moment I had seen the circumstance transplanted from India and the tropics to the Adirondack wilderness and the stringent cold of the Canadian border . . . And while I was groping for the fable and the character required, behold I found them lying ready and nine years old in my memory . . . A story conceived in Highland rain, in the blend of the smell of Athole correspondence and the Memoirs of the Chevalier de Johnstone . . . [160]

This passage, with its description of a mind and imagination that flits across, and links, cultures and landscapes, 'savagery and civilization'; that pieces together old documents; and that transplants the story of a resurrected man, seems to provide a powerful analogy to the way in which J. G. Frazer's mind worked, to demonstrate why we should link Stevenson, one of the descendants of Scott in the Scottish Romantic novel, and J. G. Frazer, one of the descendants of Scott in the Scottish tradition of anthropological assemblage—or even epic.

Both Frazer and Stevenson can be accused of an inhibiting nostalgia. Frazer, no doubt, suffers from a philosophical nostalgia for

[160] Lilly M. Grove, *Robert Louis Stevenson: sa vie et son œuvre*, Thèse de doctorat, Faculté des Lettres de l'Université de Paris (Paris: Bonvalot-Jouve, 1908), 95–6.

Enlightenment Edinburgh and Glasgow, a nostalgia fostered by the decisive contact which he had as a Glasgow student with the Hamiltonian John Veitch. Frazer studied with Lord Kelvin at Glasgow in 1872–3, yet later he could not accept Einstein; in 1873–4 he studied with Edward Caird, yet he could never accept Hegel or much of modern philosophy. Ernest Gellner's insight that Frazer 'adapted the philosophy of David Hume to the great new tidal wave of ethnographic evidence about human unreason' has been supported in detail by Robert Fraser.[161] Frazer's nostalgia for Scottish philosophy corresponds to Stevenson's nostalgia for Scottish history, with so many of his tales being set in the world of eighteenth-century Scotland. Perhaps the drag-weight of Scott was too strong in this regard. Yet in both writers there is also a deep personal and familial nostalgia, sharpened by exile. If Frazer's "June in Cambridge' suggests that scholarship by the Cam could replace and reinstate his delight in the Scottish landscape, then Stevenson used literature for the same purpose in the South Seas, returning with powerful nostalgia to the world he had left in such poems as 'To S[idney] C[olvin]' and the 1887 poem which opens:

> The tropics vanish, and meseems that I,
> From Halkerside, from topmost Allermuir,
> Or steep Caerketton, dreaming gaze again.[162]

Stevenson's writing about, and filial homage to, his own ancestors in *Memories and Portraits* should be set beside Carlyle's *Reminiscences* and Frazer's 'Memories of My Parents' and 'Speech on Receiving the Freedom of the City of Glasgow' in 1932—an honour which, Ackerman contends, was especially dear to the Cambridge don.[163] Frazer's memories of the scenes of his childhood are vivid and nostalgia-laden, as when he describes the Helensburgh he grew up in, looking westward across the loch to 'the wooded peninsulas of Roseneath' and 'the low green hills of Gareloch in the further distance, and the rugged mountains of Loch Long rising above them on the western horizon'. Here in Helensburgh, looking out to the mountains, Frazer spent much of his childhood in a house called Glenlee.

---

[161] Ernest Gellner, 'Leaves from the Golden Bough' (review of Ackerman, *J. G. Frazer*), *Times Higher Education Supplement*, 15 Jan. 1988, 18.

[162] Robert Louis Stevenson, *Collected Poems*, ed. Janet Adam Smith, 2nd edn. (London: Hart-Davis, 1971), 270.

[163] Frazer, *Creation and Evolution*, 117–28; Ackerman, *J. G. Frazer*, 5.

In the very heart of this town of gardens stands Glenlee. It is a little house with a veranda facing full south about which hops used to twine, and a sloping bank on which fuchsias with their red and purple blossoms used to grow. A burn winds through the garden, flowing for a part of its course over a pebbly bed at the foot of red sandstone cliffs.[164]

The house and garden are there today, much as Frazer described them, though the word 'cliffs' seems an exaggeration. What is most striking when one stands in that garden and thinks of Frazer, is the great exposed bell-tower of the former United Presbyterian church which looms over the grove where Frazer played. This church building was erected in July 1861, so it was there when Frazer lived in Glenlee. Another church very close, Park Church, was opened in 1863.[165] Helensburgh is a town of churches and bells as well as of gardens and groves. Of his childhood there, Frazer recalled how 'I look back to those peaceful Sabbath days with something like fond regret, and the sound of Sabbath bells, even in a foreign land, still touches a deep chord in my heart'.[166] The bells that Frazer heard in childhood from Glenlee by the Firth of Clyde underlie, one might suspect, that curious determination (for which he sought precedence in Scott) to hear the church bells of Rome by the lake of Nemi. Ackerman points out that Frazer names 'Renan as the source of the image of the "eternal bells of Rome", whose tolling symbolizes the persistence of the religious impulse in mankind, with which *The Golden Bough* ends'.[167] But long before he read Renan, the tolling of church bells had made a deep personal impression on Frazer, being intimately linked not with Rome, but with that world of Presbyterian Scotland which did so much to form him; R. A. Downie even records that Frazer attended church regularly for most of his life, despite that apparent rejection of his earlier faith which Ackerman emphasizes.[168] The bells of Helensburgh underlay the bells of Renan, just as the mountain scenery of the Firth of Clyde prepared Frazer for his reading of Scott. So much of Frazer's later interests and career seem to have roots that reach right back into his childhood. Downie's 1970 *Frazer and The Golden Bough* (which, unlike his earlier tribute of 1940, was written without Lady Frazer's supervision) records that 'An early nursemaid was responsible for his [Frazer's] only encounter with a

---

[164] Frazer, *Creation and Evolution*, 150.

[165] Anon., *Helensburgh United Presbyterian Church Jubilee, 1894* (Helensburgh: J. Lindsay Laidlaw, [?1894]), 38; George R. Logan, *Park Church Helensburgh: The First Hundred Years* (Glasgow: R. Thomson, 1964), 3.

[166] Frazer, *Creation and Evolution*, 133.   [167] Ackerman, *J. G. Frazer*, 93.

[168] R. Angus Downie, *Frazer and 'The Golden Bough'* (London: Gollancz, 1970), 21.

member of a primitive tribe, for she took him to a fairground where "The Wild Man of Borneo", dashing from his tent, set young Frazer howling with terror.'[169]

Frazer, like Stevenson, looks backwards. But Frazer, like Stevenson, also points forwards. Where aspects of Stevenson look towards Conrad or MacDiarmid, and have appealed to a variety of Modernist or post-Modernist writers from T. S. Eliot to Borges and Alasdair Gray, so Frazer's impact on Modernist writing has been well documented, most fully in John B. Vickery's study *The Literary Impact of 'The Golden Bough'* (1973), and in a recent volume of essays on *Sir James Frazer and the Literary Imagination*.[170] These studies demonstrate how Frazer's cultural assemblages, his juxtapositions of civilized and savage, and his curious combination of conscious literary style and factual encyclopaedic scope, all look towards the work of the major Modernist writers, including Eliot, Pound, Joyce, Lawrence, and Conrad. Yet these features also relate to the author of the Waverley Novels. As regards cultural and literary history, Frazer has a particular importance that, so far, has gone unrealized because he has not been seen in terms of the Scottish tradition to which he belongs. In rewriting literary history, we have to become aware that, along with Lang and Stevenson, Frazer is a crucial 'missing link' between Scott and Modernism.

Carlyle once wrote of the word 'God' that it belonged to 'the ancient dialect', the implication being that such talk can be replaced by another dialect, while still remaining within the same frame of language.[171] In *On Heroes and Hero-Worship* he writes of religious speculations in terms of changing 'dialect' (*Heroes*, 138). Dialect, like anthropology, makes one constantly aware of shifts, pluralism, and difference, of possibilities in speech or custom other than those which are taken for granted by the dominant culture. Not that either dialect or anthropology need bring with it an overthrowing of that culture's values. Frazer did not advocate that his readers became 'savages', and Wittgenstein was correct to attack Frazer's work for its assumption that the actions of 'savages' had to be 'mistakes'.[172] The very word 'dialect' implies an inferiority to 'language'; if Scott delighted

[169] Ibid. 18.

[170] John B. Vickery, *The Literary Impact of 'The Golden Bough'* (Princeton, NJ: Princeton University Press, 1973); Robert Fraser (ed.), *Sir James Frazer and the Literary Imagination* (Basingstoke: Macmillan, 1990).

[171] Carlyle, *Works*, x. 136.

[172] Ludwig Wittgenstein, *Remarks on Frazer's 'Golden Bough'*, trans. A. C. Miles, ed. Rush Rhees (Retford, Notts.: Brynmill, 1979), 1e.

in the uses of Scots, he used English as the dominant language of his narrative. Yet the terms 'language' and 'dialect' are powerfully slippery, and it is possible for the two to mix, as in the different dictions of Burns and Carlyle, to create a pertinent challenge to an established Anglocentric ethos. This chapter has drawn attention to Carlyle as a successor to Scott, the novelist with an anthropological vision and a delighter in linguistic cross-over, but, on the whole, more attention has been paid to anthropology than to dialect. That is because the idea that there was a powerful anthropological presence in nineteenth-century Scottish writing is much newer than the idea that dialect persisted in nineteenth-century Scottish writing. The most recent, valuable study of that persistence is by Emma Letley, who points to the strength of Scots in John Galt's narrative voice as well as in his characters' speeches, and to the way in which a variety of Scottish writers, from Galt to Stevenson and Douglas Brown, deployed and redeployed Scots throughout the nineteenth century as a means of both questioning and supplementing the dominant monodialectal diction of 'standard English'.[173] These linguistic challenges to Anglocentricity run through the century in parallel with the distinguished Scottish contribution to anthropology. In an era increasingly uncertain about its own fundamental values, anthropology constantly presented the possibility of cultural relativism, just as the use of non-standard diction kept alive questions of cultural relativism through language.

Sometimes the work of these writers may be viewed most fruitfully in contexts which they would not have chosen. While Scott's anthropological vision developed out of a wish to assert the validity of Scottish cultural difference within Britain, Frazer's work, enriched by Scott, was no more ostensibly concerned with Scottishness than it was with furnishing material for *The Waste Land*. Yet Frazer's work can meaningfully be seen as part of the strong currents of nineteenth-century Scottish writing which have their roots in Scott.

A devolved reading, sensitive to cultural differences within English Literature, allows a much clearer view of that problematic area, nineteenth-century Scottish Literature. It reveals a continuity between fiction and anthropology which does not constitute *the* exclusive tradition of nineteenth-century Scottish writing, but is certainly an area in which several of the strengths of Scottish culture were concen-

---

[173] Emma Letley, *From Galt to Douglas Brown: Nineteenth Century Fiction and Scots Language* (Edinburgh: Scottish Academic Press, 1988).

trated. The pattern of literary history outlined in this chapter surely includes such a novel as Galt's *The Ayrshire Legatees* (1821), which owes debts to *Humphry Clinker* in its epistolary treatment of 'national peculiarities' in Scotland and England, or Galt's *The Entail* (1822), with its linguistic and societal juxtapositions, or Susan Ferrier's *Marriage* (1818), so much of whose humour depends on the crossing of cultural boundaries.[174] The sociological attentiveness of Galt's 'theoretical histories of society', and the scope of Ferrier's fiction, which has been seen as 'a survey of human conduct in general', both indicate that these novelists might profitably be viewed as belonging to the cultural tradition outlined above.[175] The mischievous anthropological eye which uses a comparative method to juxtapose societies and their mores, provoking the reader to compare and evaluate them, is the most Scottish feature of *Don Juan*, written by the part-Scottish Byron, the great admirer of Scott's novels, who is anxious in that poem to un-English his narrative voice and who proclaims himself 'half a Scot by birth, and bred | A whole one'.[176] The Scottish writers discussed in this chapter were, without exception, pro-British; however, their work avoided or called Anglocentricity into question in various ways, and the Scottish cultural tradition of which they formed a part made available to other cultures (and to the future) examples, techniques, and materials through which the destabilizing of Anglocentricity might be taken further. In so doing, late eighteenth- and nineteenth-century Scottish writing fuelled the essentially 'provincial' phenomenon of Modernism, as well as preparing the way for twentieth-century Scottish writers' search for a post-British identity. More immediately, the Scottish cultural tradition spurred the aspirations and writings of several of the most impressive writers of nineteenth-century America.

[174] John Galt, *The Ayrshire Legatees* (1821; repr. Edinburgh: James Thin, 1978), 34.

[175] John Galt, *Autobiography* (2 vols.; London: Cochrane and M'crone, 1833), ii. 220; Herbert Foltinek, Introduction to Susan Ferrier, *Marriage* (1818; repr. Oxford: Oxford University Press, 1986), p. xiii.

[176] Lord Byron, *Don Juan* (1819–24; repr. Harmondsworth: Penguin Books, 1977), X, 17.

4

Anthologizing America

ECLECTICISM, so prominent in Scottish culture, is also of great importance in American Literature, where it involves the anthologizing of the old at least as much as (and probably more than) the gathering of the new. America was, from its beginning, not just a New World opening before the passengers from the *Mayflower*; it was a cultural anthology, the most spectacular extremes of which were those *Mayflower* 'colonials' and the native Americans. American Indians, when they do appear in literature, though, are very much an Old World construct. Roy Harvey Pearce, in his study of their depiction, pointed out that 'American theorizing about the Indian owed its greatest debt to a group of eighteenth-century Scottish writers on man and society', including William Robertson, about a third of whose popular and influential *History of America* dealt with the American Indians, and Adam Ferguson whose *Essay on the History of Civil Society* included much Indian material.[1] Some early American anthropological writers like Stanhope Smith, trained in the Scottish tradition, were keen to stress the savageness and nobility of the American Indian as seen through American eyes (as opposed, for example, to those of Lord Kames), but these were merely reactions to a dominant idea.[2] The notion of eighteenth-century America as another Scotland may seem strange today, but, even if we accept at less than face value Garry Wills's claim that Jefferson's ideas came principally 'from Aberdeen and Edinburgh and Glasgow', none the less we can see that, to Jefferson, the American colonies reproduced exactly the Scottish position.[3] In planning the Declaration of Independence, he

[1] Roy Harvey Pearce, *Savagism and Civilization: A Study of the Indian and the American Mind*, rev. edn. (Baltimore: Johns Hopkins, 1965), 82 ff.

[2] See Samuel Stanhope Smith, *An Essay on the Causes of the Variety of Complexion and Figure in the Human Species*, ed. Winthrop, D. Jordan (1787; repr. Cambridge, Mass.: Harvard University Press, 1965), pp. xli–xlii, 215.

[3] Garry Wills, *Inventing America: Jefferson's Declaration of Independence* (New York: Doubleday, 1978), 180; the most pertinent criticism is Ronald Hamony, 'Jefferson and the Scottish Enlightenment: A Critique of Garry Wills's *Inventing America: Jefferson's Declaration of Independence*', *William and Mary Quarterly*, Oct. 1979, 502–23.

recalled: 'I took the ground which, from the beginning I had thought the only one orthodox or tenable, which was that the relation between Gr. Br. and these colonies was exactly the same as that of England & Scotland after the accession of James & until the Union . . .'. This explains a section deleted from the final version of the Declaration. Jefferson complained that no one else agreed with him, and so that passage was cut out. But Scotland featured several times in the Independence discussions. In an argument about the nature of a proposed union between the states, Franklin (who had many friends in Enlightenment Edinburgh) argued that Scotland showed how a small state could come to dominate a larger one in an act of union, while John Witherspoon, the Scottish President of what is now Princeton University, argued that Scotland had suffered by just such an incorporating (as opposed to a federal) union, since 'it's inhabitants were drawn from it by the hopes of places & employments'.[4] There was a great deal of hostility to Scots in revolutionary America, because they seemed to be in control of much American trade and so to have a vested interest in the colonial status quo, but there are also important ways in which Americans looked to the Scottish model.

One reason for the fact that this continued to be the case for so long is provided by the text and title of John Clive and Bernard Bailyn's important article, 'England's Cultural Provinces: Scotland and America'.[5] Linked by a largely common language, both countries wished to assert their sense of separate cultural identity, yet were also very sensitive about appearing provincial. In each, the speech of the most distinctively indigenous groups (the Highlanders and the native Americans) appeared unintelligible to the majority. But should the language of the country and its literature be Scots or English, American or English? Jefferson, enemy of 'Purism' and friend to 'Neology', thought that the New World would remake old speech.

Certainly so great growing a population, spread over such an extent of country, with such a variety of climates, of productions, of arts, must enlarge their language, to make it answer its purpose of expressing all ideas, the new as well as the old. The new circumstances under which we are placed, call for new words, new phrases, and for the transfer of old words to new objects. An American dialect will therefore be formed; so will a West-Indian and Asiatic,

---

[4] Thomas Jefferson, *Writings*, ed. Merrill D. Peterson (Cambridge: Cambridge University Press, 1984), 9, 28, 29.

[5] John Clive and Bernard Bailyn, 'England's Cultural Provinces: Scotland and America,' *William and Mary Quarterly*, Apr. 1954, 200–13.

as a Scotch and an Irish are already formed. But whether will these adulterate, or enrich the English language? Has the beautiful poetry of Burns, or his Scottish dialect, disfigured it?[6]

This tension between standard English and a distinctively American language, as used in *Huckleberry Finn*, *The Catcher in the Rye*, or the poetry of William Carlos Williams, parallels the tension between English and Scots in Scottish Literature. But it does more than parallel it; it intersects with it. For, if we recall that Witherspoon's coining of the term 'Americanism' was based on the term 'Scotticism', and if we remember the energy with which the teaching of Rhetoric and Belles Lettres grew in American universities, often using Scottish texts, we begin to appreciate further the similarities of the pressures on language and society in early nineteenth-century Scotland and America. These similarities and the strength of Scottish literary models in America have been examined by Andrew Hook, Susan Manning, and others.[7] Andrew Hook pays most attention to the late eighteenth century, and Susan Manning to how the Calvinistic herit-age common to Scotland and early America shaped writers such as Brockden Brown, Hawthorne, and Poe. Avoiding overlap with these studies, but wishing to strengthen their sense of Scottish–American cultural connections, this chapter will pay particular attention to the way in which dialectal and anthropological elements from the Scottish tradition provided spurs to Cooper, Emerson, and Whitman in particular, with the result that these writers on the American side of the Atlantic, like the Scottish writers examined in the previous chapter, can be thought of as potent ancestors of the Modernists.

If one might suggest that the eclectic impulse behind Scottish encyclopaedism, anthologism, and anthropology, as well as behind the fictions of Scott and Carlyle, played a strong part in the building-up of a wider perspective in which Scottish culture could fulfil its role without being surpassed by England, then that was certainly a reason for some of the major American eclectic urges also. American writers were anxious to compare a wide variety of other cultures, including that of the native Americans, with English culture, so as not to be subject to the sole model offered by Old England. Yet, ironically, it

---

[6] Jefferson, *Writings*, 1295–6.

[7] Andrew Hook, *Scotland and America, 1750–1835* (Glasgow: Blackie, 1975); Susan Manning, *The Puritan-Provincial Vision: Scottish and American Literature in the Nineteenth Century* (Cambridge: Cambridge University Press, 1990); see also Richard B. Sher and Jeffrey Smitten (eds.), *Scotland and America in the Age of the Enlightenment* (Edinburgh: Edinburgh University Press, 1990), and the bibliographies in all these books.

was the Old World, often the Scottish world, which did so much to give the New an image of itself, an image which was often helpful to American writers, even as they sought to declare their cultural independence.

An excellent example of this is Thomas Campbell's *Gertrude of Wyoming* (1809), which was both popular and important in early nineteenth-century America. In New York alone it went through three editions in the year of its publication. Washington Irving wrote a short life of Campbell to go with a new edition in 1810, and organized Campbell's invitation to lecture on poetry and Belles Lettres in America in 1817, though, in the end, Campbell was unable to go.[8]

As a Scot, Campbell demonstrated how someone from a 'provincial' culture could produce work in standard English which was on equal terms with English material—he also wrote 'Ye Mariners of England'. Campbell embodied the ideal for which the teaching of Belles Lettres in Scotland and America was striving. *Gertrude of Wyoming* also helped consecrate the American landscape for literary use. Campbell, though he had various relatives in America, and in Virginia in particular, appears to have been familiar with only one Indian, a Mohawk in London, of whom he wrote to Sir Walter Scott: 'He is an arch dog, and palms a number of old Scotch tunes (he was educated in the woods by a Scotchman) for Indian opera airs, on his discerning audience.' Nevertheless, Campbell was praised by his American readership for his depiction of 'one of our native savages— "a stoic of the woods, a man without a tear" '.[9] Campbell worked eclectically from travellers' tales and printed sources, and went to considerable lengths to furnish his poem with explanatory notes on American phenomena, culled from various writers, including Jefferson, with whom he later corresponded. The notes to the poem grew after the first edition, concentrating on American Indian pecularities.

Though there had been American attempts, like Joel Barlow's *Columbiad* (1807), to produce epic poems on nationalistic themes in the eighteenth century (paralleling Scottish epics such as John Harvey's *Bruciad*), this poem, in which Campbell invented his America, was remarkably successful. What is particularly striking is the effect it had on Americans themselves. Campbell seems to have been surprised and delighted at this. One American visitor told

---

[8] Hook, *Scotland and America*, 143.

[9] *The Life and Letters of Thomas Campbell*, ed. William Beattie (3 vols.; London: Moxon, 1849), ii. 51, 185.

Campbell in 1840 of how he had been on 'a poetical pilgrimage' to the scenes of the poem. Campbell was fascinated to hear what these Pennsylvanian scenes were really like, and what especially struck him 'was the curious fact that the principal in this pilgrimage had been long blind to the beauties of natural scenery, but was moved by an inspiring influence to tread a soil which the genius of Campbell had made classic ground'.

'Every day we wandered through the primeval forests; and when tired, we used to sit down under their solemn shade among the falling leaves, and read "Gertrude of Wyoming". It was in these thick woods, where we could hear no sound but the song of the wild birds, or the squirrel cracking his nuts, away from the busy world, that I first felt the power of Campbell's genius.'
'. . . When I had finished the relation of these circumstances, Campbell, who was standing by the window, came back to the table, and taking my hand, pressed it, saying—'God bless you, sir; you make me happy, although you make me weep!'[10]

In writing *Gertrude*, Campbell, later astonished at the success of his creation, had reacted to the imaginative stimulus of an unknown world in which new combinations were possible. These combinations were not just the meeting of white and Indian, but also rearrangements of the European Old World after parts of it had taken up residence 'On Susquehanna's side', to use the opening words of *Gertrude*. Later, Campbell would be attacked by Francis Parkman, who complained in the *North American Review* of 1852 that 'jointly with Thomas Campbell, Cooper is responsible for the fathering of those aboriginal heroes, lovers, and sages, who have long formed a petty nuisance in our literature'.[11]

In 1823 James Fenimore Cooper published his third novel, *The Pioneers*, the first of the Leatherstocking Tales to be written. When he gave his book the subtitle *The Sources of the Susquehanna*, Cooper did not so much break new ground as move into the Campbellscape, where Campbell had followed Freneau in one of the many connections between Scottish and American interest in the primitive. Cooper's Susquehanna is upstream from Campbell's, but they certainly share common features, and Cooper did make use of his Scottish predecessor. *The Pioneers* is partly a sentimental pastoral set in picturesque

---

[10] Ibid. iii. 419.
[11] Cited in George Dekker and John P. McWilliams (eds.), *Fenimore Cooper: The Critical Heritage* (London: Routledge and Kegan Paul, 1973), 252.

surroundings, as well as a tale about the confrontation of Indian and white ways. The novel, as its epigraph from the American poet Paulding suggests, is set in a culturally eclectic community.

> Extremes of habits, manners, time, and space,
> Brought close together, here stood face to face,
> And gave at once a contrast to the view,
> That other lands and ages never knew. [12]

This is the New World aesthetic of eclecticism. As Cooper's novel makes clear, the American settlers bring with them older cultural patterns and speech habits; they also rub up against, fight, and mix with the older world of the Indian. Cooper is fascinated not so much by the New World's newness, as by its juxtapositions of the old, and this idea is reinforced by an epigraph from *Gertrude of Wyoming* at the head of Chapter VIII:

> For here the exile met from every clime,
> And spoke, in friendship, every distant tongue. [13]

At the head of Campbell's Pennsylvanian wilderness community was an elderly 'judge in patriarchal hall' whose only daughter fell in love with, and married, a young white man. The young man was once in the charge of an aged Indian, 'a stoic of the woods', who is later found to be 'alone . . . left on earth' of his tribe, before the poem ends with 'The death-song of an Indian chief'. [14] Cooper's novel is about a Pennsylvanian judge whose only daughter falls in love with a young white man who has, apparently, an Indian background, and is now in the company of an old Indian who is 'the last of his people who continued to inhabit this country' and who dies singing at the climax of the book. [15] There are big differences between Campbell's poem and Cooper's novel. The poem's main narrative concerns the tragic death of the young lovers in a massacre; the hero had previously survived the Indian destruction of a British fort. These elements have more in common with *The Last of the Mohicans* than with *The Pioneers*. None the less, even small details in the poem point towards *The Pioneers*. The lines from *Gertrude*,

---

[12] James Fenimore Cooper, *The Pioneers, or The Sources of the Susquehanna: A Descriptive Tale* (1823; repr. New York: New American Library, 1964), Epigraph.

[13] *The Complete Poetical Works of Thomas Campbell*, ed. J. Logie Robertson (London: Oxford University Press, 1907), 46 ('Gertrude of Wyoming', I. iv).

[14] Ibid. 47, 52, 68, 76.   [15] Cooper, *The Pioneers*, 431.

> In vain the desolated panther flies,
> And howls amidst his wilderness of fire   (I. xvii),

remind Cooper's reader of the episode where Elizabeth, the heroine,
is menaced by a panther before the eventual conflagration which
leads to Chingachgook's death. Chingachgook stoically sings himself
to death while boasting of his own exploits, recalling Campbell's
Oneyda, who sings of his own deeds as a youth, while adding,

> Then welcome be my death-song, and my death!   (III. xv),

and preaches (like Chingachgook) the Indian virtue of bloody revenge. Chingachgook laments the fact that he is the last of *his* tribe,
and Cooper's epigraph for that chapter (XXXVI) comes from *Gertrude
of Wyoming*:

> 'And I could weep'—th' Oneyda chief
> His descant wildly thus begun—
> 'But that I may not stain with grief
> The death song of my father's son.'   (III. xxxv.)

Again, in the chapter where Chingachgook dies (XXXVIII), the
epigraph—

> Even from the land of shadows, now,
> My father's awful ghost appears

—comes straight from *Gertrude of Wyoming*, III. xxxix. In writing a
book partly based on his own upbringing (his father had been a
judge), Cooper poured some of his memories into the convenient
literary mould of Campbell's American poem.

Appreciating these connections only adds to the known importance
of Scottish writing for Cooper. Like Campbell, he had little direct
knowledge of Indians, relying mainly on assembling writings about
them. As has long been recognized, his presentation of his primitives,
and the structuring of his novels, also owed a great deal to Sir Walter
Scott. Cooper was known in the nineteenth century as 'the American
Scott'; he was well read in the latter's works, not just the novels but
also the poetry, as some of the epigraphs to chapters in Cooper's
books bear witness, and as one would expect, given the remarkable
popularity of Scott's novels in nineteenth-century America.[16] Alongside the huge American impact of the Waverley Novels, we should

---

[16] On Scott's popularity in nineteenth-century America, see Hook, *Scotland and
America*, 145–52.

remember the extensiveness of Ossian's presence at the heart of American culture, an influence which ranged at least from 1773, when Jefferson stated that Ossian's works were 'to me the sources of daily pleasure' and that 'I think this rude bard of the North the greatest poet that has ever existed'.[17] Ossian was still important to Whitman when, in 1881, that creator of American epic could be inspired in crossing the Delaware to go to long passages from Ossian, where 'the Gael-strains chant themselves from the mists'.[18]

America had its own primitives and, like Scotland, its own need to unify its people (or peoples) in a national epic which would free the country from its position as an English cultural province. John P. McWilliams has pointed out that, as early as 1727, writers on the Indians could see them in terms of 'Homer's Heroes'.[19] But it was indirectly Ossian, and the immense interest in primitivism sparked off by the Ossianic poems, which led to the creation of the first *major* American epic, the Leatherstocking Tales. It was a prose epic, and it followed in the wake of the Scottish prose epic that succeeded Ossian: the Waverley Novels.

Cooper wrote of Scott that 'he raised the novel, as near as might be, to the dignity of the epic'.[20] But when, a few years later, Cooper concluded his 'Preface to the Leatherstocking Tales'—though, like Scott, he called his tales 'romances'—he implied that he, too, might have been engaged in Homeric activity.

It is the privilege of all writers of fiction, more particularly when their works aspire to the elevation of romances, to present the *beau idéal* of their characters to the reader. This it is which constitutes poetry, and to suppose that the redman is to be represented only in the squalid misery or in the degraded moral state that certainly more or less belongs to his condition, is, we apprehend, taking a very narrow view of an author's privileges. Such criticism would have deprived the world of even Homer.[21]

Both Scott and Cooper were writing sequences to entertain and to unify societies that were multicultural, and to give them pride in their distinctive national attributes. The way in which Scott formed a model for Cooper has long been recognized. The model was an

[17] Jefferson, *Writings*, 746.

[18] Walt Whitman, 'An Ossianic Night—Dearest Friends', in *CP*, 915–16.

[19] John P. McWilliams, 'Red Satan: Cooper and the American Indian Epic', in Robert Clark (ed.), *James Fenimore Cooper: New Critical Essays* (London: Vision Press, 1985), 144.   [20] Ibid. 149.

[21] James Fenimore Cooper, 'Preface to the Leatherstocking Tales', *The Deerslayer*, (1841; repr. New York: Bantam, 1982), p. x.

obvious one. By 1833 Rufus Choate in New England was comparing Scott with the *Iliad* and the *Odyssey*, and stressing, as the subtitle of his lectures demonstrates, 'The Importance of Illustrating New England History by a Series of Romances like the Waverley Novels'.[22] By 1833, though, Cooper was well on his way. And, by then, as Edwin Fussell points out in a footnote to his book *Frontier*, Scott had virtually instructed the Americans in how to imitate him. Scott had written that

He [Rob Roy] owed his fame in great measure to his residing on the very verge of the Highlands, and playing such pranks in the beginning of the 18th century, as are usually ascribed to Robin Hood in the middle ages,—and that within forty miles of Glasgow, a great commercial city, the seat of a learned university. Thus a character like his, blending the wild virtues, the subtle policy, and unrestrained licence of an American Indian, was flourishing in Scotland during the Augustan age of Queen Anne and George I. Addison, it is probable, or Pope, would have been considerably surprised if they had known that there existed in the same island with them a personage of Rob Roy's peculiar habits and profession. It is this strong contrast betwixt the civilized and cultivated mode of life on the one side of the Highland line, and the wild and lawless adventures which were habitually undertaken and achieved by one who dwelt on the opposite side of that ideal boundary, which creates the interest attached to his name.[23]

Rightly, Fussell in *Frontier* concentrates on the American-ness of his tradition, but, in doing so, he may be missing aspects of it to which we become more sensitive if we pay closer attention to some of the material hinted at in footnotes. We need to become more aware of the depth of the debt that American culture owed to Scottish modes and concepts.

Fussell emphasizes that 'the American frontier was sometimes a line and sometimes a space', but, according to the *OED* (4b), the first specific use of the American sense of this word to mean 'That part of a country which forms the border of its settled or inhabited regions' is in 1870.[24] In fact, the word which is often used in early American writing for what we now call the 'frontier' is the word 'border' as the title of the 1835 collection which Fussell mentions, James Hall's *Tales of the Border*, demonstrates. In *The Deerslayer* Cooper writes of 'frontier-

---

[22] Hook, *Scotland and America*, 162.

[23] Sir Walter Scott, 'Introduction' (1829) to *Rob Roy*, cited in Edwin Fussell, *Frontier: American Literature and the American West* (Princeton, NJ: Princeton University Press, 1965), 15–16.   [24] Fussell, *Frontier*, 17.

men' who 'pass their time between the skirts of civilized society and the boundless forests', but what these men lead is 'a border life'.[25] *The Prairie* celebrates 'that restless people which is ever found hovering on the skirts of American society', that is, it celebrates the life of 'an American borderer'.[26] Cooper's occasional use of the adjective 'American' to govern his noun 'borderers' might remind us that the best-known literary border at the time was that of Scott's *Minstrelsy of the Scottish Border*, the same border as had featured in Wordsworth's tragedy *The Borderers* (1795–6). Just as Scott's immense American popularity led to imitation of his novels, so it seems that the whole idea of the border as both a line and an area of strife and intercultural contact—that idea so crucial to Scott's novels about Scotland and to Scottish history—was an important cultural model in the United States.

Scott was far from being the inventor of the idea of the border as not merely a line but rather an area, a 'district adjoining this boundary on both sides' (*OED*, 3a). This usage goes back at least to the sixteenth century, when it seems to have originated as a peculiarly Scottish term, as seen in Sir David Lindsay's 'Baith throw the heland and the bordour' (1536). One recent writer on the Scots in America has stressed the importance of the idea of the Border in Scottish thought.

> The Border tradition is one of the central themes in Scotland's cultural history. As William Ferguson has argued, relations with England have always been of 'paramount importance' in Scottish history and were central to the forming of a Scottish national identity. In this sense, the whole of Scotland can be considered a Border region, as the perpetual presence of an ever-expanding society to the south and the legacy of national conflicts were factors of unusual importance in the creation of Scottish cultural symbols.[27]

The border, like that frontier so important to the American imagination, was a moving boundary as well as a geographical area. So, from at least the fifteenth century, that area itself was known in Scottish terms as the 'debatable land', tenanted by 'neutral men'. Though he does indicate in a footnote that Scott uses the term 'debatable land', Fussell stresses the importance 'of phrases like "neutral ground" and

[25] Cooper, *The Deerslayer*, 4–5.

[26] James Fenimore Cooper, *The Prairie* (1827; repr. New York: New American Library, 1964), 9, 113.

[27] Ned C. Landsman, *Scotland and its First American Colony, 1683–1765* (Princeton, NJ: Princeton University Press, 1985), 50.

"debatable land", which originated in the actual conditions of early American history'.[28] In the case of 'debatable land', though, this is clearly a Scottish term, taken into American usage to go with that notion of a 'border' which, from the Scottish model, developed along with, and into, what is now the more familiarly American 'frontier'.

The reason for stressing these Scottish connections is that they alert us to an important aspect of Cooper's writing which has received little detailed attention. Cooper is seen, rightly, as a novelist concerned with cultural eclecticism and, most specifically, with the coming-together of the cultural extremes of Indians and white colonists. So Leslie Fiedler claims of *The Last of the Mohicans* that 'Miscegenation is the secret theme of the world's favourite story of adventure, the secret theme of the Leatherstocking Tales as a whole', and points to the epigraph from *The Merchant of Venice* which Cooper chose for his novel —'Mislike me not for my complexion.'[29] Concern about miscegenation was prominent in American thought in the early nineteenth century, and Cooper is clearly anxious about the topic.[30] His protagonist Natty Bumppo protests time after time that he is 'a man without a cross', and though Cooper seems to subscribe to the idea of one human nature, he also stresses the separation of red and white 'gifts'. Perhaps the most striking passage on the theme of miscegenation in the Leatherstocking Tales comes in *The Last of the Mohicans*, when 'the Scotsman' Colonel Munro reveals that his wife was the remote descendant of a black West Indian slave. Munro sees the Scots' connection with slavery as resulting from the Union of Scotland and England, and, for the prejudice associated with his own mixed marriage, he apportions the blame to a 'curse entailed upon Scotland by her unnatural union with a foreign and trading people'. As Fiedler puts it, this is 'miscegenation between the Scottish noble savages and the mercantile pale-faces to their south!'[31] Cooper is reading back into the borders of his Scottish cultural model a racial dimension which is not present there. He is Americanizing Scotland. But, in the texture of his language, Cooper reveals that his America is already strongly Scotticized.

[28] Fussell, *Frontier*, 17–18.

[29] Leslie A. Fiedler, *Love and Death in the American Novel*, 3rd edn. (1982; repr. Harmondsworth: Penguin Books, 1984), 205, 202.

[30] See William Stanton, *The Leopard's Spots: Scientific Attitudes towards Race in America 1815–59* (Chicago: University of Chicago Press, 1960).

[31] James Fenimore Cooper, *The Last of the Mohicans* (1826; repr. New York: Bantam Books, 1981), 164; Fiedler, *Love and Death*, 208.

The borders which Scott investigated were not racial but cultural and linguistic ones, across which he delighted to translate his characters, such as the Waverley who learns to use the various names of Mr Mac-Ivor/ Glennaquoich/ Vich Ian Vohr and of the Prince/ Chevalier/ Pretender. Waverley finds himself 'the last of that race' (of Waverley), crossing into the 'primitive' territory of the 'new world' of Scotland, with its inhabitants' 'skins . . . burnt black'.[32] Yet one can see just how Scott provided a splendid model for Cooper's cultural eclecticism. Like Scott, Cooper gives us caves and waterfalls, native costume and civilized dress, footnotes on peculiar customs and details that reinforce the factuality of his text. The need to try and accommodate the 'wild men' to the law and the law to the 'borderers' is a constant theme in which Cooper follows Scott in *The Pioneers*, through *The Prairie*, to *The Deerslayer*. Cooper, too, is trying to further national unity while maintaining cultural identity. So he, like Scott, is attracted to mediating figures; the most prominent by far is Bumppo, in whom Indian and white meet, and who is able to interpret the one community to the other.

If Bumppo is a cultural translator, he is also translated. Even more than Scott's Jacobites, Bumppo is multi-named. He is, when he tells Hetty '*all*' his names in *The Deerslayer*, variously 'Nathaniel, or Natty' Bumppo, Straight-tongue, The Pigeon, Lap-ear, and Deerslayer.[33] But he is also 'La Langue Carabine', the Long Rifle, Hawkeye, the Leatherstocking, or the Pathfinder, depending on the company he is in. In *The Prairie* Natty makes fun of the comic Jeffersonian (or Bradwardinish) figure, the pedant Dr Battius, who is trying to systematize and tabulate nature: 'it is the man who wanted to make me believe that a name could change the natur' of a beast! Come, friend, you are welcome, though your notions are a little blinded with reading too many books.'[34] Just as he seems to stay outside the confines of the legal system, so Natty has little time for classification. He is the plain man, as a passage in *The Prairie* makes clear.

'Lord, lad, I've been called in my time by as many names as there are people among whom I've dwelt. Now the Delawares named me for my eyes, and I was called after the farsighted hawk. Then ag'in, the settlers in the Otsego hills christened me anew from the fashion of my leggings; and various have been the names by which I have gone through life; but little will it matter

[32] *W*, 72, 74–5, 434.   [33] Cooper, *The Deerslayer*, 48–9.
[34] Cooper, *The Prairie*, 102.

when the time shall come that all are to be mustered, face to face, by what titles a mortal has played his part! I humbly trust I shall be able to answer to any of mine in a loud and manly voice.'[35]

But part of the reader's pleasure in following this 'plain man', as in following figures in Scott's novels, is to watch the way in which he slips from name to name, as from language to language, with a shape-changer's ease. Though written in English, Cooper's Leatherstocking Tales make us continually aware of an implicit linguistic eclecticism. The predominant trend in criticizing Cooper has been to stress that he formed an epic myth but to ignore the language he used, since 'The Leatherstocking tales are not in any ordinary sense great art', and 'Cooper had, alas, all the qualifications for a great American writer except the simple ability to write.'[36] To say this, though, is to miss the constant fascination of the slippage between various textures of language which themselves represent different languages. There is a latent eclecticism in Cooper's language, as there is in his cultural mythology, which continually enlivens the surface of the text, and is perhaps most clearly seen in the matter of naming, with which it is closely associated.

'Though no fish, The Long Rifle can swim. He floated down the stream when the powder was all burnt, and when the eyes of the Hurons were behind a cloud.' . . .
'Can the Delawares swim, too, as well as crawl in the bushes? Where is le Gros Serpent?'
Duncan, who perceived by the use of these Canadian appellations, that his late companions were much better known to his enemies than to himself, answered, reluctantly, 'He also is gone down with the water.'
'Le Cerf Agile is not here?'
'I know not whom you call "The Nimble Deer",' said Duncan, gladly profiting by any excuse to create delay.
'Uncas', returned Magua, pronouncing the Delaware name with even greater difficulty than he spoke his English words. ' "Bounding Elk" is what the white man says, when he calls to the young Mohican.'
'Here is some confusion in names between us, Le Renard,' said Duncan, hoping to provoke a discussion. '*Daim* is the French for deer, and *cerf* for stag; *élan* is the true term, when one would speak of an elk.'[37]

[35] Ibid. 178.
[36] Fussell, *Frontier*, 68; Fiedler, *Love and Death*, 191.
[37] Cooper, *The Last of the Mohicans*, 90.

This is a striking example of what happens throughout the Leather-stocking Tales. The transition from one language to another is usually crudely indicated, frequently in ways prejudicial to the Indians, but often also reflecting badly on the 'superior' whites. This simply makes the reader all the more aware of crossing linguistic boundaries, pioneering into language, and eclectically gathering languages together in the text. Such an effect is central to the texture of Cooper's writing, as varieties of English are brought together to do duty for different languages and dialects.

'Look you here, old graybeard,' said Ishmael, seizing the trapper and whirling him round as if he had been a top; 'that I am tired of carrying on a discourse with fingers and thumbs, instead of a tongue, are a natural fact; so you'll play linguister and put words into Indian without much caring whether they suit the stomach of a redskin or not.'[38]

Time and time again, Cooper makes us aware of linguistic differences. When Hurry Harry speaks of ' "we'pons" . . . in his uncouth dialect', or Oliver Effingham asks 'with a pronunciation and language vastly superior to his appearance, "With how many shot did you load your gun?" ' Cooper is acknowledging standards of correctness and propriety as much as Blair and the other Scottish rhetoricians whose works were so widely used in the United States' educational system to strive to avoid the English of the provincial.[39] But when Cooper's greatest hero speaks, linguistic and other 'proprieties' are delightfully disrupted: 'I took him on his posteerum, saving the lady's presence, as he got up from the ambushment, and rattled three buckshot into his naked hide, so close that you might have laid a broad joe upon them all.'[40]

The linguistic boundaries are sensed most clearly when the Indians speak, whether in their native tongues, 'the original distinctions between these [Indian] nations' being 'marked by a difference in language', or in their distinctive Indian English: 'Mohegan now spoke, in tolerable English, but in a low, monotonous, guttural tone: "The children of Miquon do not love the sight of blood; and yet the Young Eagle has been struck by the hand that should do no evil!" '[41] In *The Prairie* characters move not only on the frontiers of white civilization, but also on the borders of various languages as they are

---

[38] Cooper, *The Prairie*, 307.
[39] Cooper, *The Deerslayer*, 68; Cooper, *The Pioneers*, 22.
[40] Cooper, *The Pioneers*, 24.   [41] Ibid. 78, 82.

involved in tribal conflicts. Cooper keeps us well aware of these linguistic borders and the points where he gathers them together:

he gave the usual salutation in the harsh and guttural tones of his own language. The trapper replied as well as he could, which it seems was sufficiently well to be understood. In order to escape the imputation of pedantry we shall render the substance and, so far as it is possible, the form of the dialogue that succeeded into the English tongue.[42]

The show which Cooper makes of translating his Indian languages is a device adapted from Scott, who, for example, in the chapter of *Waverley* entitled 'Highland Minstrelsy', uses the highlighting of a language barrier to increase the alien-ness and romantic fascination of his subject. Flora is introduced to Waverley as 'a translator of Highland poetry' for the benefit of Captain Waverley, who (with one of those canny satiric touches which Scott uses to send up hyper-Romanticism) 'is a worshipper of the Celtic muse; not the less so perhaps that he does not understand a word of her language'. So, after long protestations about the difficulty of translating Gaelic poetry into English, Flora obliges, as does Scott, with the apology that 'The following verses convey but little idea of the feelings with which, so sung and accompanied, they were heard by Waverley.'[43] As suggested in the previous chapter on 'Anthropology and Dialect', Scott can also emphasize language boundaries to stress confusion or babel. Cooper often uses the device of language change to stress his Indians' otherness, or their sinisterness, but there is also a Romantic fascination, because both the author of *The Last of the Mohicans* and the author of *The Lay of the Last Minstrel* were fascinated by the Romantic theme of lastness. Scott's notes and Cooper's are memorials of folklore: 'These practices have nearly disappeared . . .'. Their texts gather together passing cultural fragments and Romantically memorialize them. Cooper hardly has decaying castles, but he has decaying cultures to be frequently lamented. The few ruins Cooper does have are tied to ruined cultures. Assuring us of the accuracy of his material in a Scott-like note, Cooper takes his characters and readers to both inanimate and animate noble signs of ruin.

While Heyward and his companions hesitated to approach a building so decayed, Hawkeye and the Indians entered within the low walls, not only without fear, but with obvious interest. While the former surveyed the ruins, both internally and externally, with the curiosity of one whose recollections

---

[42] Cooper, *The Prairie*, 45.   [43] *W*, 171–2, 178.

were reviving at each moment, Chingachgook related to his son, in the language of the Delawares, and with the pride of a conqueror, the brief history of the skirmish which had been fought, in his youth, in that secluded spot. A strain of melancholy, however, blended with his triumph, rendering his voice, as usual, soft and musical.

The melancholy comes from the circumstance which Hawkeye points out to his companions with reference to Chingachgook: 'you see before you all that are now left of his race'.[44] Cooper's best-known book is a memorialization of a culture he sees as vanishing, and it ends with a funeral oration spoken in 'Indian English' by the centenarian wisdom-figure, Tamenund:

'It is enough,' he said. 'Go, children of the Lenape, the anger of the Manitou is not done. Why should Tamenund stay? The pale-faces are masters of the earth, and the time of the redmen has not yet come again. My day has been too long. In the morning I saw the sons of Unamis happy and strong; and yet, before the night has come, have I lived to see the last warrior of the wise race of the Mohicans.'[45]

The Pioneers also memorializes, and ends with the erection of an actual memorial. One of the most effective passages in the book is when the illiterate Natty asks Effingham to read the inscription on Chingachgook's gravestone to him, and, through a highlighting of certain small linguistic details and Natty's desire for their accurate preservation, Cooper makes the reader aware of a language and an entire culture whose memorializing is also the sign of its loss.

Effingham guided his finger to the spot, and Natty followed the windings of the letters to the end with deep interest, when he raised himself from the tomb, and said:
'I suppose it's all right; and it's kindly thought, and kindly done! But what have ye put over the Redskin?'
'You shall hear—
"This stone is raised to the memory of an Indian Chief, of the Delaware tribe, who was known by the several names of John Mohegan; Mohican—"
'Mo-hee-can, lad, they call theirselves! he-can.'
"Mohican; and Chingagook—"
'Gach, boy;—gach-gook; Chingachgook, which, intarpreted, means Big-sarpent. The name should be set down right, for an Indian's name has always some meaning in it.'

[44] Cooper, *The Last of the Mohicans*, 21, 127, 128.   [45] Ibid. 374.

'I will see it altered. "He was the last of his people who continued to inhabit this country; and it may be said of him that his faults were those of an Indian and his virtues those of a man." ' [46]

This passage, a passage to do with linguistic boundaries, is one of the greatest moments in Cooper's fiction. It also alerts us to the central dilemmas of his cultural and linguistic eclecticism. For Cooper, as has been pointed out, knew little of Indian languages; he learned most of his Indian knowledge from books. Like the speech of Scott's Gaels, Cooper's Indian language, when it seems most removed from English, is largely concocted 'translatorese'. The temptation to do this in constructing a work to give nobility to a dying, yet nationalistically important, culture is the same impulse as the one that produced the Ossianic epic. The problem with Cooper is that his bid to represent Indian life is also a misrepresentation.

Even in his lifetime Cooper was attacked for having Indians who were 'not of the school of nature' but were drawn from books. This criticism may seem simplistic, given Roy Harvey Pearce's demonstration of the extent to which the prevalent ideas about Indian nature were themselves heavily dependent on Scottish Enlightenment versions of Indians. [47] None the less, when we read the criticism that 'this bronze noble of nature, is then made to talk like Ossian for whole pages . . . practising for a poetic prize', it has uncomfortable bite. [48] Natty Bumppo, uniting white and Indian traits, 'who had imbibed, unconsciously, many of the Indian qualities, though he always thought of himself as a civilized being, compared with even the Delawares', is not just a unifying figure; he is also a take-over bid. [49] For, just as Cooper takes over the Indian's speech by giving us the illusion of that speech while expelling the actual language, so Natty takes from the Indian those qualities which appeal particularly to the whites, but refuses other red 'gifts', including any union with redness itself. For all the heroic pathos associated with the solitary Natty, constantly on the move from the forces of regularization, and for all his tangles with settlements (whether those of Dr Battius or of Judge Temple's vehement supporters), Natty is himself a dispossessor. He takes its 'virtues' from the weak culture, and, having incorporated these into the white world, he leaves the Indian culture all the more expendable.

[46] Cooper, *The Pioneers*, 431.   [47] Pearce, *Savagism and Civilization*, 82 ff.
[48] Grenville Mellen (1828) quoted in Dekker and McWilliams, *Fenimore Cooper*, 142.
[49] Cooper, *The Pioneers*, 432.

The same questions about cultural imperialism which need to be asked about the eclecticism of Modernist texts need also to be asked about the Leatherstocking Tales, whose eclecticism looks towards much later writing. Cooper's anxiety about miscegenation shows that he was worried about the nature of eclecticism, but it may also have heightened his imperialism. One recent critic has attempted to make much of a passage in *The Pioneers* where Elizabeth confuses Delaware with Latin and Greek. Eric Cheyfitz suggests that this is a far-reaching upset of cultural paradigms.[50] But Cooper seems to me simply to be joking, and it is debatable how much we should build on such a remark. Certainly, the joke is repeated elsewhere, as when Ellen in *The Prairie* mistakes Dr Battius's pedantic Latin for 'the Indian languages'.[51] In Chapter XIX of *The Pathfinder* Mabel tells the traitorous Scotsman, Mr Muir, that Natty 'may not understand Latin, but his knowledge of Iroquois is greater than that of most men, and it is the more useful language of the two, in this part of the world'. This certainly upsets ideas about cultural dominance, but it is equally part of the constant stress on Natty as illiterate but knowledge-able, an American scholar of nature rather than a European pedant. Though he comes to hanker after the Bible, Natty has never opened a book, but he can read the land like an Indian, and he has gathered various cultures to himself. In this eclecticism, this increasingly Romantic stress on nature as the best book, and, in *The Pathfinder* and *The Deerslayer* particularly, on God as being in the solitude of the woods, Cooper looks towards Emerson, as well as drawing on Wordsworthian Romanticism.[52]

The theme of nature versus formal scholarship parallels the theme of border freedom versus law in the Leatherstocking Tales. But accompanying both of these is the delight in slipping from register to register, from language to language (as from name to name), in a way which shows Cooper to be a Janus figure, looking back towards Scott but also looking forward to the cultural and linguistic eclecticism of Whitman, Pound, and Eliot. The meeting of 'every distant tongue', as Campbell put it, mixes with humour the French, German, Cornish, Scots, and English registers of the settlement in *The Pioneers*, just as the buildings of that settlement display a strange eclecticism in their improvisatory architecture. But when this interlinguistic

---

[50] Eric Cheyfitz, 'Literally White, Figuratively Red: The Frontier of Translation in *The Pioneers*', in Robert Clark, *James Fenimore Cooper*, 55–95.
[51] Cooper, *The Prairie*, 131.   [52] See e.g. Cooper, *The Deerslayer*, 383.

slippage is seen as extending throughout the Leatherstocking Tales as cultures meet and converse, it can be seen as one of the central pleasures of Cooper's text. Often it may be a melodramatic and compromised pleasure, even an act of racism or cultural imperialism, but it is a method of writing which places Cooper firmly in the United States' eclectic tradition. Fussell wrote of the Leatherstocking Tales that 'the rest of American writing through Whitman is a series of footnotes on them'.[53] But Fussell was referring mainly to the mythic plotting of the novels. What we must see now is that, in terms of their *linguistic* as well as their general cultural eclecticism, these novels are of great importance in the development of an American tradition.

If Cooper followed Scott in bringing together various registers and types of English in a linguistic eclecticism that paralleled his drawing-together of various cultures, then, from Emerson's American perspective, the English language has already done this job for itself, being already a splendidly eclectic construction: 'it is never my practice to read any Latin, Greek, German, Italian, scarcely any French book in the original which I can procure in an English translation. I like to be beholding to the great metropolitan English speech, the sea which receives tributaries from every region under heaven, the Rome of nations . . .'.[54] Translation partly levels down cultural difference, but Emerson may even have found this attractive, since it manifested the unity of all peoples. English was a tongue uniting not just Britain and America, but innumerable other countries and cultures also. The English were 'many-headed' and 'many-nationed', since 'their colonization annexes archipelagoes and continents, and their speech seems destined to be the universal language of men'.[55] English as 'our commercial and conquering tongue' was initially the speech of a culture whose strengths, in Emerson's view, derived from its being composite (*EL*, 670). 'The best nations are those most widely related . . . The English composite character betrays a mixed origin. Everything English is a fusion of distant and antagonistic elements. The language is mixed; the names of men are of different nations,—three languages, three or four nations . . .' (*EL*, 793). What was more, the immense, world-wide impact of the English model meant that 'The Russian in his snows is aiming to be English. The Turk and Chinese also are making awkward efforts to be

    53 Fussell, *Frontier*, 68.
    54 Ralph Waldo Emerson, journal entry cited in Joseph G. Kronick, *American Poetics of History* (Baton Rouge, La.: Louisiana State University Press, 1984), 28.
    55 *EL*, 931.

English.' (*EL*, 785.) In writing about the English, Emerson some-times seems to be holding up both a model and a warning to his American countrymen. Referring to the English as 'tribes', he seems to hint that they, too, partake of the barbaric. He makes the explicit point that 'The American is only the continuation of the English genius into new conditions.' (*EL*, 785.) In making such statements, Emerson reminds us of those eighteenth-century Scots who, wishing to profit from English models, made their society sound as close to English culture as possible.

At the same time, though, the English language's reception of international 'tributaries' to 'the Rome of nations', as the 'conquer-ing tongue' of 'colonization', makes it quite clear that linguistic dominance creates a cultural empire. Emerson's American, who, despite having won a war for independence, still finds Britain 'over-powering', must be particularly strong in resisting English cultural imperialism. The technique which Emerson suggests, and uses, is to be *more* eclectic oneself, so as to achieve a wider comparative per-spective which will serve to contain England's power.

England has inoculated all nations with her civilization, intelligence, and tastes; and, to resist the tyranny and prepossession of the British element, a serious man must aid himself, by comparing with it the civilizations of the farthest east and west, the old Greek, the Oriental, and, much more, the ideal standard, if only by means of the very impatience which English forms are sure to awaken in independent minds. (*EL*, 785.)

Emerson's is an independent mind not only because he comes from a nation where one of the key texts is the Declaration of its own Independence, but also because he is the exponent of self-reliance, anxious to attack 'worship of the past', since 'The centuries are conspirators against the sanity and authority of the soul.' (*EL*, 270.) In terms of cultural production, this means that the American, as 'the new man must feel that he is new, and has not come into the world mortgaged to the opinions and usages of Europe, and Asia, and Egypt'. He needs to sense his 'spiritual independence' (*EL*, 97).

However, while Emerson appears to be sloughing off older models, he is actually incorporating them in a strategy for being at once new and yet containing the past, a strategy which post-Emersonian American writers also adopt. The New Man is the ultimate gatherer as well as the ultimate gathering. As Emerson puts it: 'a man is the whole encyclopedia of facts', and 'can live all history in his own person' (*EL*, 237, 239).

In his attempt to develop an American Literature, Emerson pursues parallel and apparently contradictory strategies. On the one hand, he calls repeatedly for a distinctively new art for the New World; on the other hand, he fills his text with cultural fragments from the Old, and proclaims that 'the inventor only knows how to borrow' and that 'Every book is a quotation' (*EL*, 634). In using these two strategies, Emerson makes it clear that at the heart of his work is the major American concern with the question of tradition and the individual talent, of originality and genealogy. Emerson also wants to make it new. 'What is a man born for but to be a Reformer, a Re-Maker of what man has made . . .' (*EL*, 146). Every personal character reacts on an old form and 'makes it new' (*EL*, 183). In presenting the new, then, Emerson, like Pound and Eliot after him, is simply re-presenting the old. He makes it clear in the opening sentence of 'The Transcendentalist' that 'The first thing we have to say respecting what are called *new views* here in New England, at the present time, is, that they are not new, but the very oldest of thoughts cast into the mould of these new times.' (*EL*, 193.) Making it new means making it old. In this New World we are 'new-born' in 'a country of beginnings', but Emerson's essay on 'The Young American' ends by pointing out that 'This land, too, is as old as the Flood.' (*EL*, 217, 230.) 'America is a poem' in Emerson's eyes, but much of her language is old and English (*EL*, 465).

Emerson's ideal poet is far less of an instigator than an integrator: 'He whose eye can integrate all the parts, that is, the poet.' (*EL*, 9.) This view of the poet as integrator, a bringer-together, puts him in Orphic harmony with the unity of nature seen as 'unity in variety' (*EL*, 29). So Emerson can pass easily from astronomy, gravitation, and chemistry, through 'such poets as Newton, Herschel and Laplace', to quotations from, and accounts of, Zoroaster and Heraclitus, then on through Christianity to the passages from Brahmins about Vishnu, before touching down in 'Concord, or Lexington, or Virginia'—all in the space of four pages from an essay entitled 'The Method of Nature' (*EL*, 126–29). 'The whole of nature is a metaphor of the human mind.' (*EL*, 24.) 'Life is our dictionary', then, all the more so because 'the biography of one foolish person we know is, in reality, nothing less than the miniature paraphrase of the hundred volumes of the Universal History' (*EL*, 61, 422). Small wonder, then, that Emerson's prose ranges so far and wide, and that a stress on beginning afresh is bound up with a huge gathering of fragments of quotation and allusion, an encyclopaedic eclecticism.

In his endeavours, Emerson is strengthened by the example of Goethe, the 'philosopher of . . . multiplicity' who shared with Emerson the position of being 'provincial', yet managed to write in such a way that 'there is no trace of provincial limitation in his muse' (*EL*, 751). This preoccupation with the predicament of the provincial who sought to transcend provincialism was Emerson's American obsession, and was crucial in his admiration of Carlyle.

Carlyle's importance to Emerson is a theme as undeniable as it is hackneyed. The record of the 'bibliopoly' in which Emerson engaged on Carlyle's behalf, and, to a lesser extent, Carlyle on Emerson's, is remarkable. Yet, arguably, much of Carlyle's importance for Emerson has gone unnoted. To see this, and to see why it matters, it is worth looking first at another of Emerson's major cultural models, Wordsworth, and at the virtues which Emerson identified in him. Joel Porte has suggested that maybe Wordsworth, rather than Bronson Alcott, is the Orphic poet made famous by Emerson.[56] Certainly, Emerson accords Wordsworth the highest praise, stressing that 'he has done more for the sanity of this generation than any other writer' (*EL*, 1254). Wordsworth's contribution comes from his having forsaken 'the styles, and standards, and modes of thinking of London and Paris, and the books read there, and the aims pursued', devoting himself instead to 'Helvellyn and Winandermere'. He called into question 'the conduct of life . . . on wholly new grounds', and appeared to move away from conventional religion to 'the lessons which the country muse taught a stout pedestrian climbing a mountain' (*EL*, 1254–5). Wordsworth was important to Emerson as a provincial as well as a poet, and as one interested in examining the natural world for its spirit in a new, unconventionally religious way. Emerson himself wrote on *The Conduct of Life*, and much of his work might be regarded as a vast gloss on the opening poem of *Lyrical Ballads*, 'Expostulation and Reply', in which Wordsworth defends his lying alone in the country as conversing with nature: 'The eye it cannot choose but see.' Emerson celebrated himself as a huge transparent eyeball, and, as a new man keen to free himself from the weight of the past, would have known, like Wordsworth, how to reply to the expostulation:

[56] Joel Porte, *Representative Man: Ralph Waldo Emerson in his Time* (New York: Oxford University Press, 1979), 73.

> 'Where are your books?—that light bequeathed
> To beings else forlorn and blind!
> Up! up! and drink the spirit breathed
> From dead men to their kind.'[57]

If Emerson, like Wordsworth, could minimize his debt to literature for strategical purposes, then surely it was the other side of him, the compulsive literary eclectic, that warmed to Carlyle. In a sense it was, but that sense needs to be looked at carefully. In 1847 Emerson wrote to his wife from London, making a rather surprising comparison. He wrote that Carlyle was like 'a very practical Scotchman, such as you would find in any saddler's or iron dealer's shop', and then he went on to make a more detailed comparison between Carlyle and the Emersons' gardener: 'Suppose that Hugh Whelan had had leisure enough in addition to all his daily work, to read Plato, & Shakespeare, & Calvin, and, remaining Hugh Whelan all the time, should talk scornfully of all this nonsense of books that he had been bothered with,—and you shall have just the tone & talk & laughter of Carlyle.'[58] Carlyle appeals here not only as the polymath, the man with the encyclopaedic mind and conversation, but also as the man who can dismiss all that and be, literally, down to earth—like a gardener. Such a Carlyle reflects both Emerson the encyclopaedic integrator, and the Emerson who grew his own produce while emphasizing the Wordsworthian need to abandon books for nature. It is this combination that he relished in Carlyle from the first, when, in 1833, attracted to Britain, as he put it, largely through the *Edinburgh Review*, Emerson set off to locate the author of those articles on German Literature and thought which had been of great importance to himself and to his circle (*EL*, 767). Again, what is noticeable in Carlyle is not just the Emersonian quality of self-reliance, but the combination of determined, isolated provincialism among the wilds of nature, linked with an insatiable appetite for knowledge. Emerson's account stresses the wide-ranging conversation he had with Carlyle, a conversation whose topics moved from Nero, to pig-keeping, to America, to Plato, to the immortality of the soul. Emerson writes that Carlyle 'was already turning his eyes towards London with a scholar's appreciation', yet

---

[57] William Wordsworth and Samuel Taylor Coleridge, *Lyrical Ballads*, 1805 edn., ed. Derek Roper, 2nd edn. (London: MacDonald and Evans, 1976), 50 ('Expostulation and Reply').

[58] Quoted in *The Correspondence of Emerson and Carlyle*, ed. Joseph Slater (New York: Columbia University Press, 1964), 33.

Carlyle, who lives within sight of 'Wordsworth's country', is introduced not as the polymath, but as the provincial (*EL*, 775, 773).

From Edinburgh I went to the Highlands. On my return, I came from
Glasgow to Dumfries, and being intent on delivering a letter which I had
brought from Rome, inquired for Craigenputtock. It was a farm in Nithsdale,
in the parish of Dunscore, sixteen miles distant. No public coach passed near
it, so I took a private carriage from the inn. I found the house amid desolate
heathery hills, where the lonely scholar nourished his mighty heart. Carlyle
was a man from his youth, an author who did not need to hide from his
readers, and as absolute a man of the world, unknown and exiled on that hill-
farm, as if holding on his own terms what is best in London. He was tall and
gaunt, with a cliff-like brow, self-possessed, and holding his extraordinary
powers of conversation in easy command; clinging to his northern accent with
evident relish; full of lively anecdote, and with a streaming humor, which
floated every thing he looked upon.

Yet, in this remote setting and in only twenty-four hours, Emerson
and Carlyle held a conversation which ranged, as the latter put it,
'thro' the whole Encyclopedia'.[59]

At once provincial and encyclopaedic in his cultural scope, Carlyle
resembled his own Teufelsdröckh in the *Sartor Resartus* which Emerson
so eagerly devoured and saw through the American press at his own
expense. Teufelsdröckh's 'six considerable PAPER-BAGS', with their
supremely eclectic blend of knowledges, appear like an anticipatory
caricature of Emerson's essays.[60] Carlyle stressed 'the highest of all
possessions, that of Self-help'; Emerson would stress 'self-reliance' as
well as 'self-help'. Teufelsdröckh apostrophizes Voltaire, pointing
out that it is not enough to make clear

That the Mythus of the Christian Religion looks not in the eighteenth century
as it did in the eighth. Alas, were thy six-and-thirty quartos, and the six-and-
thirty thousand other quartos and folios, and flying sheets or reams, printed
before and since on the same subject, all needed to convince us of so little! but
what next? Wilt thou help us to embody the divine Spirit of that Religion in a
new Mythus, in a new vehicle and vesture, that our Souls, otherwise too like
perishing, may live?[61]

For Emerson, writing in a New England where the Puritans' 'creed is
passing away, and none arises in its room', the great need and search
is also for a new mythus, rather than a mere lamenting of the old
which is becoming increasingly less relevant (*EL*, 87). 'Why should

[59] Ibid. 12.   [60] *SR*, 61.   [61] Ibid. 92, 154.

not we also enjoy an original relation to the universe? Why should not we have a poetry and philosophy of insight and not of tradition, and a religion by revelation to us, and not the history of theirs?' (*EL*, 7.) Like Teufelsdröckh, Emerson looks at the stars from the city and feels their grandeur. Where Carlyle prophetically recasts the Delphic Oracle's *Know thyself* as 'Know what thou canst work-at', Emerson stresses that 'A man is fed, not that he may be fed, but that he may work', and he, too, returns to Delphi. 'And, in fine, the ancient precept, "Know thyself," and the modern precept, "Study nature," become at last one maxim.' Elsewhere he rewrote it as 'Obey thyself' (*EL*, 13, 56, 81).[62] In confronting his 'Everlasting No', Teufelsdröckh rebelled by stressing his own freedom.[63] In facing fearful matters, Emerson stated that his scholar must have 'self-trust' and be 'free and brave' (*EL*, 65). Discussing the 'Mythus' of Christianity, Emerson went on to stress all nature as a miracle, as Teufelsdröckh had done in *Sartor Resartus*, in the chapter headed 'Natural Supernaturalism' which tended to Transcendentalism and which saw, as did the New England Transcendentalists, nature as a huge book, 'a Volume written in celestial hieroglyphs' (*EL*, 80). Where Teufelsdröckh stated that 'Custom . . . doth make dotards of us all', and so sought for his 'new Mythus', he was followed by the searcher Emerson, who complained that 'Our age is retrospective' (*EL*, 7).[64]

Carlyle's hope, like Emerson's, is that a new prophet will arise to bind together the fragments of culture and so herald a new religion. Carlyle even has his Professor name this master eclectic, who turns out to be Emerson's encyclopaedist, Goethe.

'But there is no Religion?' reiterates the Professor. 'Fool! I tell thee, there is. Hast thou well considered all that lies in this immeasurable froth-ocean we name LITERATURE? Fragments of a genuine Church-*Homiletic* lie scattered there, which Time will assort: nay fractions even of a *Liturgy* could I point out. And knowest thou no Prophet, even in the vesture, environment, and dialect of this age? None to whom the God-like had revealed itself, through all meanest and highest forms of the Common; and by him been again prophetically revealed: in whose inspired melody, even in these rag-gathering and rag-burning days, Man's Life again begins, were it but afar off, to be divine? Knowest thou none such? I know him, and name him—Goethe.[65]

For various reasons, Emerson was unable to see Goethe quite as the poet-prophet he desired; yet there are signs that he considered

62 Ibid. 132.
63 Ibid. 135.   64 Ibid. 206.   65 Ibid. 201–2.

Carlyle for the post. Where Cooper had admired the Waverley Novels as a modern epic, Emerson is attracted to Carlyle's work for just the same reason. Emerson writes of *Sartor Resartus* as an 'Epical Song', a 'noble philosophical poem', and again speaks of Carlyle as a poet in his review of *Past and Present*, which he greets as 'Carlyle's new poem, his Iliad of English woes, to follow his poem on France, entitled the History of the French Revolution' (*EL*, 1263).[66] Earlier, linking poetry and history, Emerson had seen Carlyle as a remaker of history. 'Since Carlyle wrote French History, we see that no history, that we have, is safe, but a new classifier shall give it new and more philosophical arrangement.' (*EL*, 103.) For Emerson, Carlyle not only arranged history, he managed, in an effort of supreme eclecticism, to bring about

the first domestication of the modern system with its infinity of details into style. We have been civilizing very fast, building London and Paris, and now planting New England and India, New Holland and Oregon,—and it has not appeared in literature,—there has been no analogous expansion and recomposition in books. Carlyle's style is the first emergence of all this wealth and labor, with which the world has gone with child so long. London and Europe tunnelled, graded, corn-lawed, with trade-nobility, and east and west Indies for dependencies, and America, with the Rocky Hills in the horizon, have never before been conquered in literature. This is the first invasion and conquest. (*EL*, 1268–9.)

Emerson sees Carlyle's achievement in terms of 'expansion and *re*composition' (my emphasis), that is, a new placing-together, suitable for the New World. It is such a recomposition that Emerson seeks, and his relations with Carlyle have been perceived in terms of a 'long debate', one in which the Scot sees Emerson as increasingly 'moonshiny' in his indefatigable optimism, while Emerson sees Carlyle's increasingly gloomy intolerance.[67] Significantly, Emerson's best-known account of a meeting with Carlyle begins with an encyclopaedic conversation that took place as the two men journeyed to Stonehenge, moving from Goethe and Schiller to Confucius and America, with Emerson speculating 'that England, an old and exhausted island, must one day be contented, like other parents, to be strong only in her children', i.e. the Americans (*EL*, 916). Ever keen to leap across history and to see humanity as one, Emerson considers the construction of Stonehenge in terms of a Boston builder's yard. 'I chanced to see a year ago men at work on the substructure of a house in Bowdoin

[66] Emerson and Carlyle, *Correspondence*, 16.   [67] Ibid. 42.

Square, in Boston, swinging a block of granite of the size of the
largest Stonehenge columns with an ordinary derrick.' (*EL*, 920.)

What is comically shocking here is Emerson's abolition of the difference between prehistoric Britain and nineteenth-century Boston.
Emerson is in the position of the archaeologist Belzoni, whom he
praises for trying to 'see the end of the difference' between a remote,
ancient culture and himself (*EL*, 241). Emerson's eclectic universalism often tends to bring cultures together and dismiss what sets them
apart. So, in a small instance, Emerson censures poetic admirers of
Campbell and Scott who seek simply to replace Scottish peculiarities
by Indian peculiarities. 'The most Indian thing', Emerson contends,
'about the Indian is surely not his moccasins, or his calumet, his
wampum, or his stone hatchet, but traits of character and sagacity,
skill or passion; which would be intelligible at Paris or at Pekin, and
which Scipio or Sidney, Lord Clive or Colonel Crockett would be as
likely to exhibit as Osceola and Black Hawk.' (*EL*, 1222.) The
problem with such a view, when taken to extremes, is that it ends up
abolishing differences, and reducing all to one dominant, all-pervading
ethos. This parallels Emerson's enthusiasm for English as a language
in which all other languages come together, in an eclecticism which
can also be seen as an imperialism. Emerson seems to champion individual self-reliance, and even sides with Scott in a wish to maintain
cultural distinctiveness, when he maintains that: 'Vich Ian Vohr
must always carry his belongings in some fashion, if not added as
honor, then severed as disgrace.' (*EL*, 520.) But more often, it seems,
Emerson is concerned with a universality of manners which leads
him, for instance, to approve of the oddest gatherings as an apparent
proof of the universality of human nature, which, for him, means that
'good-breeding and personal superiority of whatever country readily
fraternize with those of every other. The chiefs of savage tribes have
distinguished themselves in London and Paris, by the purity of their
tournure.' (*EL*, 519.) So, the Indians are really just like us, at their
best. The gathering-together of all cultures may simply mean a
remaking of them in your own image. 'Self-Reliance' proclaims that
'to believe that what is true for you in your private heart is true for all
men—that is genius' (*EL*, 259). But this is also a form of egotistical
imperialism, the imposition of an individual view over all, so that the
constituting differences of other people and cultures are done away
with, or can be manipulated to suit a megalomaniac ordering, as 'All
history becomes subjective.' (*EL*, 240.)

In Emerson, such a view is clearly an attempt to beat the burden of the past, and particularly of cultural imperialism, as he urges that the reader 'must transfer the point of view from which history is commonly read, from Rome and Athens and London to himself' (*EL*, 240). Yet it also represents an attempt to set up a dictatorship over history, by denying its exteriority. The new man may cut himself off from the old world by incorporating it into himself, and so negating its independent existence. Fact becomes fiction, and the whole is gathered in the all-powerful mind of the writer.

No anchor, no cable, no fences, avail to keep a fact a fact. Babylon, Troy, Tyre, Palestine, and even early Rome, are passing already into fiction. The Garden of Eden, the sun standing still in Gibeon, is poetry thenceforward to all nations. Who cares what the fact was, when we have made a constellation of it to hang in heaven an immortal sign? London and Paris and New York must go the same way. 'What is History', said Napoleon, 'but a fable agreed upon?' This life of ours is stuck round with Egypt, Greece, Gaul, England, War, Colonization, Church, Court, and Commerce, as with so many flowers and wild ornaments grave and gay. I will not make more account of them. I believe in Eternity. I can find Greece, Asia, Italy, Spain, and the Islands,— the genius and creative principle of each and of all eras in my own mind. (*EL*, 240.)

The stress on the primary and absolute nature of the self leads to a realization that social and moral codes are merely impositions on history: 'Good and bad are but names very readily transferable to that or this.' (*EL*, 262.) Such 'scepticism' in Emerson appealed particularly to Nietzsche, who knew Emerson's work from his childhood.[68] Other aspects, too, are likely to have appealed to him. For Emerson, 'men's prayers are a disease of the will', since the strong individual is able to have a mind which 'imposes its classification on other men' (*EL*, 276, 277). 'Life' for Emerson 'is a search after power', and 'The only sin is limitation.' (*EL*, 971, 406.) In such pronouncements, the eclectic Emerson who learned from Carlyle is both excitingly and sinisterly modern. His project (like Carlyle's) anticipates developments in twentieth-century theories of the way in which language shapes and even constitutes the world. Yet Emerson can also be seen as wishing to collect and order material in a megalomaniac and proto-Fascist way that recalls some of the darker pro-

---

[68] Friedrich Nietzsche, *On the Genealogy of Morals* and *Ecce Homo*, ed. Walter Kaufmann (New York: Vintage, 1969), 339 (from a discarded draft for section 3 of 'Why I Am So Clever').

nouncements of the aged Carlyle, as well as looking towards the huge eclectic project of Pound. Mediating between Pound and Emerson, however, is the poet who developed Emerson's Carlylian eclecticism, and who himself drew sustenance from other Scottish models— Whitman.

To call Whitman an eclectic is to state the obvious. Where Emerson had seen America itself as a poem, Whitman followed, proclaiming that 'The United States themselves are essentially the greatest poem', and that 'Here is not merely a nation but a teeming nation of nations.' (*CP*, 5.) Like Emerson, Whitman likes to quote himself, as he does later when (following his frequent practice) he puts his own prose into verse to celebrate his view that 'These States are the amplest poem' because they are 'not merely a nation but a teeming Nation of nations' (*CP*, 471). What most attracts Whitman here is the way in which America is 'composite', a word which he uses several times to describe the country (*CP*, 181, 524). It is also a 'melange' or an 'ensemble' (*CP*, 511, 1259). This notion is at the heart of Whitman's wish for a new national literature. For him, too, the United States are a cultural anthology.

Ensemble is the tap-root of National Literature. America is become already a huge world of peoples, rounded and orbic climates, idiocrasies, and geographies—forty-four Nations curiously and irresistibly blent and aggregated in ONE NATION, with one imperial language, and one unitary set of social and legal standards over all—and (I predict) a yet to be National Literature. (*CP*, 1259.)

To be in tune with the eclectic nation, Whitman's poems must themselves be eclectic. Again he follows Emerson (who hailed him with confident praise) in his view of the poet as integrator, writing in an English which is itself already an amalgam, and, to pick up Whitman's word from the passage just quoted, an 'imperial' amalgam.

VIEW'D freely, the English language is the accretion and growth of every dialect, race, and range of time, and is both the free and compacted composition of all. From this point of view, it stands for Language in the largest sense, and is really the greatest of studies. It involves so much; is indeed a sort of universal absorber, combiner, and conqueror. (*CP*, 1165.)

Many of Whitman's attitudes are Emersonian, leaving Whitman also open to the charge of imperialism. This was an accusation brought against him by D. H. Lawrence.[69] Whitman was extremely ambitious. To him, it did seem 'As if the Almighty had spread before this nation charts of imperial destinies' (*CP*, 990.) In practical terms, this meant —among other things—that Whitman saw the United States as incorporating Canada (*CP*, 881, 952), but, ultimately, his grand designs had much less to do with obtaining political control of other countries than with the development of an Emersonian identification between the self and the world.

The Whitman who finds all history and geography in himself can be seen as an imperialist, reducing the world to Walt Whitman, and so celebrating his own greatness as he simultaneously abolishes difference in the world, levelling all down in his own version of the Orphic poet.

> Every existence has its idiom, every thing has an idiom and
>   tongue,
> He resolves all tongues into his own and bestows it upon
>   men, and any man translates, and any man translates
>   himself also,
> One part does not counteract another part, he is the joiner,
>   he sees how they join.   (*CP*, 315.)

In this act of translating, though, Whitman follows on not only from Emerson, but also from that earlier American cultural translator, Natty Bumppo. Like Cooper (whom he several times proudly recalls having seen) (*CP*, 701, 1190, 1200), Whitman celebrates pioneering and the crossing of boundaries as a pleasure in itself and as part of an effort to unite the 'collation' that is his country.

> I troop forth replenished with supreme power, one of an
>   average unending procession,
> We walk the roads of Ohio and Massachusetts and
>   Virginia and Wisconsin and New York and New
>   Orleans and Texas and Montreal and San Francisco
>   and Charleston and Savannah and Mexico,
> Inland and by the seacoast and boundary lines . . . and we
>   pass the boundary lines.   (*CP*, 71.)

Like Bumppo, Whitman brings together white and Indian as part of his wider eclecticism. He liked to think that there entered into him the essence of an America which included 'the tribes of red aborigines'

---

[69] D. H. Lawrence, *Studies in Classic American Literature* (1923; repr. Harmondsworth: Penguin Books, 1971), 172.

(*CP*, 7). So effective was he, that Lawrence called him 'the first white aboriginal', a title which might be more suited to Bumppo.[70] Whitman's Indians, like Cooper's, could be elegiacally Ossianic. Whitman had grown up declaiming Ossian by the seashore, and, as an old man, was still attracted to that creator of national epic who enabled him, in 1884, to eulogize a dead Iroquois orator as

> Product of Nature's sun, stars, earth direct—a towering
>   human form,
> In hunting-shirt of film, arm'd with the rifle, a half-ironical
>   smile curving its phantom lips,
> Like one of Ossian's ghosts looks down.   (*CP*, 622.)

In aiming to celebrate and help construct 'that composite American identity of the future', Whitman follows Emerson in alerting his countrymen to the need for an eclecticism, one of whose aims is to free America from the status of an English cultural province, and, in so doing, like so many of his compatriots, he profits from Scottish models (*CP*, 1147). Whitman describes his first 'serious acquaintance with poetic literature' as the time when he became 'in my sixteenth year . . . possessor of a stout, well-cramm'd one thousand page octavo volume (I have it yet), containing Walter Scott's poetry entire —an inexhaustible mine and treasury of poetic forage (especially the endless forests and jungles of notes)—has been so to me for fifty years, and remains so to this day' (*CP*, 664). Whitman tells us the precise edition which he used, with its 'various Introductions, endless interesting Notes, and Essays on Poetry, Romance, &c.', and makes the point that 'All the poems were thoroughly read by me, but the ballads of the Border Minstrelsy over and over again.' (*CP*, 665.)

At first sight, there may seem little connection between Scott's *Minstrelsy of the Scottish Border* and Whitman's *Leaves of Grass*. The ballads are strictly controlled rhyming verse-forms, whereas Whitman is poor with rhyme and is interested in how 'to essentially break down the barriers of form between prose and poetry' (*CP*, 1056). Whitman, however, suggests a connection elsewhere.

I say it were a standing disgrace to these States—I say it were a disgrace to any nation, distinguish'd above others by the variety and vastness of its territories, its materials, its inventive activity, and the splendid practicality of its people, not to rise and soar above others also in its original styles in

---

[70] Ibid. 182.

literature and art, and its own supply of intellectual and esthetic master-pieces, archetypal, and consistent with itself. I know not a land except ours that has not, to some extent, however small, made its title clear. The Scotch have their born ballads, subtly expressing their past and present, and express-ing character. (*CP*, 980–1.)

The important thing about the Scottish *Minstrelsy* is that it represents a people's poetry, a national literature. As Scott put it in one of those pieces which Whitman found so 'interesting', the 'Introductory Remarks on Popular Poetry', what we see in the ballads is 'the national Muse in her cradle'. Scott writes of the bardic poetry of 'aboriginal poets'. He relates his discussion to the 'rude poetry' of 'nations in their early state' and, in particular, to Homer, the poet whose lines, with Ossian's, Whitman recited by the sea.[71] Whitman believed that 'really great poetry is always . . . the result of a national spirit' (*CP*, 672). For all its material advances, Whitman's America at the time of the first *Leaves of Grass* (1855) still appeared primitive in terms of the lack of a national literature.

Whitman liked to quote a remark of Margaret Fuller's about America's lack of a literature (*CP*, 1074, 1262), and actually enjoyed identifying himself, as 'one of the roughs', with the supposedly primitive, unrefined nature of America. However, what Scott had also stated in following the Enlightenment views of poetry which Thomas Blackwell had helped develop was that 'The more rude and wild the state of society, the more general and violent is the impulse received from poetry and music.'[72] It is hardly accurate to see Whitman as a savage in a savage country, but he does stress America's raw 'primitiveness'. He was at the beginning, as he saw it, of a national literature, and the Border Ballads, a collection of poems written in a non-standard form of English in a country which had the status of an English cultural province, provided him with a first model for the form of his own national anthology, *Leaves of Grass*, and of a future United States Literature.

In the ballads, as in his own country, Whitman learned to treasure peculiarities of diction and thought, which were liberatingly popular, rather than belonging to what Whitman called 'the thin moribund and watery, but appallingly extensive nuisance of conventional poetry' (*CP*, 1003). 'Sometimes the bulk of the common people (who are far more 'cute than the critics suppose) relish a well-hidden

[71] Sir Walter Scott, *Poetical Works* (12 vols.; Edinburgh, Robert Cadell, 1833), i. 15, 31, 11–12.   [72] Ibid. 213.

allusion or hint carelessly dropt, faintly indicated, and left to be disinterr'd or not. Some of the very old ballads have delicious morsels of this kind.' (*CP*, 1261.)

Scott's work was important not only for collecting and presenting a people's poetry, but also for rooting itself firmly in region and in factuality; hence Whitman's appreciation of all those notes, most of which serve to ground the text in precise locality and history. Whitman incorporated the factuality of such notes into his own texts. Scott's national poetry was a spur. All his life Whitman continued to read not only Scott's poetry, but his novels also, where he found further examples of varieties of non-standard English set down as part of a celebration of nationality (*CP*, 699). However, in the imaginative leap to a democratic poetry, Scott's work, for Whitman as for Twain, was also dangerously 'feudal', and so demanded an act of liberating rejection (*CP*, 1015).

Scott was a spur, but a dangerous one; another Scottish poet acted for Whitman as an incitement towards a more democratic poetry. This poet appealed to Whitman the verbal artificer, the lover of pun whose songs are 'omniverous' [*sic*] (*CP*, 76). Yet at the same time he is the recorder of folk-speech who protests (though we may be tempted to doubt it) that one of his poems is '*Noted verbatim after a supper-talk out doors in Nevada with two old miners*' (*CP*, 652). He delights in the peculiarities of his own language that are outside the standard English tongue, in 'foofoos', 'powowing', 'koboo', and in 'gab', as he plays the role of national poet as wildman, sounding his 'barbaric yawp over the roofs of the world' (*CP*, 48, 77, 87), at the same time as knowing his Scott, Emerson, and Carlyle. As a highly literate poet who is also a celebrator of vernacular and the people who speak it, conscious of the modes and barriers of civilized gentility, as national poet and spokesman of democracy who asks 'Where is the real America?', and suggests that it is to be found among 'laboring persons, ploughmen', Whitman looks towards the poet to whom he devoted his most detailed critical essay, whom he thought 'has been applauded as democratic . . . with some warrant', and who 'forms to-day, in some respects, the most interesting personality among singers' (*CP*, 1310, 1158, 1152). This poet was Robert Burns.

F. O. Matthiessen, in his classic study *American Renaissance*, indicates some aspects of Burns which were important to Whitman, though, each time, Matthiessen is reluctant to extend the comparison. He points out that, though he praised the 'neglect, unfinish, careless nudity' of Burns's 'Scotch patois', Whitman himself did not use a folk-speech, but a synthetic construction made out of vulgar speech and wide reading.[73] Yet Whitman, as in his supposedly *'verbatim'* restatement of Nevada miners's speech, was interested in using something *like* a folk-speech, and in presenting his work as such. Whitman would not have known that Burns's 'Scotch patois' was a synthetic language, but he would have known from Scott's *Minstrelsy* that Burns produced a 'brilliant rifacimento' of folk-songs.[74] Matthiessen also points out that Whitman criticized Burns for a lack of spirituality. This should not be allowed to obscure how useful Whitman found Burns as a model; he even wrote an essay on him in *November Boughs*, 'Robert Burns as Poet and Person'. As the essay's title suggests, what particularly interests Whitman is the way in which Burns's poetry and personality are mysteriously connected while, at the same time, both have come to represent their nation's persona, 'after nearly one century, and its diligence of collections, songs, letters, anecdotes, presenting the figure of the canny Scotchman in a fullness and detail wonderfully complete' (*CP*, 1152). It is in his dual nature of Scottish individual and universal man that Burns is of most interest to American Literature and so (as Whitman's language reveals) to Whitman's own deepest concerns with the creation of a 'manly' and democratic literature. Burns is seen as 'essentially a Republican' who would have been 'at home in the Western United States', as well as being typical 'of the decent-born middle classes everywhere' (*CP*, 1152–3).

These may be misreadings of Burns, but they are misreadings which allow Whitman to situate Burns as a usefully empowering ancestor. Some of the poems which Whitman identifies as his favourites in Burns's *œuvre* do have explicitly American elements, like the 'Epistle to John Rankine', with its joking references to 'Bunker's hill' and 'Virginia'. Burns celebrated 'The man of independant mind', and was a prophet of universal brotherhood

> For a' that, and a' that,
> Its comin yet for a' that,
> That Man to Man the warld o'er,
> Shall brothers be for a' that.[75]

---

[73] F. O. Matthiessen, *American Renaissance: Art and Expression in the Age of Emerson and Whitman* (1941; repr. New York: Oxford University Press, 1979), 531.

[74] Scott, *Poetical Works*, iv. 24.

[75] *PS*, 46–7 ('Epistle to J. R——'), 603 ('Song—For a' that and a' that').

Yet, for all his universalism, he was, as Whitman saw him, 'the Ayrshire bard' (*CP*, 1157). The 'habitan of the Alleghanies' marvelled at the provincial nature of the achievement of a Burns far from the centres of great events. 'And while so much, and of the greatest moment, fit for the trumpet of the world's fame, was being transacted—that little tragi-comedy of R. B.'s life and death was going on in a country by-place in Scotland!' (*CP*, 1153.) Burns's combination of provincialism and universalism appeals strongly to Whitman, but he is attracted at least as much by his perception that 'no poet on record so fully bequeaths his own personal magnetism . . . (no man ever really higher-spirited than Robert Burns)'. The Burns myth and the perceived personality that is bonded to the poems is as important to Whitman as the poems themselves; indeed, it is part of them. 'At any rate it has come to be an impalpable aroma through which only both the songs and their singer must henceforth be read and absorb'd. (*CP*, 1154–5.) It is just such personal magnetism that Whitman, singing the body electric and identifying himself continually with his own poems and his own country, strove to achieve and succeeded in obtaining. Whitman, who writes of 'I and my book . . . My Book and I' (*CP*, 656, 660), wishes his own personality (or his created persona) to be firmly, warmly, and inextricably bound to his work. He even writes of 'my other soul, my poems' (*CP*, 916). Whitman relates how 'no man that ever lived . . . was so fondly loved, both by men and women, as Robert Burns', and stresses Burns's flesh and blood warmth (*CP*, 1154).

Just as importantly, Whitman was aware of Burns as a man with the reputation of a great lover, whose poems included 'a certain cluster, known still to a small inner circle in Scotland, but, for good reasons, not published anywhere', namely, *The Merry Muses of Caledonia* (*CP*, 1154, 1161). In the 'Calamus' poems Whitman, too, had a cluster of intensely, contentiously erotic pieces, but he also loved to project himself as a universal lover, both of men and women, in his poems, generating his own magnetism of flesh and blood warmth, incarnating himself in the people and gathering the people to himself. Burns (whose letters Whitman also knew) was celebrated as a great inseminator and a great patriot at the same time, again a Whitman-like poet who has come to be part of the national identity and the national myth.

Precious, too—fit and precious beyond all singers, high or low—will Burns ever be to the native Scotch, especially to the working-classes of North

Britain; so intensely one of them, and so racy of the soil, sights, and local customs. He often apostrophizes Scotland, and is, or would be, enthusiastically patriotic. (*CP*, 1159.)

Whitman continually apostrophizes America as well as longing for intense union with it, union which will make his own poems part of the *patria*.

> Take my leaves America, take them South and take
>   them North,
> Make welcome for them everywhere, for they are your
>   own offspring,
> Surround them East and West, for they would surround
>   you,
> And you precedents, connect lovingly with them, for
>   they connect lovingly with you.   (*CP*, 177–8.)

More literally, too, Whitman celebrates his powers as inseminator of America. He writes of the phallus as a poem (*CP*, 260). As phallic poet, he draws on Burns, but looks also towards the Pound who would be sculpted by Gaudier-Brzeska as a giant phallus and who wrote that Whitman '*is* America . . . The vital part of my message, taken from the sap and fibre of America, is the same as his.'[76] Whitman, to whom all moisture tends to be seminal, dreams of (and tries to be) the mighty poet who will inseminate America, 'Plunging his seminal muscle into its merits and demerits' (*CP*, 472).

Where Whitman celebrates Burns's fertility, his embodying 'Nature's masterly touch and luxuriant life-blood, color and heat', so he himself time after time presents his poems as life-blood or semen (*CP*, 1161). As with Burns, so with Whitman, literature and sexuality are intimately connected. 'Sex contains all.' (*CP*, 258.) 'Any one and every one is owner of the library who can read the same through all the varieties of tongues and subjects and styles, and in whom they enter with ease and take residence and force toward paternity and maternity, and make supple and powerful and rich and large . . .' (*CP*, 19). The reader is placed in the position of the lover, such is the forceful insistence of Whitman's literary 'personalism' (*CP*, 961). Whitman, too, makes us continually aware of himself as 'Poet and Person' (to use again the title of his essay on Burns), so that reading becomes orgasmic. In Whitman's verse the 'book' becomes the man

[76] Ezra Pound, *Selected Prose 1909–1965*, ed. William Cookson (London: Faber and Faber, 1973), 115 ('What I Feel about Walt Whitman').

who will 'spring from the pages into your arms' (*CP*, 611), and he delights particularly in the Burns whose verses combine a strong, sometimes censored 'indelicacy', with a personality so strong that the poem may 'call out pronouncedly in his own voice, "I, Rob, am here",' (*CP*, 1161). This device of coming forward, using not his 'proper name' but its intimate and familiar form, is one which Whitman took from Burns. Where Burns announces his enduring, comradely presence in the second 'Epistle to John Lapraik'—'I, Rob, am here'—Whitman transforms himself from 'Walter Whitman, Jr.', his early name as a writer, to the familiar 'Walt' known to readers of his poetry.[77] So, in the most famous instance, he suddenly appears in the first version of *Leaves of Grass* in a Burnsian manner:

> Walt Whitman, an American, one of the roughs, a
>   kosmos,
> Disorderly fleshy and sensual . . . eating drinking and
>   breeding,
> No sentimentalist . . . no stander above men and women
>   or apart from them . . . no more modest than
>   immodest.   (*CP*, 50.)

Whitman likes repeating the device elsewhere:

> Speech is the twin of my vision, it is unequal to measure
>   itself,
> It provokes me forever, it says sarcastically,
> *Walt you contain enough, why don't you let it out then?*   (*CP*, 213.)

Burns is 'Rab' or 'Rob' in his poems, just as Whitman is 'Walt'. 'What a relief', Whitman writes in his essay on 'Slang in America', 'most people have in speaking of a man not by his true and formal name, with a "Mister" to it, but by some odd or homely appellative.' (*CP*, 1167.) It is the 'homely' nature of Rob or Walt that is so effective, and what attracts Whitman to slang is its being 'close to the ground' (*CP*, 1166), just as Burns is 'very close to the earth' in the way in which 'he pick'd up his best words and tunes' (*CP*, 1154). Whitman is attracted to Burns's 'use of the Scotch patois' as he is attracted to the American slang which finds its way into his own work not just because it is a *people's* language, but also because it is connected with sexuality as 'the lawless germinal element, below all words and sentences, and behind all poetry, and proves a perennial rankness and protestantism in speech' (*CP*, 1165). Burns's deliberate

---

[77] *PS*, 71.

breaking of the restraints of 'proper English' is given a sexual dimension in Whitman's restraint-breaking work. Burns's physicality meant much to Whitman, and the way in which he concludes his essay on Burns means that 'the aforesaid "odd-kind chiel" remains to my heart and brain as almost the tenderest, manliest, and (even if contradictory) dearest flesh-and-blood figure in all the streams and clusters of by-gone poets' (*CP*, 1161). For Whitman, who replied with gleeful affirmation to the question 'Do I contradict myself?', that mention of Burns's contradictory nature is a final gesture of identification with the Scottish poet.

Other such revealing gestures of identification with his Scottish subject-matter are found in Whitman's piece on the 'Death of Thomas Carlyle'. Where Whitman, in Emersonian terms, longed to be the representative man of America, even of his age, he saw Carlyle in such a way. 'As a representative author, a literary figure, no man else will bequeath to the future more significant hints of our stormy era, its fierce paradoxes, its din, and its struggling parturition periods, than Carlyle.' (*CP*, 886.) Carlyle, like Whitman, had a dual nature, being on the one hand 'a cautious, conservative Scotchman', yet, on the other, knowing that 'his great heart demanded reform'. Carlyle, as Burns (and Whitman), matters not simply as a writer but as a 'man-book', a great *personality* of the age, who enters into its spirit and its passion with, for all his sophistication, what Whitman depicts as a sort of Scottish 'barbaric yawp'—'His rude rasping, taunting, contradictory tones' (*CP*, 887, 898). As Whitman imagined Burns in America, so he imagines what Carlyle would have done had he come and turned his attention to America's 'limitless air' and its 'facts' (*CP*, 891). In a rather awkward footnote, Whitman even tries to counter the suggestion that he and Carlyle might be tarred with the same brush (*CP*, 892).

Paul Zweig has written well on Carlyle's importance to Whitman, the Carlyle who had written such rhythmic poetic prose, celebrated worker-poets, advocated a new 'Mythus', and who wrote (as Whitman would continually) about natural miracles. Particularly significant for Whitman was *Sartor Resartus*, where Carlyle directs attention to the fact that 'a *Naked World* is possible', and even envisages 'a naked House of Lords', notions which would certainly appeal to Whitman's love of nudity![78] Other aspects of *Sartor Resartus*

[78] Paul Zweig, *Walt Whitman: The Making of the Poet* (1984; repr. Harmondsworth: Penguin Books, 1986), 169; *SR*, 49–50.

have a much more wide-ranging link with Whitman's eclectic compositional methods. Zweig connects the book's six 'paper-bags' of
eclectic information with the trunk that Whitman the newspaperman
kept crammed with cuttings and magazine articles, but the connection
might extend further, to Whitman's poems, which are put together
eclectically, gathering their energy from the way in which diverse
facts, details, and tones rub up against one another; his prose works
are also collections of short pieces—diary entries, short appreciations,
preserved speeches, aphorisms, even bits of commonplace-books.
Zweig points out that Whitman privately stressed the disjointed
nature of his material. But his own public work, too, revels in its
Teufelsdröckhian scrappiness, as at the very beginning of *Specimen
Days*:

Incongruous and full of skips and jumps as is that huddle of diary-jottings,
war-memoranda of 1862–'65, Nature-notes of 1877–'81, with Western and
Canadian observations afterwards, all bundled up and tied by a big string,
the resolution and indeed mandate comes to me this day, this hour,—(and
what a day! what an hour just passing! the luxury of riant grass and blowing
breeze, with all the shows of sun and sky and perfect temperature, never
before so filling my body and soul)—to go home, untie the bundle, reel out
diary-scraps and memoranda, just as they are, large or small, one after
another, into print-pages, and let the melange's lackings and wants of
connection take care of themselves. (*CP*, 689.)

The footnotes here emphasize that the material came from 'dozens of
such little note books . . . just as I threw them by after the war'
(*CP*, 689).

    Whitman was a collector of other people's work, but he was also a
self-anthologist. His national collection is not like Scott's *Minstrelsy* or
Burns's contributions to the *Scots Musical Museum*, a gathering of
existing folk compositions. Rather, Whitman (like 'Ossian' Macpherson) is faced with the need to invent the national songs which he
can then collect: his *Leaves of Grass*. *Sartor Resartus* is an invented
anthology, since Carlyle himself has to make up the scraps and
fragments which he then pieces together to represent his supposedly
quintessentially Germanic philosophy; *Leaves of Grass* is a composition
of quasi-Carlylian eclecticism, an anthology of grass-roots American
verse, all of whose parts have to be invented before they can be
collected. Part of Scott's *Minstrelsy* (the part containing the modern
imitations) had been like that. But Whitman's collection is entirely
invented and obsessively eclectic. A vital component of it, as of

American and Scottish Literature as a whole, is an in-built pleasure in eclecticism and juxaposition. Whitman brings together not just his United States, their place-names and their inhabitants, but even talks of singing all the gathered facts and figures of America in his 'bundle of songs' (*CP*, 603).

The Scottish presence in American Literature is far more extensive than has been realized, and Scott's impact was a crucial one, traceable even in the work of writers who execrated his name. Mark Twain, celebrated both for his hostility towards Scott (whose name is applied to a grand wreck in Chapter XIII of *Huckleberry Finn*) and for his cataloguing of 'Fenimore Cooper's Literary Offenses', is a novelist whose bringing-together of dialect and standard English, and whose crossings of the boundaries between communities and between the forces of wildness and those who try to 'sivilize' Huck Finn, surely make him a border-crosser, collector of languages, and cultural anthologist who owes something to the predecessors whom he disowns; perhaps it is not just Huck Finn, but also his creator, who takes 'Honest Loot from the "Walter Scott"'.[79] Again, while no writer might seem further removed from Twain than his American contemporary Henry James, James is also a delighter in broker-figures, and a novelist whose famous 'international theme' involves the repeated juxtaposition of subtly different cultures, as well as the often uncertain efforts of protagonists who try to interpret one culture to another, and then to cross the borders between them. It is a long way indeed from the naïve Edward Waverley to the racy Huck Finn or the naïvely sophisticated Lambert Strether in *The Ambassadors*, but all these characters operate in novels constructed as cultural anthologies. If Scotland was a collection of dialects, languages, and cultures, then it was one which had much to offer the anthological urges to be seen in American writing and in the literary upheavals of Modernism.

[79] See *The Portable Mark Twain*, ed. Bernard DeVoto (New York: Viking, 1946), 541–56, 539, 272. For further links between Scott and American fiction see George Dekker, *The American Historical Romance* (Cambridge: Cambridge University Press, 1987), *passim*.

5

Modernism as Provincialism

PROVINCIALISM in the nineteenth century was not an exclusively un-English phenomenon. The northwards movement of the imagination, spurred by Ossian, Burns, and Scott, was given impetus by Wordsworth and the Lake Poets. It bore fruit later in the work of the Brontës. One need only think of the opening of *Wuthering Heights* to see how the confident northern novelist could articulate a cultural identity to which the supposedly smooth southerner was a stranger. Whatever else he is, Emily Brontë's Mr Lockwood is a slightly comical cultural broker between the north of England and the metropolitan reading public, just as Lucy Snowe in *Villette* is a cultural and linguistic broker between the societies of Protestant England and Catholic Belgium. The Brontës learned from Scott, and their own, distinctive focus on women's emotional and intellectual lives, as well as on provincial communities like that of Luddite Yorkshire in *Shirley*, nourished the George Eliot whose *Middlemarch* is significantly sub-titled *A Story of Provincial Life*. A Warwickshire woman partly educated in Coventry, George Eliot deploys a Wordsworthian emphasis on the 'ordinary' in her fiction, portraying provincial lives through which may be enacted dramas every bit as vital as the supposedly important affairs of the metropolis. Again, in the work of Charlotte Brontë's biographer Elizabeth Gaskell, a similar provincial confidence is crucial. Gaskell shows a strong concern with figures who can interpret one England to another, a manifest aim in *North and South*. We may tend too easily to think of the Victorian novel as fundamentally metropolitan, rooted in the London of *Vanity Fair* or *Oliver Twist*, but the work of the provincial women novelists ought to counter and complicate any such unexamined attitude.

The genetics of nineteenth-century fiction are enriched by the way in which George Eliot enters American Literature when Dorothea Brooke becomes a model for Isabel Archer in James's *The Portrait of a Lady*. Moreover, while one might tend to think generally of the 1880s and the 'naughty' 1890s as centred increasingly on a decadent London, this period is also a heyday for provincials such as the

Irishmen Yeats, Shaw, and Wilde, the Cornishman Arthur Symons, and Scots like John Davidson, Andrew Lang, Robert Louis Stevenson, and 'Fiona Macleod'. Even among native English writers, the culminating voice of the Victorian novel is not metropolitan but resolutely provincial. It is as important for Hardy to demonstrate that the full panoply of Greek tragedy can be played out in provincial Wessex as it is for Joyce to revoice the *Odyssey* in Dublin. One need only set Hardy's view of Oxford in *Jude the Obscure* beside Matthew Arnold's earlier paean to Oxford in his Preface to the 1865 *Essays in Criticism* to see the strain between the essentially angry provincial voice and the admirer of the traditional cultural centre. Arnold may have patronized the notionally Celtic in literature, but Hardy could enjoy the dialect poetry of William Barnes as part of his authentic local culture. Barnes's other admirers included G. M. Hopkins and the Tennyson whose work contains such experiments with dialect as 'Northern Farmer, Old Style' and 'Northern Farmer, New Style'. To ignore the voice of the provincial in nineteenth-century writing in England is to distort and oversimplify the development of that country's literature. There are hidden currents between these provincial voices and their twentieth-century inheritors which have yet to be fully explored. Nevertheless, it seems undeniable that it was the un-English provincials and their traditions which contributed most to the crucially provincial phenomenon which we now know as Modernism.

If the Enlightenment and nineteenth-century Romanticism had categorized America, Ireland, Scotland, and other regions distant from the imperial metropolis as provincial, wilder places, then the writers who came from those zones learned to trade on their 'provincial' identity. At the same time, they needed to face up to, or evade, the difficulties caused by such a situation. Scott, Cooper, and Twain had all turned the inhabitants of the cultural provinces into a lively focus of attention, valorizing the primitive. The provincials who developed Modernism would also utilize the 'primitive' and 'savage' aspects of selected cultures. Sometimes they appropriated what the provinces offered them; sometimes they supplemented this by assuming and constructing identities at once primitive and cosmopolitan—so cosmopolitan that they outflanked the English cultural centre. The artist, wrote T. S. Eliot, was at once 'the most and the least civilized and civilizable' of people. This was the quintessential provincial strategy.[1]

---

[1] T. S. Eliot, 'War-Paint and Feathers', *Athenaeum*, 17 Oct. 1919, 1036.

The ways in which the provincial energies detailed earlier in this book fuelled an emergent Modernism are legion. 'I was a denizen of the "Leatherstocking" world', recalled Wyndham Lewis, whose first drawings were of redskins, ancestors of the figures in his totemic mature works. The 'overwhelming question' pondered by Prufrock is a term taken from the first of Cooper's Leatherstocking Tales, which Eliot read in childhood. Eliot later stressed the distinctive Americanness of Cooper at a time when he was emphasizing the fundamental American roots of his own work.[2] 'I should like to drive Whitman into the old world . . . "His message is my message"', wrote the American Pound in Europe in 1909, adding in 1912: 'Whistler and Whitman—I abide by their judgment.'[3] Yeats had already called Whitman 'the greatest teacher'.[4] If Pound, convinced that 'The romantic awakening dates from the production of *Ossian*', longed for a new, equally powerful awakening, then Joyce, celebrating in 1909 the way in which the signature 'Oscar Fingal O'Flahertie Wills Wilde' linked that writer to 'a savage Irish tribe' and to 'Oscar, nephew of King Fingal and the only son of Ossian in the amorphous Celtic *Odyssey*', was to begin his own Celtic *Odyssey* of *Ulysses* five years later.[5] 'You are an Irishman and you must write in your own tradition', Joyce told Arthur Power in 1930, adding that all the great writers 'were national first'.[6] Yeats would have agreed with that. His father read him *Redgauntlet*, with its celebration of the persistence of folk culture, when he was 11, and, just over a decade later, the young poet's folklore-collecting was fuelling his own work and his nationalist aspirations. In turn, Yeats's achievement would, like the work of Joyce, enthuse the young Scottish nationalist poet Hugh MacDiarmid, about whose writing Yeats made admiring remarks.[7]

A cursory account of Modernism stresses its cosmopolitanism and internationalism, presenting it as a facet of 'high' metropolitan

[2] *The Letters of Ezra Pound and Wyndham Lewis*, ed. Timothy Materer (London: Faber and Faber, 1985), 63; Robert Crawford, 'T. S. Eliot and Fenimore Cooper', *Notes and Queries*, 232/4, (Dec. 1987), 506–7.

[3] Ezra Pound, *Selected Prose 1909–1965*, ed. William Cookson (London: Faber and Faber, 1973), 116.

[4] W. B. Yeats, *Collected Letters*, i. *1865–1895*, ed. John Kelly and Eric Domville (Oxford: Clarendon Press, 1986), 9.

[5] Ezra Pound, *Literary Essays*, ed. T. S. Eliot (1954; repr. London: Faber and Faber, 1974), 215; James Joyce, *Critical Writings*, ed. Ellsworth Mason and Richard Ellmann (London: Faber and Faber, 1959), 201.

[6] Richard Ellmann, *James Joyce*, New and Revised Edition (New York: Oxford University Press, 1982), 505.

[7] Yeats, *Letters*, i. 4; Hugh MacDiarmid, *Lucky Poet: A Self-Study in Literature and Political Ideas* (1943; repr. London: Cape, 1972), 66.

culture. But there is also another, equally important, side of Modernism that is demotic and crucially 'provincial'. Only by looking in more detail at the way in which the modernists were, to use Terry Eagleton's phrase, 'Exiles and Emigrés', can we comprehend this fully. There are rewards in seeing how Modernism may be decentred to realign it with some of the 'provincial' drives set out earlier in this book and with more recent developments in 'barbarian' writing.[8] Such a decentring relies not simply on paying attention to Modernism's demotic registers and un-English ancestries, but also on attending to its most assertive cosmopolitanism. For these features are not opposites; they complement one another.

Nowhere is this clearer than in the case of Modernism's greatest *animateurs*, the apparently more-English-than-the-English T. S. Eliot, and the Ezra Pound who, in old age, was pleased to be described with a phrase which recalled not his forays into Chinese or Egyptian culture, but his own American point of origin: 'the Idaho kid'.[9] From his time in London onwards, Pound's strategy was to appear both more primitive and more sophisticated than the dominant culture expected. Emerson had admired Carlyle for seeming at once a rude gardener and a polymath, and had written of the need for Americans to acquire knowledge of a range of cultures so as to avoid English domination. Pound did just this:

Mentally I am a Walt Whitman who has learned to wear a collar and a dress shirt (although at times inimical to both). Personally I might be very glad to conceal my relationship to my spiritual father and brag about my more congenial ancestry—Dante, Shakespeare, Theocritus, Villon, but the descent is a bit difficult to establish. And, to be frank, Whitman is to my fatherland (*Patriam quam odi et amo* for no uncertain reasons) what Dante is to Italy and I at my best can only be a strife for a renaissance in America of all the lost or temporarily mislaid beauty, truth, valour, glory of Greece, Italy, England and all the rest of it.[10]

Here Pound stakes his claim both to the 'barbaric yawp' and to the formal dress of sophisticated culture. Even those who made fun of him recognized the combination. In 1911 *Punch* noted the arrival of a 'new poet' called 'Boaz Bobb, a son of the Arkansas soil' who had come to London to pursue one of the more arcane high-cultural pursuits ('studying Icelandic literature'), with a view to producing something with a primitive edge to it—'a new saga of the Wild

[8] Terry Eagleton, *Exiles and Emigrés* (London: Chatto and Windus, 1970), 9–19.

[9] Rex Lampman, 'Epitaph', in Ezra Pound, *Pavannes and Divagations* (London: Peter Owen, 1960), p. vii.　　　[10] Pound, *Selected Prose*, 115–16.

West'.[11] Though Pound did not write that Wild West saga, he was pleased to borrow a phrase from Kipling later, calling his *Cantos* 'a Tale of the Tribe'. During his early London years Pound delighted in linking the artist with the witch-doctor, maintaining in 1914 that 'The artist recognises his life in the terms of the Tahiytian savage.'[12] Pound's 'savage' pronouncements and interest in the primitive have been examined by Ronald Bush, and may be set beside Eliot's statement that the artist was at once the most and the least civilized of people, or the same poet's admiration for the artist as Tarzan of the Apes.[13] These pronouncements were part of the contemporary interest in the primitive that had been energized by the work of Frazer and other anthropological writers, but they were also useful statements for American writers to make, turning their supposedly 'wild' provenance to advantage.

No one caricatured the young Eliot as a 'Wild West' poet, yet he did describe himself later as having been born at the geographical 'beginning of the Wild West', and he was preoccupied throughout his life with comparing the uncivilized and the civilized, a characteristically American obsession worthy of Cooper, but also of Henry James. Having dealt with this in *The Savage and the City in the Work of T. S. Eliot*, I wish to concentrate here on the way in which Eliot's persistent American-ness operated throughout his career, prompting the perception that his un-Englishness was crucial to his poetic and cultural stance.[14]

In the *Christian News-Letter* of 21 March 1945 Eliot used the pseudonym 'Metoikos' ('Resident Alien') to respond to an article signed 'Civis' ('Citizen') which had impatiently contended that all Christians had to support one particular scheme for full employment. The retort of 'Metoikos' is that this particular scheme seems to him to be naïve, and that it is always dangerous to equate Christianity with the political arrangements of any one state. Eliot's use of the pseudonym 'Metoikos' was partly a piece of rhetorical cheek. Like the letter he

[11] *Ezra Pound and Dorothy Shakespear: Their Letters 1909–1914* ed. Omar Pound and A. Walton Litz (London: Faber and Faber, 1985), 46.

[12] See Michael André Bernstein, *The Tale of the Tribe* (Princeton, NJ: Princeton University Press, 1980); Ezra Pound, 'The New Sculpture', *Egoist*, 16 Feb. 1914, 68.

[13] Ronald Bush, *The Genesis of Ezra Pound's Cantos* (Princeton, NJ: Princeton University Press, 1976), 87–141; Eliot, 'War-Paint and Feathers', 1036; T. S. Eliot, 'Contemporanea', *Egoist*, June–July 1918, 84.

[14] T. S. Eliot, *To Criticize the Critic and Other Writings* (London: Faber and Faber, 1965), 44; Robert Crawford, *The Savage and the City in the Work of T. S. Eliot* (Oxford: Clarendon Press, 1987).

had just written to the *New English Weekly* (8 March 1945), taking issue with a contributor who thought that the British should be protected from all Germanic intellectual influences, Eliot's 'Metoikos' piece stresses a readiness to take up a position outside that of the comfortable group norms and unearned sympathies invoked by the writer who adopts the patriotic signature 'Civis'. Eliot's descriptions of himself tend to be tongue in cheek: Metoikos writes as 'a small business man'. But when we know that Metoikos is T. S. Eliot, the pseudonym seems to have several layers, and reveals more than simply a desire to find an appellation suitable to an opponent of Civis. It may come to be the most appropriate description of the man who, born in St Louis in 1888, spent thirty-nine years of his life as an American citizen, and thirty-eight years of his life as a British subject. Nationality for Eliot was certainly an important way of making statements, but it did not constitute cultural identity.

The young Eliot who came to England in 1914, when he was 26, was aware of not understanding the English very well (*L*, i. 57). With awkward but winning honesty he put this down to his own snobbishness, and felt (as he wrote that same year to his American cousin Eleanor Hinkley) that, though he liked the English very much, he would never feel at home in their country (*L*, i. 61, 65). At Oxford, which he found very 'foreign', he spoke in a college debate against a motion abhorring the threatened Americanization of Oxford (*L*, i. 70). Eliot's contribution was rather ironical, and the subject does not seem to have mattered to him a great deal, for it was he who, during his year as a graduate student at Merton, wrote, with a touch of the outsider's bravado, that Oxford was very pretty, but he did not like to be dead (*L*, i. 74). The following year he still felt alien, finding it easier to get on with Indians whom he met in Oxford than with most of the English people in the University (*L*, i. 92). Though Eliot grew more accustomed to the English and their ways, and more accomplished at hiding his *metoikos* sensation, this clearly persisted. By 1919 he felt he was getting used to some aspects of English society, and was convinced that he could make more of a mark there than in America (*L*, i. 285). Yet, particularly after his father's death that year, Eliot's anxiety to be published in America, to prove himself both to his mother and to his own culture, increased (*L*, i. 296). His feeling of being alienated persisted. He complained to Ottoline Morrell of being especially depressed by having lost contact with Americans and their mores (*L*, i. 307). To his brother Henry in America he wrote of how hard it was to live in, and cope with, a foreign nation. Ezra Pound, living in Kensington in 1912, had felt 'like a painted pict with a stone war club'.[15] Eliot was ill at ease with English social nuances, however hard he tried to catch them. He wrote of how he would always remain a 'foreigner' (*L*, i. 310). In that same year, 1919, he expressed this feeling much more strongly to his English friend Mary Hutchinson, asserting that he might somehow remain an American barbarian or, to use his own word, a 'savage' (*L*, i. 318). This was the year of 'Gerontion', that potential preface to *The Waste Land*. 'Gerontion', like *The Waste Land*, is a poem of cultural uprootedness, and this is signalled most blatantly by the honest, yet incongruous, homage of 'Hakagawa, bowing among the Titians'. It is the poem of a man obsessed with a tradition into which he was not born, of a poet who described himself again that year as 'a *metic*, a foreigner' (*L*, i. 318). In 1921, shortly before plunging into the composition of much of *The Waste Land*, Eliot wrote to a fellow American writer that he had 'got used to being a foreigner everywhere' (*L*, i. 431).

Though we may have grown accustomed to thinking of him in terms of Hugh Kenner's book title, *The Invisible Poet*, Eliot actually presents versions of himself quite frequently in his verse. Some knowledge of his biography is likely to tempt us to believe that aspects of the life of T. S. Eliot, American academic lecturing in Yorkshire about contemporary France (*L*, i. 152), form a 'Mélange adultère de tout', and that, as eminent man of letters, committee man, patron of artists, and director of Faber and Faber (if not as 'small business-man'), he is himself among those in 'East Coker' who 'all go into the dark'. More provocatively, he actually makes use of his own name in some poems. As a title, 'Mr Eliot's Sunday Morning Service' gener-ates an air of propriety and respect. It is as if 'Mr Eliot' has become one of those names, like 'Miss Helen Slingsby . . . my maiden aunt', or 'Mr Apollinax', which Eliot manipulates to such formal effect. Should we wish a further gloss on the particular effect which that name 'Mr Eliot' might produce, we could do worse than turn to the 'Mr Eliot' of the later 'Five-Finger Exercises':

> With his features of clerical cut,
> And his brow so grim
> And his mouth so prim
> And his conversation, so nicely

---

[15] Pound and Shakespear, *Letters*, 163.

Restricted to What Precisely
And If and Perhaps and But.

Here is another suitably clerical 'Sunday Morning Service' figure—
Virginia Woolf's Eliot in his four-piece suit; 'Pope Eliot', as Dylan
Thomas later called him—'The Pope of Russell Square'.[16]

This sober, responsible figure seems to be 'Civis' with a vengeance,
and I suspect that, for many, it represents a view of T. S. Eliot which
persists, a view which sees him as a solemn, rather cadaverous figure,
at times tormented, but always of rigidly clerical cut. Yet the poem
which follows the title 'Mr Eliot's Sunday Morning Service', like the
poem which follows the title 'Mr Apollinax', is anything but proper
and respectful of conventional pieties; the 'Mr Eliot' of 'Five-Finger
Exercises' is being sent up. The writer who is doing the sending-up
can also step outside that label, making fun of what it is seen to stand
for. Throughout his career, that writer (the poet who learned from
Laforgue) is easily able to assume a double part and hear another
voice speaking his own name, 'Knowing myself yet being someone
other', as he puts it later in 'Little Gidding'. In the matter of his own
name, Eliot appears to have utilized the status of a *metoikos*, at once an
insider and an outsider, a resident alien.

The ability simultaneously to maintain two opposite, mutually
reinforcing stances, to be an insider and an outsider at once, runs
throughout Eliot's work and career. It is particularly apparent at
moments of decision in his life. So, for instance, in 1927, a week
before he was baptized and confirmed in the Church of England,
Eliot promised his friend Bonamy Dobrée that he would soon be
sending him detailed information about the theology of the Bolovians
—the primitive tribe which inhabited a scurrilous part of Eliot's
imagination, but also one closely linked to his creativity. One month
after Eliot's confirmation those details were sent, revealing, among
other things, that the Bolovian's great god Wux numbered among his
attributes four duck feet, four arms, four testicles, two penises, and a
speedometer.[17] This information is part of a longer correspondence

[16] All quotations from Eliot's verse (unless otherwise stated) will be found in T. S.
Eliot, *The Complete Poems and Plays* (London: Faber and Faber, 1969); Dylan Thomas,
The Collected Letters, ed. Paul Ferris (London: Dent, 1985), 174, 186, 218, 222 (letters
of the 1930s); Desmond Hawkins, 'The Pope of Russell Square', in Tambimuttu and
Richard March (eds.), *T. S. Eliot: A Symposium* (London: Tambimuttu and Mass,
1948), 44–7.

[17] T. S. Eliot, letters to Bonamy Dobrée, 22 June, 29 July 1927 (Brotherton
Collection, University of Leeds).

course which Eliot was giving Dobrée on the Bolovians, and particularly on Bolovian theology. Eliot's behaviour should in no way be
taken as contradicting his sincere and fervent Christianity, but as con-
firming it, just as the apparent blasphemy of 'The Hollow Men' and
Sweeney Agonistes are steps in his conversion, a way into Christianity
through 'the back door', the door (Eliot argued) used by Baudelaire.[18]
The virulence of the potentially offensive humour, and the longing
for a Christian life, like the way up and the way down for Eliot's
cherished Herakleitos, are one. It was Eliot the *metoikos* who, apparently on a course which led far from Christianity, actually made
this course his point of entrance. One way of reading the earlier poem
'The Hippopotamus' (which contains some of Eliot's most blasphem-
ous lines) is to see it as a poem about the stone which the builders
rejected. As the hippopotamus takes wing towards heaven while 'the
True Church remains below', outsiders and insiders are dramatically
reversed.

Clearly Eliot felt something of a *metoikos* from childhood. Peter
Ackroyd, who points out Eliot's liking for describing himself as a
'resident alien', reminds us of Eliot's sense of being a southerner in
Massachusetts, and yet, when in the south, being aware that he came
of northern stock.[19] The move to England was both an escape from,
and a heightening of, such cultural differences. The living writer to
whom Eliot was closest during his early years in England was the
fellow *metoikos* Pound, who identified one of the basic bonds between
them when he recorded their first meeting and communicated to
friends his excitement at coming across 'a young American, T. S.
Eliot'.[20] Though Eliot knew his Whitman, albeit sometimes subliminally, he did not wish 'to drive Whitman into the old world', as
did the Pound of 1909.[21] But Eliot, too, had an American predecessor
to whom he clearly paid homage, and to whose example he surely
owed a great deal. This master was that most famous of American
resident aliens in England, Henry James, whose own writings, so
different from those of both Cooper and Twain, are also grounded on

[18] T. S. Eliot, 'Baudelaire', in *Selected Essays*, Third Enlarged Edition (London:
Faber and Faber, 1951), 421.

[19] Peter Ackroyd, *T. S. Eliot* (London: Hamish Hamilton, 1984), 24.

[20] Pound to H. L. Mencken, 3 Oct. 1914, in *The Selected Letters of Ezra Pound
1907–1941*, ed. D. D. Paige (London: Faber and Faber, 1950), 40–1. Pound makes it
clear several times that he is excited by Eliot as an *American* who has modernized his
verse. See Noel Stock, *The Life of Ezra Pound* (1970; repr. Harmondsworth: Penguin,
1974), 208–9.   [21] Pound, *Selected Prose*, 116.

the idea of cultural comparison, on a constant setting of America beside Europe, culture against lack of sophistication, 'primitive' beside 'civilized'.

Eliot's 1918 piece 'In Memory of Henry James' is particularly interesting because it is written at a time when Eliot is having to come to terms with his decision to remain in England and pursue a literary career, a decision which (despite Ezra Pound's attempts to explain and justify it to the business man who was Eliot's father) caused considerable strains between Eliot and his American family.[22] No less, the James piece is important because it dates from the period when Eliot was evolving the crucial critical ideas of 'Tradition and the Individual Talent'.

In his tribute Eliot is concerned to emphasize that James (who had taken British nationality in 1915, and who has always tended to be assimilated too easily into 'English Literature') is important as an American writer. Pound, in his 1912 series of articles 'Patria Mia', had recommended that anyone interested in American idiosyncrasies should read James, and had commended his subtle presentation of the quintessentially American.[23] Only an American, Eliot opines, can properly appreciate James, who receives high praise from the poet as a novelist who portrayed American life, yet who did not descend to crude stereotypes of Americans as square-jawed economic pirates, stereotypes which were so useful and comforting to readers on both sides of the Atlantic. It was just such easy angles that Eliot, like Pound, as man and poet was striving to avoid. Eliot's own early work is soaked in Henry James. Prufrock follows in the faltering footsteps of Lambert Strether, while the sex war and the oppressive atmosphere of *The Bostonians* are carried on in Eliot's ironically titled 'Portrait of a Lady', which also engages with the Jamesian themes of politeness and obsessive collecting. As an anatomizer of polite Bostonian mores, Eliot is James's only rival. After all, Eliot the visitor to Isabella Stewart Gardner's museum house in Boston, Eliot the American in Paris and throughout Europe, Eliot of Harvard whose ancestors had helped hang the Salem witches, had grown up in the James-scape.[24] He knew it not only because he had read it, but because he had lived it too.

[22] T. S. Eliot, 'In Memory of Henry James', *Egoist*, Jan. 1918, 1–2.
[23] Ezra Pound, 'Patria Mia, VIII', *New Age*, 24 Oct. 1912, 611–12.
[24] Crawford, *The Savage and the City*, 30–2.

Eliot's letters emphasize the importance of James to him. For an American, winning recognition in the English literary world, he wrote in 1920, was like breaking open a safe (*L*, i. 392). James was the most important of these American safe-crackers; he was the American whom Eliot most wanted to emulate. Writing frankly to his mother, attempting, as so often, to justify his decision to remain abroad, Eliot wrote in 1919 that a small, select section of the English public regarded him as the best living poet in England, and that he thought that, as poet, critic, and editor, he had more influence on the English literary scene than any other American had had, with the possible exception of Henry James (*L*, i. 280). As a critic, Eliot, like Pound, rightly saw James as a very American writer, and one who was descended from the New England milieu of his own family and of Hawthorne. Documents of New England civilization mattered a great deal to Eliot, as his correspondence reveals (*L*, i. 274). He was particularly impressed with James as a writer on America, thinking *The American Scene* extremely well written and filled with acute criticism (*L*, i. 233). Eliot's own criticism of the New England scene grew particularly cutting in his 1919 review of Henry Adams's *Autobiography*, which fed into 'Gerontion'.[25] His criticism of this book was so sharp because it was partly directed at himself and at the society which had produced him. We might draw parallels with Pound's '*Patriam quam odi et amo*', or Joyce on Ireland, or some of MacDiarmid's vituperations against Scotland here. James, for Eliot, was 'unique' as a critic of America, and had headed in the direction which Eliot wished to take, having become more European than almost any English person or American (*L*, i. 227).

Certainly, Eliot had read James carefully, and with an American eye. Details from the novelist permeate the poet's work. It is hard not to think of 'Portrait of a Lady' when we read in *The Ambassadors* that Parisian 'late sessions . . . when men dropped in and the picture composed more suggestively through the haze of tobacco, of music more or less good and of talk more or less polyglot, were on a principle not to be distinguished from that of the mornings and the afternoons' (Book IV. ii). Madame de Vionnet's farewell to Lambert Strether, 'we might, you and I, have been friends' (Book XII. ii), underlies the Eliotic Lady's wistful 'wondering . . . | Why we have not developed into friends'. The wish to 'tell you all' (Prufrock's wish), the bocks

[25] Grover Smith, *T. S. Eliot's Poetry and Prose: A Study in Sources and Meaning*, 2nd edn. (Chicago: University of Chicago Press, 1974), 62.

and false note of 'Portrait of a Lady', and *The Waste Land*'s fascination with what James calls 'the *femme du monde*—in these finest developments of the type—like Cleopatra in the play, indeed various and multifold', the fascinating woman seated near the fire burning under marble—all those Eliotic details have their precedents in *The Ambassadors*.[26] But other, more generally suggestive aspects of this book about resident aliens stayed with Eliot the American reader; such as that strange passage, leading on to talk of Red Indians, when Little Bilham, the American in Europe who has just used the word 'Michelangelesque' with un-Prufrockian confidence, is seen ironically as a missionary in the most primitive America.

'Oh you, Mr Bilham,' she replied as with an impatient rap on the glass, 'you're not worth sixpence! You come over to convert the savages—for I know you verily did, I remember you—and the savages simply convert *you*.'

'Not even!' the young man woefully confessed: 'they haven't gone through that form. They've simply—the cannibals!—eaten me; converted me if you like, but converted me into food. I'm but the bleached bones of a Christian.'[27]

As has been pointed out by George Dorris, this is proto-*Sweeney Agonistes*.[28] It is also a passage that was bound to have struck a man like Eliot, whose own career was involving him in the often difficult questions of cultural assimilation which he answered in one way, like Henry James, by taking British nationality. There is an interest in painful cultural shifts in *The Ambassadors* that might be related to *Heart of Darkness*. The wider themes of this passage are taken up elsewhere in *The Ambassadors* and in James's work as a whole. Questions of knowledge and identity, particularly pertaining to rootlessness, fill James's work as they permeate 'Gerontion' and *The Waste Land*. 'I'm just as English as I can be . . . Oh but I'm almost American too', says Jeanne de Vionnet, in words that are worth thinking about in connection with the T. S. Eliot who both left behind and carried with him 'the roots that clutch'.[29] Strether's position in *The Ambassadors*, the man brought up in a strict New England atmosphere who, as a *metoikos*, finds himself both free from and held by 'his old tradition, the one he had been brought up on and which even so many years of life had but little worn away', has to be of interest to readers of

---

[26] Henry James, *The Ambassadors* (1903; repr. Harmondsworth: Penguin, 1973), 94, 136, 157–8, 173, 307.   [27] Ibid. 132.

[28] George E. Dorris, 'Two Allusions in the Poetry of T. S. Eliot', *English Language Notes*, Sept. 1964, 54–7.   [29] James, *The Ambassadors*, 165–6.

Eliot.[30] The issues which James, as an American novelist, continually
raises are deeply pertinent to Eliot's career precisely because James
avoids and makes fun of the crude American stereotypes on which he
can also draw. Both James and Eliot could utilize the so-called
'savage' aspect of American-ness, as well as the urbane. When the
American Isabel Archer teases Lord Warburton for explaining to her
peculiarities of English life which she is quite able to understand,
she says:

> 'He thinks I'm a barbarian . . . and that I've never seen forks and spoons;'
> and she used to ask him artless questions for the pleasure of hearing him
> answer seriously. Then when he had fallen into the trap, 'It's a pity you can't
> see me in my war-paint and feathers,' she remarked; 'if I had known how
> kind you are to the poor savages I would have brought over my native
> costume!'[31]

When Eliot was given *An Anthology of Songs and Chants from the Indians
of North America* to review for the English journal the *Athenaeum*
(17 October 1919), he revealed some important attitudes of his own
about anthropology and the artist, as well as pointing out the dangers
of turning native American Literature into ethnic kitsch. He was also
revealed as an unstereotyped, truly Jamesian American when his
review appeared under the title 'War-Paint and Feathers'. In 1923
James, along with Frazer and Bradley, was one of the masters of
prose whom Eliot singled out and to whom he expressed a great debt.
James, English yet foreign, mattered to Eliot because, like Frazer and
Kipling, he was both an insider and an outsider in Eliot's adopted
country. Like Kipling and Eliot, too, James had an acute sense of the
terror that lay beyond the borders of the conventional 'reality' of
human experience. And, for Eliot, James had one other attribute
lacking in the *metoikos* Kipling, the writer who had written (with in-
accuracies of locution) about Eliot's Cape Ann in *Captains Courageous*:
James was an American.

Just what that meant and how important it was to the Eliot of 1918
is shown when, in 'In Memory of Henry James', Eliot admires the
way in which James's fiction subordinates individual men and women
to a larger, collective idea of society. Eliot also states, with regard to
James, that the acme of American achievement is to become not an
Englishman but a European, which, Eliot adds, no national of a

[30] Ibid. 359.
[31] Henry James, *The Portrait of a Lady* (1881; repr. London: Oxford University
Press, 1947), 73–4.

European country can become. These remarks are surely a most valuable gloss on the urge which, the following year, led Eliot to stress that the poet required to show a particular form of allegiance.

He must be aware that the mind of Europe—the mind of his own country—a mind which he learns in time to be much more important than his own private mind—is a mind which changes, and that this change is a development which abandons nothing *en route*, which does not superannuate either Shakespeare, or Homer, or the rock drawing of the Magdalenian draughtsmen.[32]

No contemporary English poet would have equated Europe with 'his own country', particularly at a time when Europe had been so bloodily divided by World War One. Eliot's vision is made possible by his American-ness. He is achieving just what he had put forward in his 1918 essay on James as the ultimate American goal: to become something which no born European could become—a European. Explicitly formulated for the first time in 1919, Eliot's 'mind of Europe' is the American construction that stands behind so much of his criticism and behind his editing of the *Criterion*, which, at its most exciting, was a full-blooded attempt to internationalize English letters. The international scope of the *Criterion* sets it apart from the other English literary journals of the time. It was very much Eliot's own journal: he read every word of what was to go to print; he commissioned articles from those whose work interested him. For all its concern with the London churches, the *Criterion* is constantly looking beyond England, looking to Europe, and trying to create Eliot's image of the European mind. When World War Two broke out and the *Criterion* ceased publication, Eliot's drive to preserve what he saw as a unified European consciousness did not cease. Where World War One had immediately preceded his concentration on the mind of Europe, World War Two brought from him essays on such topics as 'Cultural Diversity and European Unity' (1945), and a series of 'Reflections on the Unity of European Culture' (1946).

If Eliot's presentation of 'the mind of Europe' can be seen as quintessentially American, his idea of tradition is more so. Pound's advice was to be, 'Make It New', though, in giving that advice, he was being as old-fashioned as possible, since he was quoting from the inscription on a Chinese emperor's bath-tub. Moreover, Pound was being solidly American in doing this, since Thoreau's *Walden* had already quoted those 'characters engraven on the bathing tub of king Tching-Thang to this effect: "Renew thyself completely each day; do it again,

[32] Eliot, *Selected Essays*, 16.

and again, and forever again.'''[33] When Eliot wrote in 'Tradition and the Individual Talent' that tradition 'cannot be inherited, and if you want it you must obtain it by great labour', he is suggesting, surely, that each generation has to work for its tradition, to make it new by its own great labour. Eliot seems to deny that tradition can comprise of taking for granted (surely the more normal sense of the word); awkwardly, he stated earlier in that famous essay that it can be equated with 'handing down'.[34] The Eliot who argues 'that the past is altered by the present as much as the present is directed by the past' prefigures both the Pound of 'Make It New' and the Harold Bloom of *The Anxiety of Influence*, and writes in a very American grain, a grain of the New World, where each era must consciously invent or reinvent its own traditions. When we think about the writers who mattered most to Eliot (such as those brought together in his 1926 Clark Lectures), we see what a strange combination they make, giving us such an odd union as Dante, John Donne, and Jules Laforgue. Whatever the links between them that Eliot may attempt to forge, it is hard not to feel that they are figures taken from very different national traditions and yoked together to form some idea of tradition which may be of individual use to T. S. Eliot. The odd assemblage of fragments from which Eliot forms his tradition seems to owe most to the pronounced breadth and variety of his reading at the eclectically structured Harvard of his student days; that is, it grows from distinctively American roots. The Eliot who wrote on Henry James is just the person in whose writings we can see that this particular 'mind of Europe' is an American invention.

Surely only a 'provincial', in this case a rather Jamesian American from that large country no one wants to visit (as Eliot described the land of his birth), would have felt he had to acquire and exhibit traditional 'high' culture with such 'great labour'.[35] And surely only an outsider could have taken such liberties with the map of European traditions, and with the English poetic tradition in particular. Reading Eliot's criticism, especially his early criticism, gives us a view of the English poetic tradition that finds no place for Wordsworth, little place for Chaucer or Milton, and where even Shakespeare seems downgraded in comparison with John Donne. As a late nineteenth-

---

[33] As well as being the title of Pound's 1934 prose collection, the injunction to 'MAKE IT NEW' appears in Canto LIII of the *Cantos*. See also Thoreau, *Walden, or Life in the Woods* (1854; repr. London: Dent, 1974), 77.

[34] Eliot, *Selected Essays*, 14.   [35] Ibid. 14.

century American, Eliot was in apparent reaction against what would have been handed down to him as his own tradition: Longfellow, Whittier, Whitman. But his reaction against the inheritance of the turn-of-the-century Englishman is at least as virulent. Eliot's early criticism is remarkable not least for how little it has to say about Tennyson, the Pre-Raphaelites, and the Brownings. Even when we think of the often lesser-known seventeenth-century writers to whom Eliot devoted so much attention, his view of that century would seem to have no place for the then widely admired, and now unjustly ignored, Robert Herrick. The view of the English poetic tradition put forward by Eliot is one which could have been produced by none of his English contemporaries, not even by Lytton Strachey. What is most shocking today is how soon the provincial Eliot's invention of European and of English traditions became so acceptable. In actuality, it was one of the most 'making new' of American productions. Moreover, Eliot's most striking poem in its opening and its cyclic nature also revoices the Whitmanian lament, 'I mourn'd, and yet shall mourn with ever-returning spring', while its hermit thrush and lilacs have clear precedents in the same Whitman poem, 'When Lilacs Last in the Dooryard Bloom'd'. *The Waste Land*, like Whitman's 'Song of Myself', reads like a huge anthology, and the figure of Tiresias performs the function of Whitman's 'self', claiming: 'I am the man, I suffered, I was there'.[36] Eliot's impersonal and ventriloquial techniques here are a form of large-scale editing, linking the anthological and the creative drives as prefigured in the works of Scott and Carlyle and manifested in the vast assemblage of the *Cantos*. *The Waste Land*, as much as Pound's *magnum opus*, provides evidence for Pound's 1912 contention in the first part of the 'Patria Mia' series that, when you pin an American down on any issue of fundamental importance, 'you get—at his last gasp—a quotation'.[37]

The more time that passes, the clearer it is becoming that, as well as being the product of Eliot's first marriage, *The Waste Land* is the product of his American upbringing and inheritance. Not least, it draws heavily on the Harvard education which had given him the cultural apparatus which fuels the poem: the reading of Dante, of anthropology, and philosophy, the Indic materials as well as the classical ones, the Roycean interest in communities of interpretation and the transmission of cultural values. The deeply characteristic

[36] *CP*, 459, 225.
[37] Ezra Pound, 'Patria Mia, I', *New Age*, 5 Sept. 1912, 445.

eclecticism of *The Waste Land*, its mixture of cultures and languages that formed what Whitman called the 'polyglot construction stamp' of American culture, points back most decisively to Eliot's American education.[38] After all, it was in America, drawing on that culture which, to Anglocentric sensibilities, was still provincial, that the poem was begun.

To some, such arguments may seem perverse in the light of Eliot's later complaints about the Harvard, even about the America, of his youth. But Joyce made far more bitter remarks against the country from which he had fled, remarks which in no way made him other than an Irishman. The MacDiarmid who cursed Scotland's intellectual lethargy was no less Scottish for that. What is most striking about Eliot's cultural allegiances is that it is often those moments when he seems most firmly to assert his Englishness which remind us how American he remained.

Chief among these must be his declaration, in 1928, that his general point of view could be described as classicist in literature, royalist in politics, and Anglo-Catholic in religion. This, coming so soon after Eliot's taking out of British nationality and joining the Church of England, may appear the quintessence of Englishness. But no normal Englishman would have made such a public declaration, for England is a country in which individuals, like governments, prefer to avoid a written constitution. Besides, no Englishman would have said that he was making such a declaration 'to refute any accusation of playing "possum" '.[39] Eliot is using a distinctively American expression here. At the very moment of apparently denying the appropriateness of his American nickname, Old Possum, he confirms its suitability. The point at which he pronounces his English allegiances is the very moment when, in locution and attitude, he demonstrates himself most American. Eliot may be taking out a nationality. He is also revealing himself as a *metoikos*.

With this we might parallel another of Eliot's characteristic gestures: 1928, the year which brought the famous Preface to *For Lancelot Andrewes*, also saw Eliot writing a Preface to a book by Edgar Ansell Mowrer called *This American World* which demonstrated how the whole world was becoming more and more Americanized. What particularly interests Eliot here is that Mowrer does not crudely simplify

---

[38] *CP*, 1075.

[39] T. S. Eliot, Preface to *For Lancelot Andrewes* (1928; repr. London: Faber and Faber, 1970), 7.

American-ness, but sees it as a version of European qualities; Europe, in becoming Americanized, is reabsorbing something of its own making. We can see this Preface as a reconsideration of the themes that interest Eliot in Henry James, and we can note how he stresses the hold that his own family tradition had on him, as well as the sense he had of being part, yet not part, of various communities. To write this Preface at the time that he ostensibly became British was a characteristically balancing gesture on Eliot's part. He once explained in a letter to Dwight Macdonald that he felt that his own part in class politics was always to lean against the prevailing wind.[40] In his cultural allegiances, too, we can see him striving after this balance: at the very moment when he seems to be assimilating himself to a cultural centre, he performs a balancing gesture which decentres him, reasserting that Eliot the insider is also significantly (and perhaps fundamentally) an outsider. It was the English patriot Eliot, fire-watcher on the London roof-tops, who, at the end of 1940 (the year of the Battle of Britain and the London Blitz), finished 'The Dry Salvages', which draws on many memories of his youth in the United States and stands as one of his most explicitly American poems. In my experience, readers often compare the *Four Quartets* with *The Waste Land*, and see Eliot's later work as staid, formally conservative, and, in a vague way, much more 'English' than the earlier poetry. Yet, so many of the themes of the *Quartets* are developed from the earlier work. The Jamesian, Bradleian, and Indic theme of reality/unreality, which, in *The Waste Land*, gives us a City whose streets we can trace on the map yet which we are told is unreal, is also the theme of the opening of 'Burnt Norton', which leads us through a vividly perceived garden yet tells us we never walked there, and offers us a reality born from deception. Again, few poems are more formally daring than the 'East Coker' which gives the reader a passage at the opening of the second section only to place it under erasure with the words:

> That was a way of putting it—not very satisfactory:
> A periphrastic study in a worn-out poetical fashion . . .

Normally, a poet either cancels a passage in a draft, revises it, or lets it stand. Eliot tries for all three options simultaneously here; it is hard to think of a comparable effect in any other poem. At least the can-

[40] T. S. Eliot, letter to Dwight Macdonald, 29 Dec. 1956 (Yale University Library).

celled sections of *The Waste Land* were cancelled, even if they have achieved a potent afterlife. Whether or not we feel that Eliot is entirely successful in the use of such a strange device in 'East Coker', we cannot deny him an innovative sense of formal daring.

The end of that poem, like the poet's decision to be buried in the Somerset village of East Coker from which his ancestor Andrew Eliot had left for America, marks both a conclusion and a departure. On a literal and a spiritual level, its last lines embody a Columbus-motion, an exploratory putting to sea which appropriately leads towards the New World of 'The Dry Salvages', that American poem which ends with the body's descent at last to a burial in 'significant soil'. This marrying of ends and beginnings which permeates the *Four Quartets* and which culminates in 'Little Gidding' is typical of Eliot's work and literary career as a whole, which can be thought of as a continual longing and search for a home. It is also typical of his own end in East Coker, that place which is both a goal and a point of departure, an English end inseparable from the voyage to an American beginning. The Eliot who pronounced himself the descendant of pioneers complicated, but did not renounce, his relations with his family and his heritage. He brought those relationships to fruition through practising the poetics of the *metoikos*.

If some criticism has put forward the picture of an Eliot bonded to the English cultural centre, it seems time to correct that picture by emphasizing the ways in which Eliot decentres himself, and also the ways in which later writers profited from his position as a resident alien and from the wide cultural sympathies which this status conferred on his writing. The Eliot who introduced criticism to the unorthodox notion of seeing Byron as a Scottish poet, an 'intelligent foreigner' in England, was sensitive to (or, some would argue, created) this aspect of Byron because it was similar to part of himself.[41] Possibly, he had Hugh MacDiarmid in mind also. Significantly, a large number of Eliot's greatest legacies have been to figures working outside, often far outside, the London-centred culture in which Eliot himself spent much of his working life—figures such as MacDiarmid or Ralph Ellison.[42]

[41] T. S. Eliot, 'Byron' (1937), in *On Poetry and Poets* (London: Faber and Faber, 1957), 205.

[42] Robert Crawford, 'A Drunk Man Looks at The Waste Land', *Scottish Literary Journal*, 14/2 (Nov. 1987), 62–78; Steven Helmling, 'T. S. Eliot and Ralph Ellison: Insiders, Outsiders and Cultural Authority', *Southern Review*, 25/4 (Autumn 1989), 841–58.

Where Eliot succeeded as a *metoikos*, Pound remained more of a constant outsider, one whose essential American-ness remained clear. Pound valued James for the same reason as Eliot did, but he linked James frequently with another American master, Whistler. Pound's adolescent sweetheart and later literary ally, H. D., recalled how, when he presented her with Whistler's *Ten O'Clock Lecture*, Pound 'scratched a gadfly, in imitation of Whistler's butterfly, as a sort of signature in his books at that time'. H. D. recalls how Whistler was part of Pound's 'composite' personality during this period, and that the gadfly motif came from the title of a novel about a tragic revolutionary poet of the Risorgimento who takes the insect as his signature.[43] Later, in his London years, Pound was to write articles on American Literature in which he acted as a gadfly to his countrymen as he stated his frustration with American letters, but also his belief in the imminence of an American Renaissance, one of whose heralds was Whistler.[44] This artist was so important for Pound because he was at once American, a cultural provincial, and, like Henry James, confidently cosmopolitan. Pound returns to Whistler (via Whitman) at the culmination of his 'Patria Mia' series, using him as proof that being born in the United States is no excuse for accepting a provincial standard, and going on to write of how the artist sometimes needs exile for the sake of his work.[45] Setting out his own determination to win cultural riches of which his country may be proud, Pound is producing one of his most vital manifestos in the 'Patria Mia' series, and one for which Whistler's example is crucial.

Later, Pound was to place Whistler alongside Dante and Leonardo.[46] In 1912 it was 'To Whistler, American', that he addressed a verse tribute, presenting himself as one of those 'Who bear the brunt of our America | And try to wrench her impulse into art'.[47] In 'Patria Mia' he wrote of the courage that Whistler had given him, and went out of his way to show how the 'American' Whistler could work in Greek, Spanish, and Japanese modes. Pound places this culturally eclectic artist with Abraham Lincoln at 'the beginning of

---

[43] H. D., *End to Torment: A Memoir of Ezra Pound*, ed. Norman Holmes Pearson and Michael King (Manchester: Carcanet New Press, 1980), 23, 64.

[44] Pound, 'Patria Mia, I', 445.

[45] Ezra Pound, 'Patria Mia, XI', *New Age*, 14 Nov. 1912, 34.

[46] Ezra Pound, *Gaudier-Brzeska: A Memoir* (1916; repr. London: Laidlaw and Laidlaw, 1939), 6.

[47] Ezra Pound, *Collected Shorter Poems*, Second Edition (London: Faber and Faber, 1968), 251.

our Great Tradition'.[48] Later, Pound would describe one of his best, most characteristic poems, 'In a Station of the Metro', as being based on an experience which, if he had been a painter, would have led to the founding of a new school of painting, one which would communicate only by 'arrangements in colour'.[49] When we realize that there was already an American artist celebrated for his 'arrangements' of colour, we see how strong Pound's identification with Whistler remained. In 1914 he linked 'Whistler and the Japanese', and reminded readers of the *Egoist* that it was from Whistler and the Japanese or Chinese that English-speaking audiences learned to take pleasure in 'arrangements' of colours. The young poet knew well that Whistler was an American in Europe who had attempted to synthesize his own work with Eastern models; and where Whistler led, Pound followed.[50]

Like his exhortation to 'Make it New', Pound's invention of China for his time may have appeared Oriental, but it was fundamentally American, deriving from a masterly resetting of the work of Ernest Fenollosa, who was so important to Pound because he had been another American cultural synthesizer.[51] Eliot's admired Henry James, the greatest American novelist to apply a comparative cultural perspective, had realized how the American's provincial anxieties could be turned to cosmopolitan advantage. Both Fenollosa and Pound can be viewed profitably in the light of James's 1867 comments that

We are ahead of the European races in the fact that more than either of them we can deal freely with forms of civilization not our own, can pick and choose and assimilate and in short (aesthetically, etc.) claim our property wherever we find it. To have no national stamp has hitherto been a regret and a drawback, but I think it not unlikely that American writers may yet indicate that a vast intellectual fusion and synthesis of various National tendencies of the world is the condition of more important achievements than any we have seen.[52]

[48] Ezra Pound, 'Patria Mia, VIII', *New Age*, 24 Oct. 1912, 612.

[49] Pound, *Collected Shorter Poems*, 119; quoted by Michael F. Harper, 'The Revolution of the Word', in Daniel Hoffman (ed.), *Ezra Pound and William Carlos Williams: The University of Pennsylvania Conference Papers* (Philadelphia: University of Pennsylvania Press, 1983), 91–2.

[50] Ezra Pound, 'Edward Wadsworth, Vorticist', *Egoist*, 15 Aug. 1914, 30.

[51] On Pound's reading of Fenollosa and versions from the Chinese, see Wai-lim Yip, *Ezra Pound's Cathay* (Princeton, NJ: Princeton University Press, 1969); see also n. 33 above.

[52] Henry James, cited in Lawrence W. Chisolm, *Fenollosa, The Far East and American Culture* (New Haven, Conn.: Yale University Press, 1963), 3.

Fenollosa had been a poet, a patriot who had written a poem on 'The Discovery of America' as well as a long poem on 'East and West', read at Harvard in 1892 and published the following year. The poem is of poor quality, but, introducing it, Fenollosa speaks of its importance as an attempt to condense his experience of two hemispheres and his study of their history, so as to herald a great coming 'synthesis' between East and West. This, again, foreshadows Pound.[53] For Pound, what mattered about Fenollosa was that he constantly pondered parallels and comparisons between Oriental and Western art, and that he also 'looked to an American renaissance'.[54]

When Pound came to publish Fenollosa's 'The Chinese Written Character as a Medium for Poetry' in 1920, it served as a poetic manifesto written by an artist who could be both an American ancestor of Ezra Pound and who could be grafted on to the aims of Imagisme, as he was in Pound's 1915 piece on 'Imagisme and England'.[55] Fenollosa is Romantic, even Emersonian, in his attitude to language, which, for him, echoes the book of the world. 'Nature furnishes her own clues. Had the world not been full of homologies, sympathies, and identities, thought would have been starved and language chained to the obvious.' What is particularly valuable about Chinese, though, is the way in which (as Fenollosa and Pound mistakenly see it) the written characters in the languages reveal their own etymologies. Language is built up with metaphor, 'the revealer of nature' and 'the very substance of poetry', piled on top of metaphor 'in quasi-geological strata'. But, because the forms of the Chinese characters have remained unchanged for centuries, and the technique of word formation (according to this view) is to impose one character on top of another, we can still see, even in complex characters, the simple metaphors beneath. Such language, revealing its own etymology, is at once sophisticated and simple. It bears its metaphor on its face, and so is more inherently concrete and poetic. In striving for the concrete image, modern 'poetry only does consciously what the primitive races did unconsciously'. This harks back to what Pound said in 1914 about modern artists being the heirs of the witch-doctor and the voodoo, and reinforced his reading of Upward, Frazer, and

[53] Ernest Francisco Fenollosa, *East and West, The Discovery of America, and Other Poems* (New York: Thomas Y. Cromwell, 1893), p. v.

[54] Ezra Pound, head-note to Ernest Fenollosa, 'The Chinese Written Character as a Medium for Poetry', in *Instigations* (New York: Boni and Liveright, 1920), 357.

[55] Noel Stock, *The Life of Ezra Pound* (1970; repr. Harmondsworth: Penguin Books, 1974), 217.

other synthesizing anthropological writers. The artist had to be a primitive, but a self-conscious primitive, and language was bound up with poetry and the care of myth, since they had evolved together. As Pound's Fenollosa put it when he contrasted the clear, metaphorical effectiveness of the Chinese characters with the condition of 'anemic' modern speech which lacked deep metaphorical input: 'Only scholars and poets feel painfully back along the thread of our etymologies and piece together our diction, as best they may, from forgotten fragments.'[56] This would be attempted in the *Cantos*.

In that vast poem-sequence we see at work a 'comparative method' in the juxtaposition of cultures. This is most apparent on a macroscopic level in Cantos LII–LXXI, where an outline of the development of Chinese culture is followed by the 'Adams' Cantos, with their account of the mores of a maturing America. The reader is invited to compare and 'rhyme' cultural patterns, just as the reader of *Ulysses* is encouraged, by the book's structure and its very title, to align the culture of Homeric Greece with that of modern Dublin. Joyce was pleased with Eliot's essay on 'Ulysses, Order, and Myth' (1923), which pointed out that 'Psychology (such as it is, and whether our reaction to it be comic or serious), ethnology, and *The Golden Bough* have concurred to make possible what was impossible even a few years ago. Instead of narrative method, we may now use the mythical method.' This 'mythical method' involved 'manipulating a continuous parallel between contemporaneity and antiquity'.[57] Eliot saw Yeats as the method's adumbrator, and many critics have pointed out that *The Waste Land* also utilizes it. It has been shown elsewhere how this technique was essayed by Eliot in his quatrain poems, and derived from his reading and reflecting on the work of a number of anthropological writers.[58]

If we think of Modernist texts as employing a 'comparative method', analogous to (and often deriving from) that of anthropology, rather than simply a 'mythical method', this allows us to see that the 'parallels' which the writers utilize are not simply between present and past, but are more generally cross-cultural. *The Waste Land*, for instance, does not simply set the modern against the antique, but (like

[56] Fenollosa, 'The Chinese Written Character', 377–9; Pound, 'The New Sculpture', 68. For more on Fenollosa and etymology, see Sanford Schwartz, *The Matrix of Modernism* (Princeton, NJ: Princeton University Press, 1985).

[57] T. S. Eliot, 'Ulysses, Order, and Myth' (1923), repr. in Richard Ellmann and Charles Feidelson (eds.), *The Modern Tradition* (New York: Oxford University Press, 1965), 681.   [58] See Crawford, *The Savage and the City*, ch. 4.

much of Pound's work) aligns and synthesizes Eastern with Western culture. It is a comparative method that is at work in *Heart of Darkness*, where 'civilized' and 'primitive' cultures and values are juxtaposed so as frequently to erode and question one another before the eyes of the reader, who is subtly urged to compare them. A parallel comparison of cultures is encouraged in *Under Western Eyes*, whose very title alerts the reader to the implicit comparison of the mores of the Western observer and the mores of the Eastern observed, 'unrolling their Eastern logic under my Western eyes'.[59] The formal propriety of 'Lord' and the slangy familiarity of 'Jim' signal to us, through their immediate juxtaposition in the title of *Lord Jim*, that this is a novel about incongruities and awkwardnesses of classification, but *Tuan Jim*, the title used for an early draft of the novel's opening, is a reminder that the book is also about cultural juxtaposition, a sometimes uneasy rubbing-together of East and West, which might stand as both a complement to, and critique of, what goes on in the *Cantos*.[60]

Conrad's earliest fictions takes place in this territory, the domain of *Almayer's Folly* and *An Outcast of the Islands*, which is also close to the territory of Stevenson's late works *The Ebb-Tide* and *The Beach of Falesá*. Stevenson's cultural juxtapositions of East and West had taken place in the painful context of the development of imperialism. Stevenson, the literary descendant of Scott, had moved from the juxtaposition of 'primitive' Highlander and 'civilized' Lowlander to a starker version of the grating and crossing of cultures in the South Seas. Kipling, whose work may also fruitfully be viewed in relation to Scott's, had explored similar phenomena using a comparative perspective in his fiction. In particular, *Kim*, a favourite book of T. S. Eliot's, delights not only in cultural cross-overs, but also in the sort of linguistic cross-overs and assemblages that recur in *Waverley*. Like Scott, Kipling rejoices in dartingly hybrid language:

Kim rubbed his nose and grew furious, thinking, as usual, in Hindi.

'This with a beggar from the bazar might be good but—I am a Sahib and the son of a Sahib and, which is twice as much more beside, a student of Nucklao. Yes' (here he turned to English), 'a boy of St. Xavier's. Damn Mr. Lurgan's eyes!—It is some sort of machinery like a sewing-machine.

[59] Joseph Conrad, *Under Western Eyes* (1911; repr. Oxford: Oxford University Press, 1983), 381.

[60] Joseph Conrad, *Lord Jim*, ed. John Batchelor (Oxford: Oxford University Press, 1983), p. xxiii.

Oh, it is a great cheek of him—we are not frightened that way at Lucknow—No!' Then in Hindi: 'But what does he gain? He is only a trader—I am in his shop. But Creighton Sahib is a Colonel—and I think Creighton Sahib gave orders that it should be done. How I will beat that Hindu in the morning! What is this?'[61]

Kipling saw England as 'the most wonderful foreign land I have ever been in', and was, as Norman Page has pointed out, an 'outsider-insider'; Stevenson, like Conrad and the *metoikos* Eliot, was an 'outsider-insider', too, able to write of himself at times as an Englishman, but also to point out as a Scot that 'A Scotsman may tramp the better part of Europe and the United States, and never again receive so vivid an impression of foreign travel and strange lands and manners as on his first excursion into England.'[62] Conrad, whose Polishness has been emphasized acutely by Zdzislaw Najder, had the Scottish nationalist R. B. Cunninghame Graham as his closest British friend, a man who was un-English to the most marked of degrees. Conrad wrote to a Polish friend in 1903: 'Both at sea and on land my point of view is English, from which the conclusion should not be drawn that I have become an Englishman. That is not the case. Homo duplex has in my case more than one meaning . . .'.[63] Conrad lost both his parents as a result of Russian imperialism, and it is well to remember that it is his Polish roots, rather than his English residence, which do most to point him to the problems of imperialism as a literary subject. Even when he is writing his most London-centred of books, *The Secret Agent*, his readers are aware of a strange manipulation of cultural viewpoints as the Assistant Commissioner, translated from one heart of darkness to another, uses, in his invest-igation of London anarchists, skills learned 'in tracking and breaking up certain nefarious secret societies amongst the natives' of the Far East. When he writes of 'Alfred Wallace' and Wallace's 'famous book on the Malay Archipelago', Conrad makes his anthropological angle fully apparent. It operates subtly in this novel, as throughout his

61 'T. S. Eliot, 70 Today . . .', *New York Times*, 26 Sept. 1958, 29; Rudyard Kipling, *Kim* (1901; repr. New York: Bantam, 1983), 135.

62 See Norman Page, ' "The Most Wonderful Foreign Land I Have Ever Been In": Kipling's England', in R. P. Draper (ed.), *The Literature of Region and Nation* (London: Macmillan, 1989), 160–79; Robert Louis Stevenson, *Memories and Portraits* (1887; repr. as vol. xxv of the Skerryvore edition, London: Heinemann, 1925), 6–7.

63 Zdzisaw Najder, *Joseph Conrad: A Chronicle* (Cambridge: Cambridge University Press, 1983), 210–11, 295.

fiction, juxtaposing cultures in a sometimes eerie comparative perspective.[64]

This anthropological eye is shared to a degree by the Forster of *A Passage to India*, and more intensely by Lawrence, whose love–hate relationship with England makes him another outsider-insider, marked and energized by his own provincialism. Various scholars have examined the way in which Lawrence's reading of Frazer informed the writing of *Women in Love* and other works.[65] It is clear that he saw his absorption of *The Golden Bough* and *Totemism and Exogamy* in 1915 as providing intellectual backing for the intuition that 'there is a blood-consciousness which exists in us independently of the ordinary mental consciousness'.[66] Frazer appealed to the Lawrence who 'loved the Leatherstocking boots so dearly!', and saw in them 'a sort of American Odyssey, with Natty Bumppo for Ulysses'.[67] Where Joyce wrote a modern Odyssey, Lawrence wrote *about* one; his travel books are a way of effecting constant cultural comparisons, while a novel such as *The Plumed Serpent* is another aligning of 'primitive' beside 'civilized' culture, an over-excited vision of the anthropological eye.

Since the Leatherstocking Tales feature constant images of female kidnap and siege, and detail a conflict between two forces led by experienced warriors, Lawrence might have written of it in terms of the *Iliad* rather than the *Odyssey*, but his choice of the *Odyssey* is typical of the emphasis of Modernism. Many of that movement's masterworks were written during the largely static hostilities of World War One, yet it is hardly ever the *Iliad*, almost always the *Odyssey*, which is invoked. Pound, opening the *Cantos* with lines translated from the start of Book XI of the *Odyssey*, signals that his poem will be a voyage among various cultures, one that turns out to be closely related to the intellectual and physical voyages of its author. Carroll F. Terrell

[64] Joseph Conrad, *The Secret Agent* (1907; repr. Harmondsworth: Penguin Books, 1979), 87, 102.

[65] John B. Vickery, *The Literary Impact of 'The Golden Bough'* (Princeton, NJ: Princeton University Press), 280–325; see also Phillip L. Marcus, ' "A Healed Whole Man": Frazer, Lawrence and Blood Consciousness', in Robert Fraser (ed.), *Sir James Frazer and the Literary Imagination* (Basingstoke: Macmillan, 1990), 232–52; Fraser's book also contains chapters on Frazer and Conrad, Frazer and Eliot, and Frazer and Lewis.

[66] D. H. Lawrence, *Letters*, ii, ed. George J. Zytaruk and James T. Boulton (Cambridge: Cambridge University Press, 1981), 469.

[67] D. H. Lawrence, *Studies in Classic American Literature* (1923; repr. Harmondsworth: Penguin Books, 1977), 52, 55.

links the *Odyssey* to *The Golden Bough* as the sources for this first Canto.[68] The title and constant allusions throughout Joyce's *Ulysses* make that novel also a multicultural voyage, while the author's addition of the words 'Trieste–Zurich–Paris' to his text indicates that its round-Dublin Odyssey is also linked to its author's travels further afield. When Eliot alludes ironically to Homer, it is to the *Odyssey*, not to the *Iliad*, when in 'Sweeney Erect' a coming-together of incongruous partners is presented as '(Nausicaa and Poylpheme)'.[69] Again, the *Odyssey* is useful because it is a voyage among cultures, providing (as *The Waste Land* would also) material for juxtapositions as rich as those of Eliot's admired *Golden Bough*. The Modernists' use of the *Odyssey* leaves a later legacy to writers such as Derek Walcott, poet of *Omeros*.

The *Odyssey* offered the anthropologically minded Modernists not only an example of a multicultural voyage, but also a provincial home-coming. Odysseus is no Agamemnon returning to the culturally dominant Greek mainland, but a provincial returning to his island of Ithaca. In the 'Ithaca' section of *Ulysses* (Joyce's favourite part of the book), home, seen by the man returned from abroad, is presented through ceaseless questions and answers as an endlessly strange amalgam, listed and catalogued. Nowhere is the energy of this novel's eclecticism more apparent.[70] The question 'What did the first drawer unlocked contain?', brings an answer which is itself an anthology, prompting the speculation that the whole of *Ulysses* is, like *The Waste Land*, the *Cantos*, and many of MacDiarmid's long poems, a collection of songs, quotations, and innumerable cultural fragments.[71] Allusion, that favoured Modernist device, is eclecticism at its most economical. A number of acute glances towards a chosen book—be it the *Divine Comedy*, *Antony and Cleopatra*, or *Dracula*—bring into play further resonances of the work alluded to, slyly incorporating it into the modern text. At their most ambitious, Modernist works become as clotted as *The Golden Bough*, and as apparently capable of infinite expansion.

Ulysses and *The Waste Land*, like the later *Finnegans Wake*, attempt structural unity not only by internal linkages, but also by a notion of return. *Ulysses*, beginning and ending with the same letter of the

[68] Caroll F. Terrell, *A Companion to the Cantos of Ezra Pound* (Berkeley, Calif.: University of California Press, 1980), 1.

[69] James Joyce, *Ulysses*, ed. Hans Walter Gabler *et al.* (London: Bodley Head, 1986), 644; Eliot, *Complete Poems and Plays*, 42.

[70] Ellmann, *James Joyce*, 500.   [71] Joyce, *Ulysses*, 592.

alphabet, enacts the homeward return of its wanderers; *The Waste Land*'s first section opens with talk of 'stirring | Dull roots with spring rain', and its conclusion is also dominated by the image of rain as a potential agent of revival, one that would painfully return the poem's occupants (and its readers) to that point in the endless cycles of creation at which the eclectic voyage of the text began; *Finnegans Wake* is celebrated for its linking of end and beginning.

All these devices are attempts to form a potentially endless anthology into a single work of art. In the *Cantos* the effort to do this is uncertain. The original version of the first Canto confronts the problem of how to present the breadth of the modern consciousness in a work of art, and sees Browning's *Sordello* as a potential model, since 'the modern world | Needs such a rag-bag to stuff all its thought in'. But the idea of a bag stuffed with multifarious ideas also strongly recalls Carlyle's Teufelsdröckh.[72] William Irving and Park Honan have pointed out that, as Browning presents it, 'Sordello's view of history resembles Carlyle's', and the whole 'rag-bag' assemblage of Pound's *Cantos* also recalls Carlyle not just as a summit of the Scottish eclectic tradition, but also as a presence behind the Whitman who drew on a vast chest of cuttings and fragments.[73]

Sometimes Pound's debts to Carlyle are as plain as his praise of Whitman; his first book of criticism declares baldly that 'The study of literature is hero-worship', a contention reminding us that the hero-worship of both Carlyle and Pound had its darker, Fascist implications.[74] But Pound's debt to Carlyle is usually far less obvious, more fully mediated. On a formal level, it is clearest when we realize that the *Cantos*, like *Sartor Resartus*, is a fantastically eclectic work that revels in 'translatorese', and, as such, represents the effort of a 'provincial' to outflank both English culture and the 'proper' language of 'standard English'. Pound's eclecticism both outstrips Carlyle's and, in so doing, falters with the worry over structure— 'I cannot make it cohere.' (Canto CXVI.) Similar anxieties attend the reading of MacDiarmid's long poems. Even *A Drunk Man Looks at the Thistle* may be seen as an anthology rather than a single unit. Such problems intensify with MacDiarmid's later work, which is

---

[72] The first published versions of Pound's first 3 Cantos are most conveniently reprinted in Bush, *The Genesis of Pound's Cantos*, 53–73; this quotation is from Canto I. 53.

[73] William Irvine and Park Honan, *The Book, the Ring, and the Poet: A Biography of Robert Browning* (London: Bodley Head, 1974), 91; Paul Zweig, *Walt Whitman: The Making of the Poet* (1984; repr. Harmondsworth: Penguin Books, 1986), 169.

[74] Ezra Pound, *The Spirit of Romance* (1910; repr. London: Peter Owen, 1970), 7.

written very much in the tradition of Scottish eclecticism that MacDiarmid inherited most immediately from the poet whom he described as 'God through the wrong end of a telescope'—the Scottish poet John Davidson.

In 1906 Davidson agitated for a 'Library of Anthologies', and he told an interviewer in 1901 that 'Dictionaries, encyclopaedias, &c., seemed' to him to be 'the proper source of a poet's vocabulary'.[75] Lauded by MacDiarmid and Eliot, Davidson's work as an urban, provincial, and encyclopaedic writer has yet to be considered fully in the context of Modernism.[76] In MacDiarmid's later poetry, particularly *In Memoriam James Joyce*, the technique is developed further, again in a poem which is a supremely eclectic 'rag-bag'. W. N. Herbert has argued eloquently that MacDiarmid's vast digressions and analogies are themselves a method of pleasurable discovery; the same might be said about those of Pound.[77] For the purposes of the present project, the important point is not to decide which Modernist texts are most successful, but, having shown the anthropological and eclectic focus of much Modernist writing, to demonstrate why MacDiarmid's work should be seen as part of the wider Modernist enterprise. This way of looking at Modernism reinforces the sense in which it is vitally 'provincial', and highlights its links with the tradition of Scottish eclecticism.

Scott and Whitman were national anthologists. Yet Whitman also came to realize that he had produced not only a national poetry, but a world poetry:

> Though from no definite plan at the time, I see now that I have uncon-sciously sought, by indirections at least as much as directions, to express the whirls and rapid growth and intensity of the United States, the prevailing tendency and events of the Nineteenth century, and largely the spirit of the whole current world, my time . . . (*CP*, 1009).

Here a sense of potential formlessness in the work, of its being bound up with the swirling nature of the modern world, anticipates Pound

[75] *MCP*, 362; John Davidson, 'October 1906' letter to Grant Richards, Davidson Archive, Princeton University Library; John Davidson, 'About Myself', *Candid Friend*, 1 June 1901, 178.

[76] *John Davidson: A Selection of his Poems*, ed. Maurice Lindsay (London: Hutchinson, 1961), pp. xi–xii ('Preface' by T. S. Eliot), 47–54 ('John Davidson: Influences and Influence', by Hugh MacDiarmid).

[77] W. N. Herbert, 'Continuity in the Poetry and Prose of Hugh MacDiarmid', D.Phil. thesis (Oxford, 1990), presently being revised for publication by the Clarendon Press.

and MacDiarmid, and is the brighter aspect of the Arnoldian worry that modernity is 'unpoetical' in its formlessness.[78] A relishing of eclecticism allows Whitman to get round this problem in his sprawling *vers libre* assemblages, whose own poetic lines are often part of an artwork which is constructed from interlinked shorter passages. This is a way of gathering the world together (seen from an American perspective) and, in a poetry of science and lyricism, of celebrating the world's imminent unity as it is welded together like railway lines or the lines of Whitman's 'Passage to India':

> The earth to be spann'd, connected by network,
> The races, neighbors, to marry and be given in marriage,
> The oceans to be cross'd, the distant brought near,
> The lands to be welded together.   (*CP*, 532.)

Here is the supreme eclectic dream, and it is one of the aspects of Whitman which most appealed to Pound, and to Hugh MacDiarmid when he proclaimed that in 'moving towards a world poetry today . . . I don't think we can ignore the best of Whitman . . .'.[79] MacDiarmid took up many themes and devices from his fellow provincial Whitman, not least of which was the urge for an eclectic and international poetry of science. For fact is one of Whitman's great Muses, and the bringing-together of facts—often the most apparently 'unpoetic'—is at the heart of his verse. He must be the only poet to attempt to construct a sort of index poem, largely made up of the titles of his other poems, where he addresses them 'by every name', listing them off (*CP*, 634). 'The true use for the imaginative faculty of modern times', Whitman proclaims, 'is to give ultimate vivification to facts, to science, and to common lives.' (*CP*, 659.) MacDiarmid's work carries forward such contentions in its development of his 'poetry of fact' in particular. MacDiarmid, like Pound, knew what to do with Whitman. This is only one of the many things they shared.

Both Pound and MacDiarmid grew up with the Celtic Twilight in provincial towns; the inheritance of these men was Romantic, and it never really left them. Pound's work drew on the late Romantic energy of Pater, whose *Studies in the History of the Renaissance*, encouraged by Andrew Lang's work, taught Pound to seek the 'virtue' of a

---

[78] Matthew Arnold, *Letters to Arthur Hugh Clough*, ed. Howard Foster Lowry (Oxford: Clarendon Press, 1932), 99.

[79] Hugh MacDiarmid, *The Thistle Rises: An Anthology of Poetry and Prose*, ed. Alan Bold (London: Hamish Hamilton, 1984), 231.

work of art among all the various 'equal' schools of taste, periods and types, something he was to do later in *The Spirit of Romance*. Pater's *Renaissance* also traced the undercurrent of pagan energies which, 'pre-eminent for light', were conducted by the Renaissance from the Middle Ages through such channels as the Albigensian cult that Pound wove into his *Cantos*.[80] Other aspects of Pound's thought owe much to late nineteenth-century Romanticism. His enthusiasm for Villon, for instance, can be traced back to Stevenson, some of whose work Pound admired, though he suspected its 'charm'. Aged about 19, it was in Stevenson that he read about Villon both in the short story 'A Lodging for the Night', with its poet stranded in the 'flying vortices' of the snow, and, in the essay in which Stevenson compares Villon with Burns, a comparison in which Pound, too, would engage in *The Spirit of Romance*.[81] Pound described his student years as 'drunk with "Celticism"', and he, like MacDiarmid, would look to the energies of the Celtic Renaissance in Ireland in order to help fuel a renaissance of his own culture.[82]

Pound's wished-for American 'Renaissance' and MacDiarmid's Scottish Renaissance have many parallels.[83] The drunken, late Romantic diction of Pound's early work can be matched by such early MacDiarmid poems as 'La Belle Terre sans Merci', with its 'Hatchments of houses multitudinous | Shine starry-white, and Eden-green | Glimmer the cypress-groves innumerous' (*MCP*, 1197). Yet each poet moved beyond such diction, because both were interested in their native variants of English as well as in constructing a synthetic language suitable for handling all themes. Where MacDiarmid read laundry baskets full of books, Pound set out to master world literature. The two poets delighted in formal variety,

---

[80] Robert Crawford, 'Pater's *Renaissance*, Andrew Lang, and Anthropological Romanticism', *ELH*, Winter 1986, 849–79; Peter Makin, *Pound's Cantos* (London: Allen and Unwin, 1985), 8–10; Walter Pater, *The Renaissance: Studies in Art and Poetry*, ed. K. Clark (1873; repr. Glasgow: Collins, 1961), 28–9, 188. See also Leon Surette, *A Light from Eleusis: A Study of Ezra Pound's 'Cantos'* (Oxford: Clarendon Press, 1979).

[81] Ezra Pound, *Guide to Kulchur* (1938; repr. London: Peter Owen, 1966), 299; Ezra Pound, *Selected Letters 1909–1941*, ed. D. D. Paige (London: Faber and Faber, 1950), 93; Pound and Shakespear, *Letters*, 118, 158; Robert Louis Stevenson, 'A Lodging for the Night', in *New Arabian Nights* (1882; repr. as vol. i of the Skerryvore edition, London: Heinemann, 1924), 251; Robert Louis Stevenson, 'François Villon', in *Familiar Studies of Men and Books* (1882; repr. as vol. xxiii of the Skerryvore edition, London: Heinemann, 1925), 161, 164; Pound, *The Spirit of Romance*, 176.

[82] Pound, *Literary Essays*, 367.

[83] Pound's essay 'The Renaissance' is reprinted from *Poetry* (Chicago, 1914) in *Literary Essays*, 214–26.

and both experienced exile (even internal exile), as their vision of
their respective native countries came to be at odds with those of the
countries themselves. Both sought to find strength in native as well as
foreign models, to transfuse the richness of international culture into
the culture of their homeland, and to form a vast and eclectic vision
which leaves them open to the charge of megalomania. MacDiarmid
and Pound were (like Whitman) aggressive self-publicists and
energizers of groups of writers. They, too, became man-myths and
'man-books' as well as compulsive editors. MacDiarmid was to be
caricatured as 'Scotland's Vortex-maker', following hot on the heels
of Vortex Pound.[84]

MacDiarmid had started writing in the *New Age* in 1909, just
before Pound began to publish there in 1910, although the Scotsman
did not become a regular contributor until the 1920s. For the young
MacDiarmid, the *New Age* was 'the most brilliant journal', and for
Pound, in the period around the First World War, it was an important organ.[85] Remembering that MacDiarmid read the *New Age* when
Pound was writing about the possibilities of an American Renaissance,
further suggests the role such thinking may have had in the fuelling of
the dream of a Scottish Renaissance. Such a dream had already been
adumbrated by one of MacDiarmid's early polymath heroes, Patrick
Geddes, who had written of a 'Scots Renascence' in his magazine the
Evergreen, which looked to the Scottish Enlightenment anthologist
Allan Ramsay. Geddes recognized Ramsay's *Ever Green* as stimulating
collections like 'Scott's "Border Minstrelsy"' and so championing
the spirit of 'Celtic Renaissance'. Geddes campaigned for a 'new
synthesis' and a 'dream of reuniting Art and Science'. MacDiarmid
paid tribute to Geddes, but the latter achieved most in fields other
than that of poetry.[86]

As far as verse was concerned, it was Pound whose work offered the
spur of synthesis. A late tribute by MacDiarmid puts succinctly the
importance of the effort which was made in the *Cantos*:

[84] This caricature of MacDiarmid by 'Coia' appeared in the *Bookman* in September
1934, and is reproduced in Gordon Wright, *MacDiarmid: An Illustrated Biography*
(Edinburgh: Gordon Wright Publishing, 1977), 147.
[85] On MacDiarmid and the *New Age*, see Catherine Kerrigan, *Whaur Extremes Meet:
The Poetry of Hugh MacDiarmid 1920–1934* (Edinburgh: James Thin, 1983), ch. 2, esp.
p. 27.
[86] See Hugh MacDiarmid, *Letters*, ed. Alan Bold (London: Hamish Hamilton,
1984), 271; Patrick Geddes, 'The Scots Renascence', *Evergreen*, Spring 1895, 136–7,
139; Patrick Geddes, 'The Sociology of Autumn', *Evergreen*, Autumn 1896, 34.

To take the whole field of knowledge and to assimilate all the diverse com-
ponents into a general view—establish a synthesis, in fact—is clearly a
Herculean undertaking scarcely to be attempted by anyone, and hardly more
likely to be understood by anyone else. Yet it is a task of paramount importance to undertake, and one (*pace* Snow and his two cultures) on which the
whole future of poetry depends. No one save Pound has yet undertaken it or
shown any sense of the need . . .[87]

This piece is correct in emphasizing Pound's attempt at synthesis,
but—and MacDiarmid probably wishes his readers to recognize this
—it is inaccurate in saying Pound is alone in his efforts. For just such
efforts characterize MacDiarmid's own nurturings of that Davidsonian
encyclopaedic Muse, with its poetic love of reference books and
eclectic diction.

Davidson (who trained as a chemist) could move from conventional
lyrics to daring explorations of language and its deployment in
strange and new combinations—in the opening passage to 'The
Crystal Palace', for example:

> Contraption,—that's the bizarre, proper slang,
> Eclectic word, for this portentous toy,
> The flying machine, that gyrates stiffly, arms
> A-kimbo, so to say, and baskets slung
> From every elbow, skating in the air.
> Irreverent, we; but Tartars from Thibet
> May deem Sir Hiram the Grandest Lama, deem
> His volatile machinery best, and most
> Magnific, rotatory engine, meant
> For penitence and prayer combined, whereby
> Petitioner as well as orison
> Are spun about in space: a solemn rite
> Before the portal of that fane unique,
> Victorian temple of commercialism,
> Our very own eighth wonder of the world,
> The Crystal Palace.[88]

The jerky, striking geometries of this verse, as well as its eclecticism
and urban liveliness, remind us why Davidson's name was often
linked to that of Whitman by early reviewers, who also argued about
whether or not Davidson's work was poetry, contending, for

[87] Hugh MacDiarmid, 'The Esemplastic Power', *Agenda*, Autumn–Winter
1970, 28.
[88] John Davidson, *The Poems*, ed. Andrew Turnbull (2 vols.; Edinburgh: Scottish
Academic Press, 1973), ii. 427.

instance, that 'Science is the enemy of poetry and all poets ought to abhor it.' In a poem such as 'Fleet Street' Davidson catalogues scientific elements, as he points to

> The carbon, iron, copper, silicon,
> Zinc, aluminium vapours, metalloids,
> Constituents of the skeleton and shell
> Of Fleet Street.

'Fleet Street' supplies a sparky eclecticism of styles and vocabularies —'gride', 'centrifugal', and 'metalloids' set against 'the dark | Consummate matter of eternity'—as Fleet Street is juxtaposed with, and linked to, the elements of the universe.[89]

Along with Pound, Davidson supplies a precedent for the eclectic gathering and swerving momentum of MacDiarmid's later poems, where the reader moves through passages on scientific and literary themes as if moving through entries in a dictionary or an encyclopaedia. The presentation of such material in poetic lineation invites us to read it with particular attention to its linguistic possibilities, its phonetic and ideational resonances, savouring an eclecticism of textures as rich as that in *Waverley* or *Sartor Resartus*, texts which also celebrate the informational, the factual, and the oddities of linguistic assembly. Seeing 'On a Raised Beach' and *In Memoriam James Joyce* against the background of Davidson, Whitman, and MacDiarmid's own previous synthetic/eclectic achievements and, behind those, Poundian Modernism and the various branches of the Scottish eclectic tradition, allows us to make far greater cultural sense of the work; it provides a nourishing context in which to read. So often MacDiarmid's 'Poetry of fact' reads like a data bank or encyclopaedia which presents a multiplicity of linguistic possibilities

> Whereby things not yet discovered are foreknown to
>   Science
> —As Meldelyev [*sic*] predicted scanium [*sic*], germanium
>   and polonium,
> As astronomers have foretold where a planet should be
> And the telescope later has found it—as blue roses
> Can never be found, but peas with yellow blossoms
> And haricot beans with red blossoms will yet be found,

---

[89] James Douglas, 'Books and Bookmen', *The Star*, 19 June 1909 (among press cuttings in Princeton University Library's Davidson archive); Davidson, *Poems*, ii. 444.

Guests not yet arrived—whose places await them.
(At the table Moseley revised—the table at which
Between aluminium and gold, masurium, rhenium
    and illinium
Have taken their places—not without question still—
While 85 and 87—alabamine and virginium—
Are still in dispute,
New places have been requisitioned
For Fermi's 9 and 94,
And whether the neutron is itself an element,
Lower than hydrogen,
And to be distinguished therefore as No. 0,
Is now in fierce debate.)   (*MCP*, 741–2.)

Part of the pleasure of this poetry is in its juxtapositions of themes and textures, its thrusting energies, impelling the reader onwards through an incredibly polymorphous text. It plants flags in innumerable territories considered by many as alien to poetry, and it goes beyond the *Cantos* on which, in part, it builds. The fullest defence of MacDiarmid's late work has been given by poet-critics whose own work falls within the line of Scottish eclecticism, namely, Edwin Morgan and W. N. Herbert. Herbert argues, validly, that MacDiarmid's late work has important post-modern affiliations.[90] Yet surely the way in which it is, like the rest of the poet's work 'assembled' (to use Edwin Morgan's words) 'out of scraps of art and life and knowledge', suggests that it may also be seen within the ambit of the eclectic Modernist texts, of Pound, Eliot, and the master memorialized in *In Memoriam James Joyce*, as well as against the background of earlier Scottish assemblages such as those of Scott or Carlyle. No aspects of MacDiarmid's work strengthen this suggestion more than his deployment of the celebrated 'Caledonian antisyzygy' and the construction of his poetic language. No writer on Modernism has yet seen these in the Modernist context. As a consequence, our views of MacDiarmid and of Modernism have been impoverished, while the forcefully provincial nature of Modernism has gone largely ignored.

It is a commonplace of MacDiarmid criticism that the poet drew on the work of one of the few teachers of English Literature to concern themselves with an extended study of Scottish writing: G. Gregory

---

[90] Herbert, 'Continuity in Hugh MacDiarmid'; Edwin Morgan, *Collected Poems* (Manchester: Carcanet, 1990), 153 ('To Hugh MacDiarmid').

Smith, whose *Scottish Literature: Character and Influence* appeared in 1919.[91] Most importantly, this study provided MacDiarmid with the concept of what Smith had playfully called 'the Caledonian antisyzygy' (GS, 4). For Smith, this was a recurring hallmark of Scottish Literature, which delighted in combining diverse, often opposing, materials and, particularly, in linking the fantastic to the densely factual. For his own work, MacDiarmid borrowed the image of the gargoyle that grins at a saint's elbow (used in *A Drunk Man Looks at the Thistle*) from Gregory Smith, as well as the title of his first book, *Annals of the Five Senses* (1923). Gregory Smith's book is really a series of essays rather than a history of Scottish Literature. Its emphasis on the Scottish Middle Ages and on the danger of continuing to imitate Burns was of use to MacDiarmid, who wished to replace Burns with Dunbar as a model for Scottish poets, though Gregory Smith's opposition to the sort of artificial 'Braid Scots' constructed by recourse to Jamieson's *Dictionary of the Scottish Language* is an opposition to exactly the sort of linguistic experimentation which MacDiarmid would practise (GS, 138–9). Where Gregory Smith was sceptical about the idea of a Scottish Renaissance along the lines of an Irish Renaissance, MacDiarmid was enthusiastic, seeing at the same time the need to utilize non-Scottish models. This impulse led him to Eliot's work, and particularly to *The Waste Land*, which he read in 1922 and which is one of the most important underpinnings of *A Drunk Man Looks at the Thistle*, where Eliot's Modernist techniques are harnessed to the long poem in Scots.[92]

MacDiarmid's disagreements with Gregory Smith were to be as fruitful as his agreements. The greatest direct help that Smith's book provided was the idea of the 'Caledonian antisyzygy', which MacDiarmid adopted as one of the main planks of his own work, and his determination always to be at the point 'whaur extremes meet' — extremes of politics, language, culture, and sensibility. Such meetings were also being explored by MacDiarmid's admired Joyce, who, in the Circe episode of *Ulysses*, had portrayed a moment where 'Extremes meet' as 'Jewgreek meets greekjew'.[93] Joyce's Daedalan wish to 'Hellenize Ireland', allowing Dublin to take her rightful place in art as 'one of the European capitals' rather than being trapped by

[91] See e.g. Nancy K. Gish, *Hugh MacDiarmid: The Man and his Work* (London: Macmillan, 1984), 63; GS.

[92] See Crawford, 'A Drunk Man Looks at the Waste Land'.

[93] See Ellmann, *James Joyce*, 395.

'Irish Ireland', is paralleled by MacDiarmid's wish to internationalize
Scottish culture and to

> ha'e Scotland to my eye
> Until I saw a timeless flame
> Tak' Auchtermuchty for a name,
> And kent that Ecclefechan stood
> As pairt o' an eternal mood.   (*MCP*, 144.)[94]

In *Ulysses*, to which MacDiarmid paid significant homage, Joyce had
demonstrated how the panorama of human existence could be presented in a modern 'provincial' town which was, like Edinburgh, a
lapsed European capital. MacDiarmid longed to unite the 'provincial'
and the international in a similar redeeming gesture, and, in the
Caledonian antisyzygy, he sought an enabling myth through which
power could be generated from meeting extremes.

In retrospect we may suspect that Gregory Smith's formulation of
a Caledonian antisyzygy was not simply an attempt to describe an
aspect of Scottish tradition, but also an outgrowth from that tradition.
Scotland had long seen itself as a cultural amalgam, and had been
presented as such by Smollett, Scott, and others. The importation of
the doppelgänger from German Romanticism was a useful one for the
Scottish writer, producing (among other works) those antisyzygical
works of fiction, Hogg's *Memoirs and Confessions of a Justified Sinner* and
Stevenson's *The Strange Case of Dr Jekyll and Mr Hyde*. Like those texts,
Gregory Smith was a product of nineteenth-century Scotland. Like
Stevenson, he grew up in Edinburgh and attended Edinburgh University, but thereafter he lived mostly out of Scotland, with whose
culture, however, he remained preoccupied. Gregory Smith (born in
1865) was a young man during the height of Stevenson's popularity;
Scottish Literature: Character and Influence, written when Smith was
Professor of English at Queen's University, Belfast, sees Stevenson as
a recent example of the antisyzygical Scottish tradition. More than
that, Gregory Smith relates his own idea of the 'jostling of contraries'
to a phrase from *Stevensoniana* about 'the "polar twins" of the Scottish
Muse' (GS, 20).

Gregory Smith was not the only Scottish academic of this period
whose concerns led him to an interest in the fusing of opposites. The
Shetlander Herbert Grierson (1866–1960), who was educated at

---

[94] James Joyce, *Letters*, ii, ed. Richard Ellmann (London: Faber and Faber, 1966),
109.

Aberdeen and Oxford, became Professor of English at Edinburgh, and who edited Scott's *Letters*, was most celebrated for his pioneering study of Donne and the Metaphysical Poets which helped bring back to prominence those poets whose work, at its most distinctive, had been condemned by Samuel Johnson as a poetry where 'the most heterogeneous ideas are yoked by violence together'.[95] Yet it is just this antisyzygical blending which appears to have appealed most to Grierson when he introduced his selection of *Metaphysical Lyrics and Poems of the Seventeenth Century*, published in 1921, two years after Gregory Smith's *Scottish Literature*. In his Introduction Grierson celebrates Donne as a poet of 'passionate thinking', linking him with Dante as combining great erudition with 'a full-blooded temperament and acute mind'. Grierson makes no direct reference to Johnson's celebrated attack on the Metaphysicals, but his own description of the finest moments in Donne's work seems to be directed at both accommodating and refuting Johnson's objections:

These vivid, simple, realistic touches are too quickly merged in learned and fantastic elaborations, and the final effect of every poem of Donne's is a bizarre and blended one; but if the greatest poetry rises clear of the bizarre, the fantastic, yet very great poetry may be bizarre if it be the expression of a strangely blended temperament, an intense emotion, a vivid imagination.[96]

'Strange blending' is what appealed to those two Scottish critics, Grierson and Gregory Smith, in their chosen subjects (Gregory Smith, too, had worked on seventeenth-century poetry, as Grierson's Introduction indicates[97]). MacDiarmid would develop this antisyzygical idea in his borrowing from Gregory Smith. And T. S. Eliot would respond to it enthusiastically in 'The Metaphysical Poets' (1921), his celebrated consideration of Grierson's anthology. Eliot's essay develops not only what Grierson had called 'the passionate fulness of [Donne's] mind', but also, in explicit refutal of Johnson's criticism, celebrates Eliot's own antisyzygical vision of the poet's business.

When a poet's mind is perfectly equipped for its work, it is constantly amalgamating disparate experience; the ordinary man's experience is chaotic,

---

[95] Samuel Johnson, 'Cowley', in *Lives of the English Poets* (1779 and 1781; 2 vols.; repr. London: Oxford University Press, 1961), i. 14.

[96] *Metaphysical Lyrics and Poems of the Seventeenth Century: Donne to Blake*, ed. Herbert J. C. Grierson (Oxford: Clarendon Press, 1921), pp. xiii, xvi, xxi, xxii.

[97] Ibid. p. xxxiii.

irregular, fragmentary. The latter falls in love, or reads Spinoza, and these two experiences have nothing to do with each other, or with the noise of the typewriter or the smell of cooking; in the mind of the poet these experiences are always forming new wholes.[98]

Eliot's own work, like that of the provincials MacDiarmid, Joyce, and Pound, would show a fascination with points of intersection 'whaur extremes meet', delighting in fusing opposites—whether ends and beginnings, 'savage' and 'civilized', or East and West.[99] MacDiarmid's 'Caledonian antisyzygy' is part of the Modernist delight in accumulating energy from the bringing-together of opposites, seen equally clearly in Yeats's ideas of self and anti-self, or his whole 'philosophy' of intersecting gyres. When *Ulysses*, *The Waste Land*, and *A Drunk Man Looks at the Thistle* juxtapose demotic, urban pub scenes with the emblems of high culture, they all, like Yeats with his intersecting self-masks, or Conrad and Lawrence with their clashing cultures, deploy the energies of antisyzygy, and rejoice in the eclecticism of Modernist cultural assembly. Similarly, all these writers, like MacDiarmid, were in some way provincial in their orientation, cultural ec-centrics who liked to confront the English cultural centre with material pointedly foreign to it.

Eliot reviewed Gregory Smith's *Scottish Literature* a month before he published the first part of his most celebrated piece of criticism, the essay on 'Tradition and the Individual Talent'.[100] In both the review and the essay Eliot is honing his own, more sophisticated ideas against the simpler perceptions of Gregory Smith. But he is also adapting some of the material from Smith's book for his own purposes. Eliot chose as the title of his review, 'Was There a Scottish Literature?' The tense of the verb is significant. Moreover, Eliot makes it clear that he sees Scottish Literature as fragmented into several periods, each heavily indebted to English Literature; since the days of Gavin Douglas, Scottish writing appears to Eliot to have been in a sometimes brilliant, but continually flickering decline.

In many ways Gregory Smith and Eliot were writers moving in opposite directions, and Eliot's reaction to Smith's study takes the form of a series of seminal disagreements. If it was just Eliot's

---

[98] Ibid. p. xxiii; Eliot, *Selected Essays*, 287.

[99] See Crawford, *The Savage and the City*, 1–4.

[100] T. S. Eliot, 'Was There a Scottish Literature?', *Athenaeum*, 1 Aug. 1919, 680–1, partly repr. in Edwin Muir, *Uncollected Scottish Criticism*, ed. Andrew Noble (London: Vision/Barnes and Noble, 1982), 88–9. The references that follow are to the original *Athenaeum* version.

American-ness that led him to construct the 'mind of Europe' and to both over-protest and undercut his allegiance to English metropolitan values, then in Gregory Smith he perceived another outsider, one who came from a culture which (whatever its status in the Middle Ages) could now be seen as 'provincial'. Eliot responded with pleasure to the glimpse of Edinburgh around 1800 which Gregory Smith offered, writing that Edinburgh at about this time was analogous to Boston around 1850, since both had momentarily challenged London as centres of cultural importance. In making this connection, Eliot links the earlier predicament of Edinburgh to the cultural milieu of his own roots. But Eliot also stressed that challenges could not be sustained in the provinces, where there was not a continuous supply of 'important men', and that it was inevitable that, after such moments passed, important men would turn to the metropolis again.[101] The Eliot who had devoted a considerable amount of his early poetry to breaking out of a now decadent and oppressively polite Bostonian milieu, and who had followed Henry James to England and to the metropolis, was committed to incorporating 'provincial' traditions into the central, metropolitan one, strengthening and subtly changing the centre. So was the Pound who wrote his 1917 series of essays on 'Provincialism the Enemy', a title with which Joyce would have concurred.[102] Yet that title is revealing, for it is the provincial, not the inhabitant of the cultural centre, who feels obliged (like the founders of Rhetoric and Belles Lettres) to worry about 'provincialism'. In his apparent rejection of his immediate American cultural inheritance, the provincial Eliot, like Gregory Smith, was also interested in searching for, and even constructing, his own idea of the tradition that mattered most to him. This Eliot was stimulated by Smith to think more deeply about what the word 'tradition' meant, and about the problems of the provincial.

Where Gregory Smith sees tradition in essentially local terms, Eliot's review swings to the opposite extreme. We can see that the ideas of 'Tradition and the Individual Talent' are already taking shape in his mind, and are being sharpened in his reaction to Smith's essays. For Eliot, 'History . . . for us is the history of Europe', and when we suppose a literature, we suppose something more than can be offered by Scottish or other 'provincial' literatures.

[101] Ibid. 681.
[102] Ezra Pound, 'Provincialism the Enemy', in *Selected Prose 1909–1965*, ed. William Cookson (London: Faber and Faber, 1973), 159–73.

We suppose not merely a corpus of writings in one language, but writings and writers between whom there is a tradition; and writers who are not merely connected by tradition in time, but who are related so as to be in the light of eternity contemporaneous, from a certain point of view cells in one body, Chaucer and Hardy. We suppose a mind which is not only the English mind of one period with its prejudices of politics and fashions of taste, but which is a greater, finer, more positive, more comprehensive mind than the mind of any period. And we suppose to each writer an importance which is not only individual, but due to his place as a constituent of this mind.[103]

This passage looks back to Pound's desire to weigh Theocritus and Yeats in one balance, but, more clearly, it looks forward to 'Tradition and the Individual Talent', the first part of which would be published the following month. Eliot appears to be articulating his ideas in total opposition to Gregory Smith, but this should not blind us to the similarities in the endeavours of the two writers. Both are attempting to articulate some sort of cultural home, a usable tradition. Eliot argues that Gregory Smith's construction is discontinuous, lacks an anchor in a single language, and will not cohere. Yet, with hindsight, it is just such charges that might be levelled at Eliot's own constructed tradition. Where Gregory Smith wishes to connect all the writers who share the common geography of his immediate homeland, the young Eliot seems equally determined to bond together a number of writers who do *not* share such a connection with the America he had left. If it is a provincial motive that drives Smith to construct, or reconstruct, a tradition which marks out a Scottish Literature, then it is the equally provincial obsession with acquiring all the trappings of cosmopolitan culture which drives Eliot to assemble his vast 'mind of Europe', whose essential cosmopolitan adjuncts allow him to outflank English-ness. Neither the tradition of Smith nor the tradition of Eliot comprises the unexamined body of material whose nature can be taken for granted by writers who conform passively to contemporary metropolitan taste. Both traditions are constructed by outsiders, and are the products of daring and labour. It is probably for this reason that Eliot is ready not only to attack Gregory Smith, but also to be stimulated by him.

When he opens 'Tradition and the Individual Talent', Eliot remarks that 'In English writing we seldom speak of tradition.'[104] Gregory Smith's book, though, had devoted particular attention to

---

[103] Eliot, 'Was There a Scottish Literature?', 680.
[104] Eliot, *Selected Essays*, 13.

the 'attitude to tradition' in Scottish Literature, and discusses recent attempts to connect the *'disjecta membra'* of Scottish culture to a coherent history and the way in which 'the historical habit rules in Scottish Literature', with the result that the literature is 'deliberately and exceptionally conservative' (GS, 57–9). This is seen most clearly in the case of Burns, whose importance comes only in part from his originality and individuality, since much of his power is drawn from the way he reproduces and 'edits' a version of the work of his traditional predecessors:

it would appear that the wider and deeper we go, the more reason do we find for the application of what we may call the 'editorial' theory. This is not suggested in the narrower sense in which we speak of Ramsay's recasting of *Christis Kirk*, for the worse, or of Burns's re-handling of the songs, to their bettering, but with the meaning that from as far back as Henryson, and notwithstanding all checks from without, the indebtedness of each poet to his predecessors, individually and corporately, is unmistakeable, and in none more so than in Burns himself. Through him the tradition passed on . . . (GS, 60).

This is a subject to which Gregory Smith returns several times in the course of his book, emphasizing how 'Literature is always looking back, shaking out the old garments, rummaging the old stores of subjects and forms. Its work is for the most part tasks of editorship; even Burns is a sublime example of the art of continuation.' (GS, 131.) For Gregory Smith, this strength of Burns is also a problem, because, while he gathered the old and gave new life to it, 'edited supremely well', he none the less marked 'the end of a process: genius is expressing itself in terms of what has gone before' (GS, 133, 134). Eliot would later feel that Milton and Shakespeare, by the peculiar supremacy of their literary achievements, became fatal models for other writers to follow, but what Eliot is able to do in 'Tradition and the Individual Talent' is to develop Gregory Smith's discussion of Burns in a positive, daring way, making Burns's type of achievement into a model for the achievement of the great poet in general. If Gregory Smith showed MacDiarmid just why Burns might be a 'barrier to emancipation' (GS, 135), and therefore not a model to be followed, he showed Eliot something different, but equally useful: the concept of the poet as editor, as a writer who matters at least as much for what he repeats from the inheritance of tradition as for what he says that is new and original. Among other things, *The Waste Land* would be Eliot's editing of the tradition of 'the mind of Europe'. 'Tradition and the Individual Talent', in its own terms, develops the concept of the poet as an editor of his tradition.

The Burns whose creative work Gregory Smith can present as a kind of editing sits firmly in that tradition of Scottish eclecticism where creative writing, anthologizing, and editing are closely connected. Eliot's response to that view of Burns shows how easily that tradition can be linked to Modernism. Eliot did not believe in a particular Scottish literary tradition, but, as well as constructing his 'mind of Europe', he did acknowledge a limited devolutionary force in English Literature, pronouncing in a lecture at University College, Dublin, in 1936 that 'there must be three literatures in the English language, English, Irish, and American'. 'In spite of the gallant efforts of Mr. Hugh MacDiarmid', Eliot was pessimistic about 'the difficult case of Scotland'.[105]

Speculations arising from the strength of 'provincial' cultures and their literatures were hardly new to the poet from St Louis. In 1919, as if the questions raised in reviewing Gregory Smith's work were still in Eliot's mind, 'Tradition and the Individual Talent' gestures, near the start of its crucial second paragraph, towards the idea that 'Every nation, every race, has not only its own creative, but its own critical turn of mind.' But that paragraph goes on to discuss no individual poet or tradition, just a universal idea of the poet, and the way in which we attend to the parts of his work which appear most individual. Eliot attacks the emphasis on 'the peculiar essence of the man',[106] just as Gregory Smith had warned against an obsession with Burns's 'personality' (GS, 229), and as MacDiarmid would savage, in *A Drunk Man*, the fatuous perversions of Burns Clubs in general (*MCP*, 84–6). Eliot's insistence on the importance of the poet's creative unoriginality is strongly reminiscent of Gregory Smith's discussion of Burns as editor. Eliot explains that

We dwell with satisfaction upon the poet's difference from his predecessors, especially his immediate predecessors; we endeavour to find something that can be isolated in order to be enjoyed. Whereas if we approach a poet without this prejudice we shall often find that not only the best, but the most individual parts of his work may be those in which the dead poets, his ancestors, assert their immortality most vigorously.[107]

[105] T. S. Eliot, 'Tradition and the Practice of Poetry', in *T. S. Eliot: Essays from The Southern Review*, ed. James Olney (Oxford: Clarendon Press, 1988), 16–17.
[106] Eliot, *Selected Essays*, 13–14.   [107] Ibid. 14.

Gregory Smith had set up Burns as 'the guardian of a literary tradition and its renewer' (GS, 226). Eliot's poet is to perform the same
functions, to become tradition's mouthpiece, as much an editor as a
poet. Eliot's own poetry (like MacDiarmid's) was to be accused of
plagiarism. Its thefts were really pieces of daringly creative and
eclectic editing—a collecting of Dante and Shakespeare, nursery
rhymes and Sanskrit—an anthologizing and expansion of the 'mind
of Europe' which is readily related to both American and Scottish
traditions of eclecticism. The Modernists' anthologizing of cultures
and cultural fragments is paralleled on the detailed level of their
poetic language. What many readers are struck by when they first
open the *Cantos* or *The Waste Land*, not to mention *Ulysses* or *Finnegans
Wake*, are the exotic elements of vocabulary—the Sanskrit, Chinese
ideograms, or Egyptian hieroglyphics—that stick out from the text.
These are only the most obvious of the foreign elements, such as the
polylingual rush of fragments at the end of *The Waste Land*, Joyce's
repeated translingual puns and hybridizations, Pound's citations and
foreign aphorisms. All these contribute to the side of Modernism
which is most off-puttingly cosmopolitan, higher than highbrow,
biblio-holic, and arcane. Yet, in that often rebarbative hyper-
cosmopolitanism is a provincial urge—the urge to let the Irish people
have 'one good look at themselves'—which propels Joyce to 'look
abroad' to Europe in order to garner its treasures for the refurbishing
of an Irish art lost in what he sees as detestable naïve insularity.[108]
There is a sense in which the deepest American-ness of Pound's
poetry comes from its seeming so un-American, so far removed from
Vachel Lindsay, *The Spoon River Anthology*, and the other poetry so
fêted in the States that Pound left behind. His work often tells
America, particularly the 'provincial' America that he had left in his
youth and that had formed him, precisely what America does not
want to hear but, Pound is convinced, needs to hear. So, first of all,
he collects for America the music of Provence, and then moves
through the classical spectrum as far as China and ancient Egypt,
gathering from the whole world those beauties which were so lacking
yet so necessary in his 'provincial' 'Patria Mia'. In 1913 Pound had
heard Laurence Binyon enthusing about Detroit's Freer Collection, a
gathering so rich both in Whistlers and Oriental art.[109] The *Cantos* are

---

[108] Joyce, *Critical Writings*, 70; James Joyce, *Letters*, ed. Stuart Gilbert (London:
Faber and Faber, 1957), 64.   [109] Pound and Shakespear, *Letters*, 178.

the Freer collection in verse. In them, as in so many other Modernist texts, mere Englishness is decisively outflanked.

That is to put it at its most noble and flattering. From another angle, the same can be said of Pound's output as can be said of MacDiarmid's: that it represents a megalomaniac attempt to take over all world culture which went hand in hand with sympathies towards those who were trying to take over the world. Uneasily, the word 'totalitarian' takes in both the writing of a poem embracing world culture and the striving for political world domination. Pound's Pater-derived search for 'virtues', which, in *The Spirit of Romance*, had led him to desire that 'literary scholarship, which will weigh Theocritus and Yeats with one balance', was, like the open-mindedness which leads MacDiarmid to 'compare Burns and Victor Khlebnikov', a liberating and exciting concept, but when Pound achieves the equivalent in the *Cantos*, we may well feel the uncomfortably strong presence of that danger always present in the eclectic tradition, the danger of levelling down, abolishing cultural difference in what is essentially a form of cultural imperialism, wishing to plunder the riches of the world for the American epic.[110]

This charge, however, is too neat. It ignores the texture of the poem and does not quite apply to Pound, as we see from an almost random glance at an example of the developed technique of the *Cantos*. What becomes apparent is that the materials are *not* fully integrated into the text, but retain something of their alien quality. The collage of Pound's poem is one of the great sources of its difficulty. This bringing-together of languages, this crossing and leaping of boundaries, is clearly the Poundian development of the provincials' eclectic tradition, but Pound does not always 'translate' his material fully. He has, one might say, the generosity to allow it to retain, even when ripped from its context, its own distinctive and often alien cultural flavour. Through repetition in the course of the poem, the foreign materials become partially domesticated. While the patient reader does achieve a certain familiarity with them, and they come to generate emotional and intellectual resonances within the work, it would be foolhardy to claim that their otherness is completely lost.

> The boat of Ra-Set moves with the sun
> 'but our job to build light' said Ocellus:

[110] Pound, *The Spirit of Romance*, 8; Hugh MacDiarmid, *Contemporary Scottish Studies* (Edinburgh: Scottish Educational Journal, 1976), 117.

Agada, Ganna, Faasa

新   hsin[1]

Make it new
Τᾶ ἐξ Αἰγύπτου φάρμακα
Leucothea gave her veil to Odysseus
Χρόνος
πνεῦμα Θεῶν
Καὶ ἔρως σοφίας
The Temple (hieron) is not for sale.   (Canto XCVIII.)

Yet it is just this 'generosity' of Pound which makes us continually wonder if the poem does cohere. The poet throws out various structural plans as his poem progresses, but it is questionable how much any of these assists the reader. Pound presents a palimpsest (another Paterian idea).[111] He reveals those 'quasi-geological strata' that Fenollosa had spoken of, inasmuch as, through his modern text, he lets show fragments of ancient cultures which he sees as part of a great whole. The problem is that the fragments appear to overcome him and his readers. The real pleasures of reading the *Cantos* are not the pleasures of a whole, but of bits, of stunning juxtapositions of cultures and languages, of piercingly lyrical textures (particularly images of light and water), with a continual and musical mixing of different cultural referents. Pound's learning is reminiscent of Baron Bradwardine's. All these juxtapositions are part of what Eliot called, in 1928, Pound's 'steady effort towards the synthetic construction of a style of speech'.[112] Such a description calls to mind the MacDiarmid who, in 1926, was involving Russian and English-language Modernist writing (including that of Joyce and Eliot) in his campaign for a move 'Towards a Synthetic Scots' of the sort which he had begun to write in 1922, but Eliot's description would also suit the work of Joyce, or the Grassic Gibbon of *A Scots Quair*, or his own language in *The Waste Land*.[113]

Modernism's delight in the construction of synthetic languages full of exotic and learned terminology can be seen as an attack on

---

[111] On the relevance of Paterian palimpsest to Pound, see Ian F. A. Bell, *Critic as Scientist: The Modernist Poetics of Ezra Pound* (London: Methuen, 1981), 190–6, to which might be added the fact that palimpsests also feature in ch. 21 of one of Pater's best-known works, *Marius the Epicurean*.

[112] T. S. Eliot, Introduction to Ezra Pound, *Selected Poems* (London: Faber and Faber, 1928), p. xiv.

[113] MacDiarmid, *Contemporary Scottish Studies*, 117 and 125.

'standard English' by writers wishing to escape the latent limitations
in their provincial origins by forging a diction so polylingual and
sophisticated that it tops and outflanks the English cultural centre. In
practice, the matter is a more complex one. For, if the cosmopolitan
linguistic drive of Modernism may have had a provincial motivation,
that élite cosmopolitan drive was matched by a demotic urge.

The Joyce who was pandied (belted) as a child for his 'vulgar
language' grew up to be concerned with 'correct English', yet as a
young man he translated Hauptmann into Irish dialect and relished
his native demotic so much that he once offered to give lessons in
Dublin English.[114] When he was constructing the history of English
prose that constitutes the 'Oxen of the Sun' episode of *Ulysses*, he
planned to end it 'in a fearful jumble of Pidgin English, nigger
English, Cockney, Irish, Bowery slang and broken doggerel'—all the
demotics, in fact. What emerges is an instructive contrast to the 1921
Newbolt Report, with its wish to use English Literature in the
moulding of an English national consciousness, its emphasis on
propriety, and its generally Anglocentric attitudes, most apparent in
the title of a book by one of the members of the committee which
produced the Report—George Sampson's *English for the English*
(1921).[115] Joyce 'un-Englishes' English and offers it to everybody:

> Waiting, guvnor? Most deciduously. Bet your boots on. Stunned like,
> seeing as how no shiners is acoming. Underconstumble? He've got the chink
> *ad lib.* Seed near free poun on un a spell ago a said war hisn. Us come right in
> on your invite, see? Up to you, matey. Out with the oof. Two bar and a wing.
> You larn that go off of they there Frenchy bilks? Won't wash here for nuts
> nohow. Lil chile velly solly. Ise de cutest colour coon down our side. Gawds
> teruth, Chawley. We are nae fou. We're nae tha fou. Au reservoir, mossoo.
> Tanks you.[116]

If Carlyle had viewed with some relish the defeat of Johnsonian
English by 'ragged battalions of Scott's-novel Scotch, with Irish,
German French and even Newspaper Cockney', then Joyce enacts
the rout with great gusto here.[117] This passage is far from being the
only one in *Ulysses* where demotic speech is relished. The continual
presence of the demotic in *Ulysses* is matched by its thorough diffusion

---

[114] Joyce, *Critical Writings*, 29; Ellmann, *James Joyce*, 30, 87, and 187.

[115] Joyce, *Letters*, 139; see Chris Baldick, *The Social Mission of English Criticism 1848–1932* (Oxford: Clarendon Press, 1983), 89–107 on the Newbolt Report, and 100 on *English for the English*.   [116] Joyce, *Ulysses*, 347.

[117] See Ch. 3 n. 66, above.

throughout the word-carnival of *Finnegans Wake*, in which the ground-rhythms of common Irish speech are given their head.

Well, after it was put in the Mercy Cordial Mendicants' Sitterdag-Zindeh-Munaday Wakeschrift (for once they sullied their white kidloves, chewing cuds after their dinners of cheeckin and beggin, with their show us it here and their mind out of that and their when you're quite finished with the reading matarial), even the snee that snowdon his hoaring hair had a skunner against him.[118]

Nora Joyce's biographer has argued that it is Nora, 'her husband's portable Ireland', whose voice is heard throughout the *Wake* and *Ulysses*.[119] This can be neither proved nor disproved, but the constant demotic note is sounded in Joyce's work, and offers (like the foreign puns and borrowings) a way of giving the text an un-English accent. The Joyce who had Stephen Dedalus complain of the English Jesuit dean that 'The language in which we are speaking is his before it is mine', would ponder the idea that 'The Irish, condemned to express themselves in a language not their own, have stamped on it the mark of their own genius and compete for glory with the civilised nations. This is then called English literature.'[120]

Exaggerated, these sentiments become those of the unsympathetic-ally chauvinistic Citizen in *Ulysses*, but the Joyce who could vituperate against Ireland could also spring abrasively to its defence: 'To me an Irish safety pin is more important than an English epic.' If he was pleased to be thought 'no Englishman', maintaining that 'every day in every way I am walking along the streets of Dublin', he also deployed the demotic to escape from the bounds of standard English propriety. The Irish demotic in particular is used to give his voice at once a local, provincial as well as an international, cosmopolitan accent.[121] His remark in 1915 that 'I cannot express myself in English without enclosing myself in a tradition', is an important one, and a development of Stephen's worries before the dean. Declaring that Ireland 'has abandoned her own language almost entirely and accepted the language of the conqueror without being able to assimilate the culture or adapt herself to the mentality of which this language is the vehicle', Joyce chose to write not in Gaelic, but in his own un-English

---

[118] James Joyce, *Finnegans Wake* (1939; repr. London: Faber and Faber, 1982), 205.

[119] Brenda Maddox, *Nora: A Biography of Nora Joyce* (London: Hamish Hamilton, 1988), 492.   [120] Ellmann, *James Joyce*, 217.

[121] Joyce, *Ulysses*, 266–7; Ellmann, *James Joyce*, 423, 456, and 704.

synthetic language, which Richard Ellmann called 'Dublin Greek'.[122] When, with regard to *Finnegans Wake*, Joyce said: 'I'll give them back their English language. I'm not destroying it for good', the third-person form of the pronoun ('their' English language) marks just that significant distancing from the English cultural centre which is apparent throughout his writings.[123]

Joyce's use of the demotic is far from unique in Modernism. Pound was eager that the language of his works should be seen to be un-English. About *Homage to Sextus Propertius*, which Pound presented as a poem against English 'imperial hogwash', he wrote in 1919 that he did not think he had consciously paid any attention to grammar anywhere in the poem, but had rendered it all ideographically, putting the entire poem 'into Chinese and then into English'.[124] If *Cathay* relies on 'translatorese' and the *Cantos* are notorious for their ideograms, we should not forget that there is also in Pound's work a strong, slangy, demotic presence mixed in with his orientalizing.

> Manchu custom very old, revived now by YONG TCHING
> An' woikinmen thought of. If proper in field work
> get 8th degree button and
>   right to sit at tea with the governor
> One, european, a painter, one only admitted
> And Pope's envoys got a melon
> And they druv out Lon Coto fer graftin'
> sent him to confino to watch men breakin' ground.   (Canto LXI.)

Here we can listen in for a moment to the constructed demotic voice of Uncle Ez's letters, the 'Uncle Rufus-speak' in which he communicated with such allies as Old Possum (whose nickname comes from Joel Chandler Harris and Mayne Reid).[125] This voice is present, too, in Pound's late translation of Sophocles' *Women of Trachis*, one of the strangest productions to emerge from Modernist demotic, and in his idea of translating Homer into slang.[126]

---

[122] Ellmann, *James Joyce*, 397; Joyce, *Critical Writings*, 212–13.

[123] Ellmann, *James Joyce*, 546.

[124] See Donald Monk, 'How to Misread: Pound's Use of Translation', in Philip Grover (ed.), *Ezra Pound, The London Years: 1908–1920* (New York: AMS Press, 1978), 82; Pound, quoted in Neda M. Westlake, 'Ezra Pound and William Carlos Williams Collections at the University of Pennsylvania', in Hoffman, *Pound and Williams*, 224.

[125] See Joel Chandler Harris, *Uncle Remus: His Songs and His Sayings* (1880; repr. Harmondsworth: Penguin Books, 1982), 59–62. While Harris talks of 'Brer Possum', Reid has an 'Old Possum'.

[126] Sophokles, *Women of Trachis: A Version by Ezra Pound* (London: Neville Spearman, 1956); *Selected Letters of Ezra Pound and Louis Zukofsky*, ed. Barry Ahearn (London: Faber and Faber, 1987), 91.

The demotic presence in Modernism, whether in Lawrence's relishing of dialect and mockery of 'The Oxford Voice' (in his poem of that title), or in Eliot's use of Cockney in *The Waste Land*, is too easily forgotten. Consideration of it makes it easier to see why MacDiarmid's synthetic work, which joined the demotic to the arcane should be viewed in the Modernist orbit. MacDiarmid's synthesizing is a further and major link between international Modernism and the cultural traditions of Scotland, where the nineteenth-century concerns with dialect and anthropology ran on together in the work of Lewis Spence and Lewis Grassic Gibbon, respectively a herald of and a literary collaborator with MacDiarmid.

Poet, encyclopaedist, writer on anthropology and myth, Fellow of the Royal Anthropological Institute, Spence is, in various respects, a minor successor to Andrew Lang. His Scottish-flavoured *Introduction to Mythology* (1921) happily quotes Scots poetry, pays tribute to Lang, and includes a survey of the 'Progress of Mythic Science' which accords acclaim to McLennan and the 'remarkable' Robertson Smith before devoting most of its space to Lang and Frazer.[127] Spence urged that the elucidation of myth should be 'eclectic', drawing on what is best in various systems, and this eclecticism is paralleled by the vocabulary of his poetry in Scots, which combined materials from various stages of the language in the 'tentative experiments in the application of older Scots phrases and syntax to the modern tongue' which critics rightly see as stimulating and reinforcing the related development of a synthetic Scots by the MacDiarmid whom Spence described in 1928 (before the two poets fell out) as 'the leader of the "Scottish" Literary Renaissance'.[128] Spence's speculative anthropological writings were also drawn on by MacDiarmid's friend J. Leslie Mitchell—'Lewis Grassic Gibbon'. Though Mitchell was later to dismiss Scott, a childhood passion for his work ('the greatest novels ever written') went hand in hand with an interest in archaeology and primitive man which matured into the preoccupation with anthropology that runs through Grassic Gibbon's fiction and non-fiction alike.[129] This combination of interests is also observable in the work of another Scottish literary descendant of Scott and

[127] Lewis Spence, *An Introduction to Mythology* (London: Harrap, 1921), 61.

[128] Ibid. 6; Spence, cited in Alan Bold, *MacDiarmid* (London: John Murray, 1988), 128; Lewis Spence, *The National Party of Scotland* (Glasgow: National Party of Scotland, [1928]), 4.

[129] J. Leslie Mitchell, *The Conquest of the Maya* (London: Jarrolds, 1934), 272; Grassic Gibbon, aged 13, cited in Ian S. Munro, *Leslie Mitchell: Lewis Grassic Gibbon* (Edinburgh: Oliver and Boyd, 1966), 18.

Stevenson (and a mentor of MacDiarmid), John Buchan, whose novels, such as *Witch Wood* (1927), show an interest in primitive survivals. Gibbon was well aware of the 'pleasing literary style' of 'Frazer, a Scotsman by birth', but he dismissed Frazer's 'gigantic compendiums' as 'crude' in *Scottish Scene*, written with MacDiarmid in 1934. Gibbon was committed instead to the Diffusionist theories of G. Elliot Smith in which T. S. Eliot took an interest and which saw the 'progress' of civilization as a progressive enslavement.[130] Grassic Gibbon's most successful works are probably those in which his anthropological interests are most subtly integrated. The trilogy *A Scots Quair* (1932–4) traces what he called elsewhere 'the codes and circumstances of a civilisation'.[131] Its three volumes set before the reader the societies of rural peasant Scotland, small-town Scotland, and capitalist urban Scotland, inviting comparisons as his protagonist Chris Guthrie moves from one to another. Such alignments of the varieties of Scottish civilization and culture recall the work of Scott, but the books are also consciously modern, not least in their synthetic language, which blends spoken and archaic Scots idiom into the fundamentally English-language narrative voice, producing a seductive constructed speech which, MacDiarmid argued, is 'not really indebted to James Joyce at all, any more than Joyce himself was to Tobias Smollett'.[132] None the less, Grassic Gibbon's work provides a further link between the native tradition of Scott and that of Modernism, lending support to the contention that the work of his admirer MacDiarmid may fruitfully be seen in *both* a Scottish and a Modernist perspective.

The Modernist, synthetic quality of his own Scots was asserted by MacDiarmid from the early 1920s, and nowhere more clearly than in a piece written in 1923, in which he linked his linguistic adventuring to Joyce's in *Ulysses*:

We have been enormously struck by the resemblance—the moral resemblance—between Jamieson's Etymological Dictionary of the Scottish language and James Joyce's *Ulysses*. A *vis comica* that has not yet been liberated lies bound by desuetude and misappreciation in the recesses of the

130 Lewis Grassic Gibbon, *A Scots Hairst: Essays and Short Stories*, ed. Ian S. Munro (London: Hutchinson, 1967), 151–2; Crawford, *The Savage and the City*, 150, 177–8, 229, 231.

131 Gibbon, quoted in Ian Campbell, *Lewis Grassic Gibbon* (Edinburgh: Scottish Academic Press, 1985), 39.

132 Hugh MacDiarmid, Foreword to Munro, *Leslie Mitchell*, p. ix.

Doric: and its potential uprising would be no less prodigious, uncontrollable, and utterly at variance with conventional morality than was Joyce's tremendous outpouring. The Scottish instinct is irrevocably, continuously, opposed to all who 'are at ease in Zion'. It lacks entirely the English sense of 'the majesty of true corpulence'. Sandy is our national figure—a shy, subtle, disgruntled, idiosyncratic individual—very different from John Bull. [133]

For MacDiarmid, the Scots dictionary became a power-source which he used to generate sharp, striking lyrics whose non-standard language is in keeping with the strange perspectives on the world which they provide. They thrill and are valuable in their eccentricity, their apartness from the English cultural centre. So, in MacDiarmid's first Scots lyric, 'The Watergaw' (*MCP*, 17), certain 'uncanny' effects are pulled off specifically because of the un-Englishness of the language. This poem is written in natural speech rhythms, and uses much normal spoken Scots vocabulary. But MacDiarmid has enriched his verbal texture by importing certain strange, sometimes obsolete, and very specific words from the Scots dictionary. 'Yowtrummle'—a word without an English equivalent (it means the time at the cold end of July when ewes tremble)—introduces the idea of cold and trembling, appropriate enough in a poem about the last look of a person on the point of death. That unique, once-in-a-lifetime, almost once-out-of-a-lifetime, look is preceded by the image of 'that antrin'—that strange or rare—'thing'—a partial rainbow. But the connotations of 'watergaw' include the idea that the fragment of rainbow is like a broken tooth. Its shivering, chattering light again suggests cold and trembling, together with the idea of teeth chattering —what teeth do when you are cold (perhaps mortally cold) or when you are frightened or horrified. The rainbow is obscured—its light is behind the falling curtain of the snow. Only then is the other subject of the poem—the look of a person on the point of death—introduced. The poem urges us to see the human and the non-human phenomena as related, but it does not spell out their relationship, any more than Ezra Pound, whom MacDiarmid greatly admired, spells out the nature of that Imagist equation which is 'In a Station of the Metro'. [134] We can see MacDiarmid's poem as being like an Imagist poem, but it is much more extreme than Pound's 'In a Station of the Metro'. MacDiarmid is fascinated by the moment in eternity when life and death intersect for an instant. 'The Watergaw' anticipates the

133 MacDiarmid, *The Thistle Rises*, 129.
134 Pound, *Collected Shorter Poems*, 119.

later poem 'At My Father's Grave', when 'A livin' man upon a deid man thinks | And ony sma'er thocht's impossible' (*MCP*, 299). MacDiarmid's first Scots poem, 'The Watergaw', where the individual confronts the stormy universe, and life meets death, marks out his determination not only to be a poet of the extreme and universal as well as the provincial and local, but also, as he later puts it, always to be 'whaur extremes meet'.

Many of the remarkable early Scots lyrics which MacDiarmid wrote in the 1920s strikingly juxtapose the human plight with the behaviour of the cosmos. In 'Empty Vessel' (*MCP*, 66) a girl whose child has died continues to sing to it, cradling its absence. Her song is described as sweeter than that of the elemental forces of the universe. Light, bending over all phenomena, gives less attention to its object than this dishevelled girl gives to hers. Another famous Scots lyric, 'The Bonnie Broukit Bairn' (*MCP*, 17), sees the earth as being like the soot-smudged face of a child, more important than the entire galaxy. Like Burns, MacDiarmid sets the small beside the great and finds in favour of the small, but, unlike Burns, he makes it perfectly clear that his linguistic strategies are linked to a left-wing Scottish nationalist strategy, are part of a search to find a post-British identity for Scotland and Scottish culture. MacDiarmid's Scots, at a challenging angle to the English language, is accompanied by remarkable slants of perception. In 'The Innumerable Christ' the world is seen from outer space, twinkling as if it were a star. The earth become unearthly. It gives off an 'unearthly licht', and Christs are seen being crucified throughout endless inhabited galaxies. Time after time these lyrics stun both by the clarity and the weirdness of their diction and by the imagination they manifest, an imagination (like Joyce's) which is both provincial and cosmological, familiar and uncanny:

> An' when the earth's as cauld's the mune
> An' a' its folk are lang syne deid,
> On coontless stars the Babe maun cry
> 　　An' the Crucified maun bleed.　　　　(*MCP*, 32.)

The odd effects of that cosmological imagination, loving both precision and eerie sweep, would continue to manifest themselves in Scots and later English poems. We see the same elements in a little poem written in the 1930s, 'The Skeleton of the Future', but in the mid-1920s MacDiarmid turned towards the long poem, writing what many consider to be his Scots masterpiece, the poem published in 1926 as *A Drunk Man Looks at the Thistle*.

Amongst many other things, *A Drunk Man* makes clear what I have suggested about the Scots poems that preceded it: being 'whaur extremes meet' involves being at the intersection of the resolutely 'provincial'—the Scots and Scottish 'other tradition'—and the international, even intergalactic. To borrow Edwin Morgan's phrase, the effect of reading *A Drunk Man* is one of being in constant rapid transit 'from Glasgow to Saturn' and back again.[135] I have pointed out elsewhere the ways in which *A Drunk Man* draws on *The Waste Land* in terms of detail, theme, and structure, and have suggested that it is part of twentieth-century Scottish Literature's search for a post-British identity.[136] What should be emphasized here is that the method of *A Drunk Man* is very much a Modernist one. Where Pound links Henry Adams to Chinese history and looks to Ovid and Dante, where Joyce's Dublin looks to Irish tradition and to Homer and Vico, so MacDiarmid's Scotland becomes the Scotland of German, Belgian, and Russian writers, the Scotland of Alexander Blok as well as the Scotland of Dante. The 'provincial' and the international are bonded, but not in the kind of cosmopolitan sense that makes a poem's speech as international as nylon. Reading *A Drunk Man*, we should not forget that we are reading a Scottish text; just as reading *Ulysses* is an experience of Irish Literature, and reading the *Cantos* and *The Waste Land* is a particularly American experience. The essential un-Englishness of these texts does not, of course, prevent them taking their place *also* in the canon of international Modernism. Rather, the provincial challenge to Anglocentric identity is a crucial part of what makes them part of that international Modernism.

MacDiarmid's later poetry of fact, with its synthetic English filled with scientific terminology, is another Modernist act of linguistic exploration, the learned counterpoint of some of his more demotic uses of Scots in, for example, satirical passages of *A Drunk Man*. In MacDiarmid's work we see the Modernist collage of language at its strongest. If we take the diagram at the start of James Murray's *New English Dictionary* (later renamed the *Oxford English Dictionary*) which shows the make-up of the English language, then we can see how the striking thing about the Modernists is that their most characteristic effects are gained by combining the materials

---

135 Edwin Morgan, *From Glasgow to Saturn* (Cheadle: Carcanet Press, 1973).
136 See Crawford, 'A Drunk Man Looks at the Waste Land'.

on the outer edges—slang, foreign, dialectal—rather than simply
rearranging the common pool with the literary and colloquial.[137]
There seems to be a geographical correlative of this, inasmuch as
most of the High Modernists did not come from the centre of English
culture. However much they 'possumed' their way into the English
centre, the roots of their creativity came from Hailey, Idaho, from
Poland, from Dublin, St Louis, Langholm, or the Nottinghamshire
pits. Their language is not the language of English gentlemen, nor is
it meant to be. If Modernism's cosmopolitanism can be seen as partly
the result of 'provincial' concerns, then so is its use of the demotic. In
this use, it brought back to the centre of high art those provincial
improprieties which the teachers of Rhetoric and Belles Lettres and
their successors had tried to banish. It is this demotic aspect of
Modernism which constitutes one of the movement's most important
legacies. Drawing so strongly on both anthropology and dialect, and
aiming to outflank the Anglocentricity of established Englishness
through a combination of the demotic and the multicultural, Modernism was an essentially provincial phenomenon. As such, it placed
various powerful stimuli at the disposal of a number of writers whom
we can characterize not simply as provincials, but as 'barbarians'.

---

[137] This diagram is reproduced with permission from James A. H. Murray (ed.),
New English Dictionary, i (Oxford: Clarendon Press, 1888; later renamed the *OED*),
'General Explanations', p. xvii.

6

Barbarians

THE most obvious legacy of Modernism is to Post-Modernism, where eclectic assemblages of linguistic textures are omnipresent. In John Ashbery's verse the polyphony of the Eliotic 'He Do the Police in Different Voices' has undergone a further development. Such an evolution has in turn contributed to the growth of L = A = N = G = U = A = G = E writing, where again there is an eclecticism of registers and textures that constantly draws attention to the materiality of language. This kind of writing is carried out, as Jerome McGann has indicated, 'under the explicit sign of what Veronica Forest-Thomson called, in her important study of 1978, *Poetic Artifice*'. Significantly, in both her critical and creative work, the Scottish poet Forest-Thomson was preoccupied with Eliot, whose work has recently been linked with that of Ashbery by various commentators.[1]

Yet there are other legacies of Modernism which are harder to perceive, perhaps, because they are found in the work of writers explicitly or implicitly opposed to the Modernist movement. These writers conceal a demotic and provincial inheritance from Modernism which continues the energies examined in the present book. The result is that we come full circle, moving to a position where the very issues fundamental to the Scottish invention of English Literature—issues of linguistic impropriety and cultural authority—are again the focus of attention in the work of a number of poets who might best be described, with a fruitfully provocative intention, as barbarians. One of the clearest bridges between the work of the Modernists and these barbarians is an unlikely but major one: the poetry of Philip Larkin.

---

[1] Jerome McGann, 'Postmodern Poetries', *Verse*, 7/1 (Spring 1990), 8. See Veronica Forrest-Thomson, *Collected Poems and Translations* (London: Allardyce, Barnett, 1990), and *Poetic Artifice* (Manchester: Manchester University Press, 1978); on Eliot and Ashbery, see Ian Gregson, 'Epigraphs for Epigones: John Ashbery's Influence in England', *Bête Noire*, 4, Winter 1987, 94; Harriet Davidson, 'John Ashbery and the Postmodern Legacy of T. S. Eliot', Conference paper given by Prof. Davidson of Rutgers University at the 1988 Conference on 'T. S. Eliot and his Legacies', University of Glasgow.

Larkin, celebrated for his opposition to Modernism and his championing of a native English tradition whose greatest modern exemplar was Hardy, stated that he thought of Eliot as American.[2] As time passes, Larkin's identification of Eliot's American-ness seems increasingly astute; but it also becomes clear just how much Larkin learned from that American. The early Eliot, writing, like the Joyce of *Dubliners*, about ignored urban corners, scraps, smells, and uninviting small rooms, consecrated, in poems like 'Rhapsody on a Windy Night', the cramped territory to be tenanted by Larkin's Mr Bleaney. The Eliot who could write of the cheap, city-clerk world of the modern urban landscape, yet who, in *The Waste Land*, could also find there an unexpected moment of respite and beauty beside a public bar, with its sounds of ordinary inelegant sociability counterpointed by inexplicable churchly splendour, prefigures the Larkin who would see in his poem 'Here' both the material desires of the cut-price crowd and the fleeting beauty of a 'Pastoral of ships up streets'. In that poem the movement from Hull to 'unfenced existence: | Facing the sun, untalkative, out of reach', parallels the movement from Lower Thames Street to the glimpsed peace at the end of *The Waste Land* 'which passeth understanding'.[3] The eye of T. S. Eliot that was trained in urban St Louis and Boston has given something to the eye that came to love Hull.

When Larkin's great poem 'The Whitsun Weddings' draws towards its conclusion with

> I thought of London spread out in the sun,
> Its postal districts packed like squares of wheat,

the effect of dense, modern, mechanical control and deadening, counterpoised by the intense fertility of the wheat, is borrowed from the effect the young Eliot achieves when

> The readers of the *Boston Evening Transcript*
> Sway in the wind like a field of ripe corn.

That Eliot poem, 'The Boston Evening Transcript', is one, like many a Larkin poem and persona, about frustration. Where Eliot in *The Waste Land*'s drafts borrowed James's title 'In the Cage', Larkin in a revealing doodle produced a series of studies in limitation when he

---

[2] Philip Larkin, interview with Thwaite (1973), cited in Geoffrey Harvey, *The Romantic Tradition in Modern English Poetry* (Basingstoke: Macmillan, 1986), 5–6.

[3] Philip Larkin, *Collected Poems*, ed. Anthony Thwaite (London: Faber and Faber, 1988), 136 ('Here'); all the Larkin poems cited are from this edition.

sketched a number of little boxes whose form linked the condemned cell to the bridal suite, a bird cage to a librarian's office.[4] Eliot's 'multifoliate rose' from 'The Hollow Men' becomes Larkin's 'million-petalled flower | Of being here' in 'The Old Fools'. Both poets were much concerned with escape and limitation. The Larkin poem 'Breadfruit'—about bored men who futilely seek escape from drab lives by dreaming of naked girls bringing breadfruit—is pure *Sweeney Agonistes*.[5] Terry Whalen has pointed out that the final poem in Larkin's early collection *The North Ship* opens with a stanza which 'reads like a pastiche of early T. S. Eliot'.[6] But very little attention has been paid to the ways in which Larkin also revised and destabil-ized that tradition as a provincial writer whose identity was in some ways as complex as Eliot's.[7]

Described as part of 'the quintessentially English tradition of equipoise', Larkin has become a key figure in the concept of English-ness in modern literature.[8] He has created one of the 'Englands of the Mind', according to Seamus Heaney, and Tom Paulin has seen his work as leading 'Into the Heart of Englishness'.[9] A famous photo-graph shows Larkin beside a large road sign that reads simply 'ENGLAND'. His clearest statement on the matter of Englishness in literature occurs in an interview with Anthony Thwaite:

I had in my mind a notion that there might have been what I'd call, for want of a better phrase, an English tradition coming from the nineteenth century with people like Hardy, which was interrupted partly by the Great War, when many English poets were killed off, and partly by the really tremendous impact of Yeats, whom I think of as Celtic, and Eliot, whom I think of as American.[10]

It is hard for critics to go on agreeing with Larkin's view of his own work when it becomes apparent that, in his poetry's language and in

[4] Larkin's doodle is reproduced in Harry Chambers (ed.), *An Enormous Yes: In memoriam Philip Larkin (1922–1985)* (Calstock, Cornwall: Peterloo Poets, 1986), 45.

[5] Larkin, *Collected Poems*, 141.

[6] Terry Whalen, *Philip Larkin and English Poetry* (Basingstoke: Macmillan, 1986), 3.

[7] See e.g. Harvey, *The Romantic Tradition*, 6, for a conventional view of Larkin's place in English tradition.   [8] Ibid.

[9] Seamus Heaney, 'Englands of the Mind', in *Preoccupations: Selected Prose 1968–1978* (London: Faber and Faber, 1980), 150–69; Tom Paulin, 'Into the Heart of Englishness', *TLS*, 20–6 July 1990, 779–80.

[10] This photograph is reproduced in Anthony Thwaite (ed.), *Larkin at Sixty* (London: Faber and Faber, 1982), opposite p. 60; Larkin, cited in Harvey, *The Romantic Tradition*, 5–6.

his defiantly 'provincial' stance at Hull, Larkin draws continually on the Modernism he sees as un-English.

In various ways, Larkin's work depends on, and develops from, Modernism. He may put forward his interest in jazz as being popular and anti-Modernist, but jazz is one of the strongest demotic elements in Picasso, Stravinsky, or, for that matter, in the work of the poet of *Sweeney Agonistes*. Larkin's poetry, obsessed with love, death, and mutability, only appears to eschew grand themes; he chooses minor characters, his protagonists live urban and suburban bit parts, and they are apparent failures, which makes them very much the kin of Leopold Bloom wandering round Dublin, or J. Alfred Prufrock, with his lust for, and dread of, society. Like these Modernist figures, the people of Larkin's verse remind us that, in the apparently provincial and marginal, great issues are enacted. Surely the last word of Molly Bloom's final monologue in *Ulysses* gives Larkin the 'enormous yes' of love in his poem 'For Sidney Becket'. Barbara Everett's examination of Larkin's debt to Symbolism, and Terry Whalen's view of Larkin in the context of Pound and Imagism, suggests that this poet's statements about his work are not to be taken at face value. In one interview he declares: 'I'm afraid I know very little about American poetry'; a moment later he makes a joke about John Ashbery; earlier he has revealed that he has been reading Frank O'Hara.[11]

Larkin demonstrates just how slippery the word 'English' is. His *Oxford Book of Twentieth-Century English Verse* might be assumed to set out his 'English tradition'. But it allocates most space to T. S. Eliot ('whom I think of as American'), while next, in terms of volume, comes W. B. Yeats, another of Larkin's masters ('whom I think of as Celtic'). Larkin's Preface to the anthology provides a rather generous definition of what he means by 'English' writers, while he also states: 'I have not included poems by American or Commonwealth writers.' Even excepting the case of Eliot, this is misleading, since an Australian (Peter Porter) and a Caribbean (Derek Walcott) are present in the book.[12] Larkin's idea of 'English' verse is actually far more wideranging than many of his pronouncements about English tradition might suggest. His 'English' tradition is appropriating and incorporating extra-English elements at exactly the same time as it asserts its

[11] Barbara Everett, 'Philip Larkin: After Symbolism', *Essays in Criticism*, July 1980, 227–42; Whalen, *Philip Larkin*, 96–100; Philip Larkin, *Required Writing* (London: Faber and Faber, 1983), 70, 64.

[12] Philip Larkin (ed.), *The Oxford Book of Twentieth-Century English Verse* (Oxford: Clarendon Press, 1973).

English purity. It may seem an anti-Modernist and narrow-mindedly English gesture to exclude Hugh MacDiarmid's Scots poems, and all synthetic Scots verse, from his *Oxford Book*, while incorporating other un-English writing. On the other hand, many readers must be surprised to find that Philip Larkin admires MacDiarmid's later work enough to include a large section of the Scottish nationalist poet's 'Lament for the Great Music'.[13] Such readers might be further stimulated if they juxtaposed the opening of MacDiarmid's 'The Glass of Pure Water' with Larkin's 'Water', a poem which at once ironizes yet fundamentally relies on what he elsewhere mocks as the 'myth-kitty', a kitty fundamental to 'The Whitsun Weddings' and much of the rest of Larkin's verse.[14]

As well as specific debts to Eliot, the demotic aspect of Larkin's diction owes a general debt to Modernist use. If Molly Bloom thinks about a 'fuck', Lawrence's Mellors follows her.[15] It is salutary to remember Eliot's annoyance when Lewis refused to print in *Blast* anything that ended with -uck, -unt, or -ugger.[16] These terms may not be an attack on standard English; but around 1920 they certainly represented an attack on standard literary English, an attack which hints at Modernism's legacy to a writer such as Tony Harrison.

Yet it was Larkin who made the word 'fuck' fully canonical. The Larkin poem which most people find easiest to remember is 'This Be the Verse', which begins with a fine pun:

> They fuck you up, your mum and dad.
> They may not mean to, but they do.

Rochester and Burns exploited this sort of language; nineteenth-century literature did not. These lines of Larkin could not have been written without Modernism, which made such language acceptable once again as the stuff of art, just as it made nursery-rhyme rhythms acceptable for serious purposes—as in 'The Hollow Men'. This point is an important one, because so often Larkin achieves his lyricism by an aggressively anti-literary opening that deploys the demotic which Modernism had brought into high art. It is Modernist demotic that allows Larkin to open an intensely lyrical poem with the line

---

[13] Ibid., p. v.

[14] Ian Hamilton, 'Four Conversations', *London Magazine*, 4/6 (Nov. 1964), 71.

[15] James Joyce, *Ulysses*, ed. Hans Walter Gabler *et al.* (London: Bodley Head, 1986), 621.

[16] *The Letters of Wyndham Lewis*, ed. W. K. Rose (London: Methuen, 1963), pp. 66–7.

'Groping back to bed after a piss' ('Sad Steps'), just as the librarian Larkin whose persona pronounces, in 'A Study of Reading Habits', that 'Books are a load of crap' owes much to the Pound persona who, in *Hugh Selwyn Mauberly*, lamented western culture as 'an old bitch gone in the teeth', represented by 'a few thousand battered books'.[17] Pound, like Joyce and the Eliot who invented his own 'Prof. Dr. Krapp', would have relished 'Jan van Hogspeuw' in Larkin's poem 'The Card-Players', while 'Old Prijck' in the same poem seems straight out of Samuel Beckett.[18] Larkin has the Modernists' demotic energy and sense of zany vulgarity, not to mention their gift of condescension. As Blake Morrison suggested in a stimulating piece called 'Dialect Does It' (*London Review of Books*, 5 December 1985), four-letter words that play off against gentilities can be seen as Larkin's equivalent of dialect.

Only by devolving the unitary notion of 'English Literature' can we come to appreciate the extent to which Larkin's work has been formed and infiltrated by those very un-English forces which, at first sight, he would appear to be resisting. We can see, too, that Larkin, like so many of the Modernist writers, is a 'provincial', rather than simply a poet of the English cultural centre. Notorious for his dislike of London, and eloquent on behalf of Hull's 'sudden elegancies', Larkin identifies himself most with the Englishness of the provincial Hardy whose attitude towards London and Oxbridge was at times sycophantic, yet generally hostile.[19] Larkin's short, elegant Foreword to Douglas Dunn's 1982 anthology *A Rumoured City: New Poets from Hull* is laden with material pertinent to his own verse. Surely it is of himself he writes when he describes 'others who come [to Hull], as they think, for a year or two, and stay a lifetime, sensing that they have found a city that is in the world, yet sufficiently on the edge of it to have a different resonance'.[20] What Hull is seen offering here is the position of insider-outsider. Hull gave Larkin a valued provincial status as a place that can only be reached, in the language of his poem 'Here', by 'Swerving' aside from the main flow of the traffic, and this provincial status of Larkin's was to be of use to such writers as Seamus Heaney, able to see Larkin as one of the modern English

[17] Ezra Pound, *Collected Shorter Poems*, Second Edition (London: Faber and Faber, 1968), 208. [18] *L*, i. 42.

[19] Philip Larkin, Foreword to *A Rumoured City: New Poets from Hull*, ed. Douglas Dunn (Newcastle: Bloodaxe Books, 1982), 9. On Hardy's attitudes towards the English cultural centre, see John Halperin, 'Hardy Rising', in *Novelists in their Youth* (London: Chatto and Windus, 1990), 57–95. [20] *A Rumoured City*, 9.

poets 'now possessed of that defensive love of their territory which was once shared only by those poets whom we might call colonial—Yeats, MacDiarmid, Carlos Williams'. By seeing Larkin as, in some sense, a 'colonial' writer, Heaney is able both to identify and compete with Larkin, the poet of 'English nationalism'.[21] If Larkin's 'English nationalism' is really very much of a provincial variety, that makes him all the closer to the Northern Ireland-born Heaney. A Janus-faced Larkin, looking in one direction towards Modernists for whom he expressed dislike, and in another direction towards un-English poets such as Heaney, may seem an unlikely phenomenon. Yet just such a Larkin was instrumental in encouraging the poetry of Douglas Dunn.

Dunn recalls that he was constantly reading and rereading Larkin's verse before he went to live in Terry Street, Hull, the area which provided much of the subject-matter for his first book of poems. Yet this reading of Larkin followed a period in Dunn's adolescence when he was 'holding imaginary conversations with W. H. Auden, Yeats and Eliot'. In both Eliot and Auden, Dunn would have met the use of the modern, often dingy, urban locations and details which Larkin's work develops. Dunn implies such a link when he writes of being particularly impressed by 'the up-to-dateness of observation in Larkin's verse', and connects this to 'the sensation contemporary readers testify to have felt when faced with Auden's poetry of the nineteen-thirties'.[22] For Dunn, then, Larkin seems to have functioned as an intermediary between the Eliot-derived Modernism of the 1930s Auden and the 1960s world of *Terry Street*, a book fascinated throughout by the provincial.

Underlying the writing of *Terry Street* is Dunn's United States experience in Akron, Ohio, where he read a great deal of American writing, admiring particularly the depiction of a small provincial community in the short stories of Sherwood Anderson's *Winesburg Ohio*.[23] But as far as English-language poetry was concerned, Dunn was working in a territory opened up by the Modernists, who had built on the work of such poets as John Davidson. The damp, litter-strewn, yellowing, lower-class urban world, where men smoke their

---

[21] Heaney, *Preoccupations*, 151, 167.

[22] Douglas Dunn, *Under the Influence* (Edinburgh: Edinburgh University Library, 1987), 1, 3. Dunn's links with Larkin, Modernism, and Scottish culture are more fully explored in Robert Crawford and David Kinloch (eds.), *Reading Douglas Dunn* (Edinburgh: Edinburgh University Press, 1992).

[23] Douglas Dunn, interview with Robert Crawford, *Verse*, 4 (1985), 28.

pipes and despondency is in the air, is a landscape of T. S. Eliot's first
book as well as Dunn's. When Dunn writes poetry that includes
masturbation and toilet habits ('Ins and Outs'), he is using subjects
that were only made possible for serious modern literature by Joyce.
When Dunn ends a love poem with the 'brushing of teeth', he is
building on Eliot's example; not only does Eliot perfect the ability to
dismiss a tender image with a harsh urban one at the conclusion of
'Preludes', he also (towards the end of 'Rhapsody on a Windy
Night') introduces what is probably the first reference in 'serious'
modern English-language poetry to the brushing of teeth.[24] Modernism extended the territory of nineteenth-century poetry by constantly
incorporating the 'unpoetic'—Eliot's 'sawdust restaurants', 'grimy
scraps', 'tooth-brush', and trousers with rolled bottoms.[25] Auden,
Larkin, and Dunn, with their industrial machinery, advertisements,
and television aerials—further this process in the area of urban landscape. Two of the *Terry Street* poems, 'After Closing Time' and
'Winter', are Poundian Imagist pieces, juxtaposing a series of shots—
'Recalcitrant motorbikes; | Dog-shit under frost; a coughing woman'
—as if to remind us that Pound's quintessential Imagist poem, 'In a
Station of the Metro', was fundamentally an urban one, of rapid city-
life perceptions.

The Modernist underpinnings of Dunn's first book are deep-
buried, like those of Larkin's work; much clearer is Dunn's debt to
Larkin, an indebtedness seen not just in general terms—the move
into an up-to-date, fashion-conscious, lower-income bracket Hull-
scape, observed in such Larkin poems as 'Here' and the despondent
'Afternoons'—but also in particular word usages. Dunn's 'Horses in
a Suburban Field' in *Terry Street* is very much a Larkinesque poem
about limitations. When, writing of children in that poem, Dunn uses
the expression 'headache-soothing absences', the unusual plural
'absences' points to the Larkin poem of that title, while 'headache-
soothing' may owe something to the milieu and suburban specifics
of the Larkin poem of childhood, 'Coming', where light 'Bathes the
serene | Foreheads of houses'. Larkin's adjective 'cheap'—'Cheap
suits' ('Here') or 'cheap clothes' ('The Large Cool Store')—is picked
up by Dunn in the 'cheap Spanish Burgundy' ('The Clothes Pit') and
'cheap wine' ('Ins and Outs') drunk in Terry Street. It marks out
both an economic poverty and a poverty of limited horizons, eroded

[24] Douglas Dunn, *Terry Street* (London: Faber and Faber, 1969), 28, 47.

[25] T. S. Eliot, *The Complete Poems and Plays* (London: Faber and Faber, 1969), 13,
22, 26, 16.

aspirations. Mr Bleaney's 'thin' curtains and the 'thin scarves' of Terry Street's 'Young Women in Rollers' function in a similar way. Mr Bleaney's 'sixty-watt bulb' environment is picked up in the Terry Street policeman's 'low-powered motorcycle', as he waits under 'a gone-out streetlamp' in 'Late Night Walk down Terry Street'. If the landscape of 'Here' is out of the way, inasmuch as it has to be reached by 'Swerving', then so is Terry Street, down which taxis 'swerve' only to use it as a short cut ('Late Night Walk down Terry Street'). Larkin offered Dunn a subject-matter, locutions, and a certain autumnal tone: if Larkin's 'Young mothers' in 'Afternoons' seem already to be prematurely ageing, the 'Young Women in Rollers' in *Terry Street* are also dated 'in last year's fashions'.

Both Larkin and Dunn can find beauties in their provincial world —whether the 'Pastoral of ships up streets' of 'Here', or the work-man's trowel that 'becomes precious' in 'On Roofs of Terry Street'. It might seem that, where Eliot and Larkin are outsiders, con-descending to their lower-class subject-matter, Dunn is living as part of his; but this is complicated by the realization that Eliot's Cockney pub drinkers are no better or worse than his Cleopatra, while Larkin's diction invites us to consider what, if anything, is wrong with 'cheap suits' rather than simply to condemn them. Dunn, while showing much sympathy with his Terry Street neighbours, can also condescend to them, like an anthropologist towards 'a lost tribe' ('The Silences'), when he points out, with a note of irony, that 'they lack intellectual grooming' ('The Clothes Pit'). The isolation of the poet-observer from his subjects is faced up to in 'A Window Affair' and 'Young Women in Rollers', where we realize that however much, like Larkin, the poet of *Terry Street* may wish to take part in the communal and sexual life of the society he watches, he is also cut off as the observing librarian-poet 'at my window, among books' ('Young Women in Rollers').

As Dunn told John Haffenden, he also felt cut off in Terry Street because he was a Scot. This does not emerge in the poems about Terry Street, though in the 'Envoi' Dunn added to the Terry Street group in 1981 he recalls Hull's 'Surprise of damp and Englishness'; in the Englishman Larkin's 'Here', the 'Englishness' could be taken for granted and did not need to be stated—Hull was just 'the surprise of a large town'.[26] Larkin physically assembled the poems of Dunn's

---

[26] John Haffenden, *Viewpoints: Poets in Conversation* (London: Faber and Faber, 1981), 15; Douglas Dunn, *Selected Poems 1964–1983* (London: Faber and Faber, 1986), 20.

first book—he may even have chosen its title—and he certainly recommended it for publication to Faber, as Dunn discovered later.[27] Dunn has written that Terry Street reminded him of the community in which he grew up and about which he had, so far, been unable to write.[28] It might seem as if, whatever private resonances it had for its Scottish author, *Terry Street* entered the world as a Larkinesque English book about provincial terrain. Yet it also contains a few significantly Scottish poems about the Clyde estuary, particularly 'Landscape with One Figure', which shows both a desire and inability to be 'part of a place' that is explicitly a Scottish place. Most important of these Scottish-accented pieces is 'A Dream of Judgment', which opens:

> Posterity, thy name is Samuel Johnson.
> You sit on a velvet cushion on a varnished throne
> Shaking your head sideways, saying No,
> Definitely no, to all the books held up to you.
> Licking your boots is a small Scotsman
> Who looks like Boswell, but is really me.

At first sight it may seem odd that Dunn moves from being an inhabitant of 1960s Hull to being an eighteenth-century Scotsman, but only at first sight. This poem is not simply about the young poet's anxiety over his first book, it is also about operating as a Scottish writer in an Anglocentric cultural milieu. Though not a poem about Larkin, it is a poem about the apparently unshakeable dominance of metropolitan Englishness. Its imagery reminds us that the poet of *Terry Street* is not simply a 'provincial'; he is also a Scot. Dunn chooses an image from the British culture of two hundred years earlier, because the issues of cultural imperialism present then in Scoto-English literary relations are just the issues which are important to him as a modern writer. It was the voice of Larkin which helped Dunn to locate his own poetic voice and to begin to enunciate those issues, combining his acute senses of class, provincialism, and Scottishness.

These senses combine most vigorously in Dunn's 1979 collection *Barbarians*, whose title announces a commitment to people and voices excluded from the dominant culture, particularly those whose speech is in some way non-standard or 'barbarous'. Dunn angled the book's blurb so as to explain that 'To the Greeks a barbarian was someone who did not speak their language. *Bar-bar, bar-bar* was their onomato-

---

[27] Dunn, *Under the Influence*, 10.   [28] Haffenden, *Viewpoints*, 14, 16.

poeic description of these un-Greek tongues.'[29] 'The Come-on', the book's first poem, with its particular attention to moments 'when the vile | Come on with their "coals in the bath" stories | Or mock at your accent', manifests a hostility towards the centres of English cultural power:

> Black traffic of Oxbridge—
> Books and bicycles, the bile of success—
> Men dressed in prunella
> Utter credentials and their culture rules us.

Against such cultural power, Dunn advocates a barbarian literary attack, but one which appropriates the weapons of the dominant culture,

> We will beat them with decorum, with manners,
> As sly as language is.
> Take tea with the king's son at the seminars—
> He won't know what's happening.

Two hundred years on, this argument parallels that of the eighteenth-century Scots who wished their countrymen to learn proper English in order to be able to defeat the English on their own cultural ground. *Barbarians* is the most markedly Scottish of Dunn's first four collections not only because its first group of poems contains 'The Student', studying at his 'Mechanics' Literary Club' in 'Renfrewshire, 1820', and wondering 'Is literature a life proved much too good | To have its place in our coarse neighbourhood', but also because the first section of the book, 'Barbarian Pastorals', is followed by a second section whose poems are set implicitly and explicitly in Scotland.

Throughout the book Dunn maintains a decorum of diction which can appear mannered ('Heartbreak and loneliness of virtue!'—'Elegy for the lost Parish'), yet at the same time he protests himself to be suspicious of, or hostile towards, the values of 'civilization', as in 'The Wealth', a poem about America, which concludes:

> In your culture, I am a barbarian,
> But I'm that here, and everywhere,
> Lulled by alien rites, lullabyed with remorse
> Here on the backstreets of the universe.

---

[29] Douglas Dunn, *Barbarians* (London: Faber and Faber, 1979), blurb on rear jacket.

The strategy of this book is simultaneously to maintain a sophisticated, cultured diction and metrics and an emphasis on the writer's
barbarian stance: to be at once sophisticated and primitive. If Dunn
has learned from the English provincialism of Larkin, an English
provincialism which itself owes a debt to Modernism, then he has
ended up adopting his own version of a common Modernist cultural
stance, being what Eliot called 'the most and least civilized and
civilizable of people'.[30] The issues which are at the centre of
Barbarians, issues of propriety of speech and cultural imperialism, are
also the issues which were at the centre of the debates surrounding the
eighteenth-century establishment of the teaching of Rhetoric and
Belles Lettres. Dunn appears to have become the uneasy countryman
of James Boswell again.

Aspects of Dunn's *Barbarians* may recall the linguistic and cultural
concerns of 'provincial' eighteenth-century Scotland, but they are
also particularly characteristic of the time the book was published. In
1980 Tony Harrison followed up his 1978 collection, *The School of
Eloquence* (whose title looks to the eighteenth century), with *Continuous*,
a book containing such poems as 'The Rhubarbarians' which, in title
and subject-matter, are very much consonant with Dunn's work.
Harrison is concerned here with the ' "mob" *rhubarb-rhubarb* ' of the
working-class North of England labourers (which is not 'poetry'), as
opposed to the 'tribune's speech' of the ruling classes, recorded by
'gaffers' blackleg Boswells'.[31] Boswell's name is used here not just to
indicate sycophantic recording, but to exemplify the provincial
speaker who appears to 'go over' to the mores of the dominant
culture. It is that culture and its speech-acts with which much of
Harrison's verse engages, asserting the strengths and independences
of those who operate outside, against, or below the rule of standard
English. If Dunn's barbarians are anxious to occupy the garden of
culture from which they are excluded, Harrison's aim is forthrightly
to 'occupy your lousy leasehold Poetry' without betrayal or com-
promise.[32] In 'Them & [uz]' Harrison, confronting his English
teacher, challenges the assumptions behind eighteenth-century ideas
of linguistic impropriety:

> 4 words only of *mi 'art aches* and . . . 'Mine's broken,
> you barbarian, T. W.!' *He* was nicely spoken.
> 'Can't have our glorious heritage done to death!'

[30] T. S. Eliot, 'War-Paint and Feathers', *Athenaeum*, 17 Oct. 1919, 1036.
[31] Tony Harrison, *Selected Poems*, New Expanded Edition (Harmondsworth:
Penguin Books, 1987), 113 ('The Rhubarbarians, I'). All Harrison poems are quoted
from this selection.   [32] Ibid. 123 ('Them & [uz], II').

I played the Drunken Porter in *Macbeth*.

'Poetry's the speech of kings. You're one of those
Shakespeare gives the comic bits to: prose!
All poetry (even Cockney Keats?) you see
's been dubbed by [ʌs] into RP,
Received pronunciation, please believe [ʌs]
your speech is in the hands of the Receivers.'

'We say [ʌs] not [uz], T. W.!' That shut my trap.
I doffed my flat a's (as in 'flat cap')
my mouth all stuffed with glottals, great
lumps to hawk up and spit out . . . *E-nun-ci-ate*!³³

Harrison uses his Leeds demotic speech to conflict with, and undermine, the credentials of the received pronunciation which has the backing of the institutional teaching of English Literature. He proclaims 'RIP RP' and determines to use 'my *name* and own voice: [uz] [uz] [uz]'.³⁴ Yet Harrison's own poetry is not, for the most part, written in demotic. Often, as in 'Book Ends', his verse makes clear the separation which the poet's book-learning has imposed between him and his cultural background. Harrison's wish is fully to take on board the language of 'high culture', while maintaining an edgy, often awkward loyalty to his provincial home. In so doing, like the Modernists, he takes in an unusually wide linguistic range. As well as setting provincial demotic and standard English, the typesetter of 'The Rhubarbarians' will need access to a Greek font and to the international phonetic alphabet. Much of Harrison's work is bound up with translation, at which he works 'the hardest in his class' ('Classics Society')—translation not just from one language to another, but from one class to another, and from provincial to metropolitan. Such themes of translation, ever powerful in provincial writers who are compelled to measure their own cultural and linguistic standards against other, dominant ones, continually lead Harrison to juxtapose materials drawn from either side of a cultural, linguistic, or class barrier. So 'The Rhubarbarians, II' carries the subtitle '(On translating Smetana's *Prodaná Nevesta* for the Metropolitan Opera, New York)', which is immediately followed by an epigraph from George Formby about '*mi little stick of Blackpool Rock!*'. The very form of the poems in *The School of Eloquence* is a strange amalgam, bringing

³³ Ibid. 122 ('Them & [uz], I').
³⁴ Ibid. 123. For more detailed comment on Harrison's language and its politics, see Neil Astley (ed.), *Bloodaxe Critical Anthologies*, i. *Tony Harrison* (Newcastle upon Tyne: Bloodaxe Books, 1991).

together the language and subject-matter of working-class Leeds with the high-cultural vehicle of the extended sonnet (sonnets being a form more normally associated with courtly eroticism or literary homage). A similar bonding had been achieved in the title and body of Edwin Morgan's 'Glasgow Sonnets' (1972), where, as Douglas Dunn points out, 'High Culture, in one of its most precious forms, encounters the issues of Glasgow' in a combination which challenges the reader to decide whether or not the title 'Glasgow Sonnets' is oxymoronic.[35]

Where Harrison uses provincial demotic to question the cultural authority of an accredited standard English, often in an anguished way, Tom Leonard's 1970s poems mounted a related 'barbarian' critique deploying comic means, for example, in his use of phonetically transcribed Glaswegian:

> this is thi
> six a clock
> news thi
> man said n
> thi reason
> a talk wia
> BBC accent
> iz coz yi
> widny wahnt
> mi ti talk
> aboot thi
> trooth wia
> voice lik
> wanna yoo
> scruff. if
> a toktaboot
> thi trooth
> lik wanna yoo
> scruff yi
> widny thingk
> it wuz troo.[36]

Leonard, like Harrison, is able to use non-standard forms as part of a gesture of solidarity with lower-class speakers of a provincial vernacular, as well as using these forms as a means of interrogating the

---

[35] Douglas Dunn, 'Morgan's Sonnets', in Robert Crawford and Hamish Whyte (eds.), *About Edwin Morgan* (Edinburgh: Edinburgh University Press, 1990), 75.

[36] Tom Leonard, *Intimate Voices: Selected Work 1965–1983* (Newcastle: Galloping Dog Press, 1984), 88 ('Unrelated Incidents (3)').

established structures of linguistic and cultural power. Aware of linguists such as Chomsky and Sapir, his work in both English and Scots once again seeks to combine sophistication with an apparently barbarian inflection. It would be as naïve to see this writing as limited in its concerns to comic or solely Glaswegian issues as it would to see Tony Harrison as simply a poet of Leeds. Leonard is aware that 'the criticism of "provincialism"' is 'an international pattern', and, like Seamus Heaney, he is preoccupied with 'governance'—with issues of language and political control.[37]

The second poem of Leonard's 1986 *Situations Theoretical and Contemporary* clearly encourages the reader to draw connections between the Scottish situation and wider instances of cultural imperialism. The poem begins with the arrival in the bay of 'The schooner *The Mother of Parliaments*', and concludes:

> Run and tell your fellow-tribesmen.
> We are going to have a referendum!
> Shall we join the British Empire?[38]

This is clearly a poem of widely applicable cultural reference, while Leonard's use of the word 'referendum' gives it a particular resonance in the context of Scotland after the 1979 Devolution Referendum. Leonard's work in verse may be related to the prose of Glaswegian James Kelman, who also uses a voice clearly differentiated from that of standard English in order to express solidarity with a particular linguistic and cultural community. The narrative voice of Kelman's *A Disaffection* is very much the voice that was repressed by the teachers of Rhetoric and Belles Lettres and their institutional descendants: 'The clothes. He was going to don a shirt and tie and generally affect the conventional appearance of an establishment sort of bloke, an ordinary upholder of the Greatbritish way of eking out this existence. He would polish the shoes. Naw he fucking wouldni.'[39] Kelman's diction, mixing high register ('don') and low register ('wouldni'), is very much an artistic construct, yet one which owes clear allegiance to working-class Glasgow speech. The voice in the novel is barbarian, but also sophisticated; it swears, and discusses philosophy. Kelman

[37] *Radical Renfrew: Poetry from the French Revolution to the First World War*, ed. Tom Leonard (Edinburgh: Polygon, 1990), pp. xxv–xxvi; cf. Seamus Heaney, *The Haw Lantern* (London: Faber and Faber, 1987), 28.

[38] Tom Leonard, *Situations Theoretical and Contemporary* (Newcastle: Galloping Dog Press, 1986) [3].

[39] James Kelman, *A Disaffection* (London: Secker and Warburg, 1989), 125.

has a strong interest in Third World writing, and his linguistic strategy is a way of combating what is perceived as class and cultural imperialism. Kelman has also encouraged Alasdair Gray to deploy phonetic representation to emphasize that 'standard English' is simply another dialect—'the main dialect of the British rich' and 'the language of Shakespia and Docta Johnson'.[40] Such a device, upending the conventions of phonetic representation, is the barbarian's revenge.

Critics have shown a general reluctance to establish connections between material produced in various English-language locations that might be described as provincial. This has meant, for instance, that Scottish and Northern Irish work tends to be seen separately. Yet there are obvious, strong shared preoccupations which make it worthwhile examining significant similarities. In 1983 Seamus Deane's Field Day pamphlet *Civilians and Barbarians* pointed out how English norms of culture and civility had meant that differing Irish phenomena were defined as barbaric both by the English and the colonized Irish. This idea of a barbarian un-English identity which may be both utilized and transcended has been of great concern not only to Joyce, but also in Irish writing.[41]

One of the texts which Deane sees as constructing the English/Irish distinction between civilians and barbarians is Edmund Spenser's *View of the Present State of Ireland* (1596); when Heaney reads the same text, he identifies with the barbarians, feeling 'closer to the natives, the geniuses of the place'.[42] Like the Dunn of *Barbarians* and the Harrison of 'The Rhubarbarians', Heaney, with his 'guttural muse', has rejoiced in a language able to enunciate cultural differences at a significant angle to standard English.[43] The first part of his 'Singing School' (whose title and subject-matter precede and parallel aspects of Harrison's *The School of Eloquence*) is dedicated to Seamus Deane, and revels in an environment where

> Those hobnailed boots from beyond the mountain
> Were walking, by God, all over the fine
> Lawns of elocution.[44]

[40] Alasdair Gray, *Something Leather* (London: Cape, 1990), 31. On the importance of Gray in 20th century Scottish writing, see Robert Crawford and Thom Nairn (eds.), *The Arts of Alasdair Gray* (Edinburgh: Edinburgh University Press, 1991).

[41] Seamus Deane, *Civilians and Barbarians* (1983), repr. in Field Day Theatre Company, *Ireland's Field Day* (London: Hutchinson, 1985), 33–42.

[42] Ibid. 33; Heaney, *Preoccupations*, 34–5.

[43] Seamus Heaney, 'Traditions', in *Selected Poems 1965–1975* (London: Faber and Faber, 1980), 68.   [44] Ibid. 130.

Where Dunn as a lower-class Scot sees his younger self, ironically, as treasuring 'a grudge as lovely as mine', Heaney the rural Ulster Catholic here relishes 'Inferiority | Complexes, stuff that dreams were made on'.[45] While it may have seemed to him that 'Ulster was British, but with no rights on | The English Lyric', he, like Dunn and Harrison, strove for a mastery of decorous English, at the same time as seeking the ability to pronounce his own cultural difference.[46] In his search for both sophisticated grace and local strength, Heaney found kin among the Modernists. He values MacDiarmid, for instance, not only for his 'local accent' in *A Drunk Man Looks at the Thistle* (a work which Heaney ranks with *The Waste Land*), but also because MacDiarmid attempts to write in an English 'out of a region where the culture and language are at variance with standard English utterance and attitudes'. Heaney rightly sees such problems as linking 'Americans, West Indians, Indians, Scots and Irish', and goes on to connect MacDiarmid with Joyce.[47]

While Yeats and the Celtic Twilight, offering 'a discovery of confidence in our own ground, in our place, in our speech', are clearly important to Heaney, it would be wrong to ignore the example of Joyce.[48] If Heaney's friend Richard Ellmann was fond of stressing Joyce's wish to Hellenize Ireland, there are few clearer examples of Joyce's success in this endeavour than the fact that Heaney opened his own account of a Northern Irish childhood in 'Mossbawn' with the word and concept of the '*omphalos*', a term that was redeployed to denote the persistence of an ideal Ireland in the poem 'The Toome Road'.[49] If Heaney is a writer who is ambitious and international in his range, yet one for whom 'the English tradition is not ultimately home', then the importance that Joyce's preoccupations hold for him is evident in the prominence he gives to that writer at the conclusion of his 1974 lecture 'Feeling into Words', where, after writing of his view of poetry, and particularly of his own Irish poetry, as 'a restoration of the culture to itself', Heaney ends with the statement that 'to forge a poem is one thing, to forge the uncreated conscience of the race, as Stephen Dedalus puts it, is quite another and places daunting pressures and responsibilities on anyone who would risk the name of poet'.[50]

---

[45] Douglas Dunn, *Love or Nothing* (London: Faber and Faber, 1974), 34 ('The Competition'); Heaney, *Selected Poems*, 130.   [46] Heaney, *Selected Poems*, 131.
[47] Heaney, *Preoccupations*, 195–8.   [48] Ibid. 135.
[49] Ibid. 17; Seamus Heaney, *Field Work* (London: Faber and Faber, 1979), 15.
[50] Heaney, *Preoccupations*, 34, 60.

Just such pressures and responsibilities are at the heart of Heaney's poetry, one of whose main themes is the constant tug between the forces which demand that he act as spokesman for a particular culture under pressure, and the forces which demand that he concentrate on producing an aesthetic pleasure freed from the attacks and propulsions of cultural politics. Innumerable Heaney poems attempt to connect or reconcile the Ulster farm-boy with the Harvard or Oxford Professor of Poetry, the aesthete with the Republican, the barbarian with the admirer of civilian achievement. Sometimes the balances are accurately, painfully spelled out, as when Heaney the civilian knows how he 'would connive | in civilized outrage' at the same time as the barbarian side of him would 'understand the exact | and tribal, intimate revenge' wreaked on a girl tarred and feathered for going out with British soldiers ('Punishment').[51] Elsewhere, the balances or resolutions have an awkwardness about them that carries its own unstated eloquence. If the poet of 'Digging' resolves to dig with his pen, so preserving a continuity between the tradition of his potato-digging forefathers and his own individual literary talent, the reader may be left reflecting that, however useful a metaphorical solution the poem's concluding 'I'll dig with it' may be, it remains the case that, on a literal level, few tools are less geared to moving earth than a pen.[52] Yet Heaney's later career, whether through his bog poems or through such titles as *Field Work*, continues to stress obsessively the bond between literary work and earth work.

Furthermore, we are made aware of the specific Irishness of both in the factor which most readily connects word and soil—the place-name. In the poem 'Anahorish' the Irish place-name is celebrated as '*Anahorish*, soft gradient | of consonant, vowel-meadow'. In 'Gifts of Rain' 'The tawny guttural water | spells itself: Moyola', while in 'Broagh' word and thing are brought almost (but not quite) to unity, and Heaney, in repeating the place-name, relishes 'that last | *gh* the strangers found | difficult to manage'.[53] This final phrase is a pun on the ideas of 'hard to pronounce' and 'awkward to govern'. The pun looks forward to the title of Heaney's second book of essays, *The Government of the Tongue* (1988), which both celebrates the power of the strong poetic voice and examines the political control of speech. Heaney's deployment of almost the same phrase in the 'Clearances' sequence, written in memory of his mother, reveals again how close

[51] Heaney, *Selected Poems*, 117.   [52] Ibid. 10–11.
[53] Ibid. 58, 65, 66.

some of his own preoccupations are to those of the Harrison who constantly memorializes his own parents and attempts to enact through language a fidelity to their world, a world from which he has, in various ways, separated himself. Heaney recalls how

> I governed my tongue
> In front of her, a genuinely well-
> adjusted adequate betrayal
> Of what I knew better. I'd *naw* and *aye*
> And decently relapse into the wrong
> Grammar which kept us allied and at bay.[54]

Here again is the wish to be faithful to the provincial language of the parent culture while inhabiting the dominant language culture of 'proper standard English'. Heaney may employ occasional dialect words like 'clabber' in his poems, but his work is written almost entirely in standard English.[55] Explicitly and implicitly, though, Irish subject-matter stresses the culture from which the verse originates, making the poems' cultural origins a necessary part of their meaning. Heaney likes to use distinctively Irish place-names—'Glanmore Sonnets', 'A Lough Neagh Sequence', 'Anahorish'—because local place-names (frequently deployed by writers such as Dunn, Harrison, Norman MacCaig, and Les A. Murray) are a way of asserting a local identity.[56] Place-names may function as dialect, asserting the bond between a particular culture and its soil; at the same time, as Burns knew, the setting of place-names in poetry is a celebration of the importance of the often denigrated provincial, an assertion that any and every place is good enough for literature—as Larkin put it (ironically parodying one of his own lines): 'Poetry, like prose, happens anywhere', whether in Hull, Glanmore, or Glasgow.[57]

Place-names may assert a link between culture and territory, a bond which Heaney, with his 'vowel of earth', may like to celebrate. Yet, like Brian Friel, whose play *Translations* examines the attempts of different cultures to impose their own names on the same territory, he must confront the problem that name and earth are never quite one.[58] The same thing may be designated by various terms, each pointing to a different culture, a different 'government of the tongue'. Just as

---

[54] Heaney, *The Haw Lantern*, 28 ('Clearances, 4').
[55] Heaney, *Selected Poems*, 26 ('Poem').
[56] Heaney, *Field Work*, 33; Heaney, *Selected Poems*, 40, 58.
[57] *A Rumoured City*, 9.
[58] Brian Friel, *Translations* (London: Faber and Faber, 1981).

Waverley's 'Chevalier' might also be 'the Prince' or 'the Pretender', so the place Heaney calls 'Derry' might also be called 'Londonderry'. The terms 'Ulster', 'Northern Ireland', and 'Ireland' can all take in the six counties, but each term conducts them in a different political and cultural direction. Provincials, always liable to be labelled at the convenience of the dominant culture, may be particularly sensitive to acts of naming, and if Heaney clearly rejoices in deploying Irish place-names in his work, he has also made a strong stand about the cultural labels attached to his verse. His 1983 Field Day pamphlet *An Open Letter* explains his clear reluctance to be described as 'British':

> As Empire rings its curtain down
>   This 'British' word
> Sticks deep in native and *colon*
>   Like Arthur's sword.[59]

Heaney is responding here to the English editors of *The Penguin Book of Contemporary British Poetry*, Andrew Motion and Blake Morrison. His poem is written in the 'Burns stanza', as if to emphasize his problematic cultural identity, and his response may be compared with Douglas Dunn's complaint that 'Nowhere in Morrotion's Introduction am I referred to as Scots, simply as "provincial." '[60] Heaney the Irish republican and Dunn the Scot who is '*looking* for a constitutional crisis' and convinced that 'what Scotland needs is much closer to an autonomous political life than anything meant by devolution' are particularly wary of having their 'barbarian' un-English identities submerged in an English-dominated 'British' context. For that context, like an undevolved monolithic English Literature, ignores the strength of their 'provincial' traditions, and uses that adjective 'provincial' without any awareness of the cultural imperialism that the term implies.[61] As Dunn puts it:

At certain points the cultures of Wales, Ireland, Scotland and England overlap. But there's too much resistance from each of them—and, quite rightly so, from England too—for these tentatively shared concerns to make a 'Britain'. Some observers see that as self-evident. Readers of poetry need

[59] Seamus Heaney, *An Open Letter*, in Field Day Theatre Company, *Ireland's Field Day*, 23.

[60] Douglas Dunn, interview with Bernard O'Donoghue, *Oxford Poetry*, 2/2 (Spring 1985), 50.

[61] Douglas Dunn, 'Interview with the Devil', in Sean O'Brien and Richard Plaice (eds.), *The Printer's Devil: A Magazine of New Writing* (Tunbridge Wells: South East Arts, 1990), 17.

only compare the underlying convictions of Welsh, Irish, Scottish and English poetry.[62]

Heaney, too, is obsessed with such issues and the way in which they intersect with language. Yet the Irish context differs from the Scottish, because contemporary Irish cultural struggles are frequently violent, and Heaney is wary of becoming too ensnared by the potentially deforming pressures of the politics of language. The clearest sign of this comes in the 'Station Island' sequence, where he engages with the Modernists in a debate about language and culture.

The concluding part of Heaney's 1984 version of 'Station Island' is written in a *terza rima* whose form and content pay homage not just to Dante, but also to the Eliot who meets his 'familiar compound ghost' in *Little Gidding*, and with whom Heaney has revealed an increasing fascination in such pieces as his lecture 'Learning from Eliot'. Heaney's Modernist ghost is a Joyce whom the poet greets warmly, yet by whom he is also reproached and counselled:

> 'Old father, mother's son,
> there is a moment in Stephen's diary
> for April the thirteenth, a revelation
>
> set among my stars—that one entry
> has been a sort of password in my ears,
> the collect of a new epiphany,
>
> the Feast of the Holy Tundish.' 'Who cares,'
> he jeered, 'any more? The English language
> belongs to us. You are raking at dead fires,
>
> a waste of time for somebody your age.
> That subject people stuff is a cod's game,
> infantile, like your peasant pilgrimage.
>
> You lose more of yourself than you redeem
> doing the decent thing. Keep at a tangent.
> When they make the circle wide, it's time to swim
>
> out on your own and fill the element
> with signatures on your own frequency,
> echo soundings, searches, probes, allurements,
>
> elver-gleams in the dark of the whole sea.'[63]

---

[62] Dunn, interview with O'Donoghue, 46.
[63] Seamus Heaney, 'Learning from Eliot', *Agenda*, 27/1 (Spring 1989), 17–31; Seamus Heaney, *Station Island* (London: Faber and Faber, 1984), 93–4.

Heaney's reference here is to the way in which the 'tundish', picked out by the English dean as a local curiosity in Stephen Dedalus's speech, comes to be a powerful, sanctified emblem of the strength of Irish culture. Yet in Joyce's own text there is an added irony, for the word 'tundish' is discovered to be an English one after all, and so may stand as both Irish and English. In advising Heaney to drop his barbarian or provincial anxieties and take confident possession of the post-imperial English language, the Modernist Joyce seems to be pointing the way to a confident artistic future, free of 'That subject people stuff'—a future of individual aesthetic satisfactions, free of political drag-weights. In its later, revised version this section of 'Station Island' considerably truncates Joyce's advice and the poet's response to it, as if the lessons involved no longer need to be spelt out in full. In either version the end of the 'Station Island' sequence does not make it clear whether or not the poet takes the Joycean advice. Heaney's next collection, *The Haw Lantern* (1987), continues to ponder the problems and celebrate the energies essayed in his earlier books. He may have become cosmopolitan in his academic affiliations and sophisticated in his powers of expression, learning from literary theory as well as from writers from Eastern Europe, yet he remains an Irish writer, contributing to, and seeking to hold aloft, Irish culture. However, in what is arguably Heaney's finest volume of poetry, *Seeing Things* (1991), the strains and toings and froings which have dominated his work are further universalized and translated. Men sawing a tree move 'backwards and forwards | So that they seemed to row the steady earth'. A basket is swung backwards and forwards by excited hands. We glimpse 'the zig-zag hieroglyph for life itself'.[64] Such to and fro or zigzag patterns, however, are seen as natural and exciting—the rocking of a boat, the releasing and returning of memories, the pulsations of sex—rather than the constant and contrary tugs of the Northern Irish Catholic sensibility. In this book filled with visionary crossings of borders Heaney, for so long nurtured on being pulled in two directions, seems to have won through to a perception that his condition may yield cause for an intensely visionary celebration that builds on, without betraying, the Irish identity so important to his earlier work and to his stance as a

[64] Seamus Heaney, *New Selected Poems 1966–1987* (London: Faber and Faber, 1990), 193; Seamus Heaney, *Seeing Things* (London: Faber and Faber, 1991), 9 ('Markings'), 17 ('Seeing Things'). On Heaney and literary theory, see the chapter on 'Heaney among the Deconstructionists', in Henry Hart, *Seamus Heaney: Poet of Contrary Progressions* (Syracuse: Syracuse University Press, 1991).

critic. It is noticeable, for instance, that the strategy in his prose books is to focus attention either on writers whose fame is internationally recognized or else who are Irish. So a subtle valorization of writers such as Patrick Kavanagh and John Montague is part of his critical project. Cunningly, in his criticism, as in his poetry, Heaney is concerned with significant realignments of the cultural map.

Heaney's fellow directors of the Field Day Theatre Company are involved in similar attempts at realignment, efforts often geared towards the reduction of British cultural (and political) power in Ireland. The first of the Field Day pamphlets, Tom Paulin's *A New Look at the Language Question*, opens by returning to the eighteenth century to glance at the cultural politics of Johnson's lexicography and to look at Swift's essay 'On Barbarous Denominations in Ireland', where, as well as attacking Irish accents, Swift 'criticised the Scottish accent and most English regional accents as "offensive"'. Paulin's pamphlet links eighteenth-century linguistic attitudes about propriety to the development of an Anglocentric ethic of Britishness. Written by a poet eager to see British hegemony over Northern Ireland come to an end, Paulin's pamphlet concludes by paying particular attention to Irish dialect words and their study, and envisages the possible development of an 'Irish English', which would be precisely the opposite of the standard English advocated in the eighteenth century. In the late twentieth century, when the recognized ruling language of Britain is standard English, Paulin's 'Irish English' would be a distinctively Irish language and

would redeem many words from that too-exclusive, too-local, usage which amounts to a kind of introverted neglect. Many words which now appear simply gnarled, or which 'make strange' or seem opaque to most readers, would be released into the shaped flow of a new public language. Thus in Ireland there would exist three fully-fledged languages—Irish, Ulster Scots and Irish English. Irish and Ulster Scots would be preserved and nourished, while Irish English would be a form of modern English which draws on Irish, the Yola and Fingallian dialects, Ulster Scots, Elizabethan English, Hiberno-English, British English and American English. A confident concept of Irish English would substantially increase the vocabulary and this would invigorate the written language. A language that lives lithely on the tongue ought to be capable of becoming the flexible written instrument of a complete cultural idea.[65]

[65] Tom Paulin, in Field Day Theatre Company, *Ireland's Field Day*, 5, 15.

As if gesturing towards such a proposition, Paulin's poems in the significantly titled *Liberty Tree* and in *Fivemiletown* draw on such dialect words as 'sheugh', 'scraggy', 'eejits', and 'dwammy'. As editor of the 1990 *Faber Book of Vernacular Verse*, Paulin continues to champion dialect and 'ungentlemanly' language. Though its effect is rather scattered, his aligning of Whitman beside Burns, and his inclusion of Dunn, Heaney, Murray, and Walcott along with MacDiarmid, Harrison, street rhymes, and the demotic Eliot of *Sweeney Agonistes*, show that, like the Douglas Dunn who in 1990 aligned Harrison, Heaney, and Tom Leonard, Paulin wishes to support his own poetic endeavours by emphasizing the broad church of vernacular writing. A sophisticated barbarian, he collects voices which 'know that out in the public world a polished speech issues orders and receives deference. It seeks to flatten out and obliterate all the varieties of spoken English and to substitute one accent for all the others. It may be the ruin of us yet.'[66]

Paulin's own dialect usages and his nationalist stance can be paralleled by the sprinkling of such Scots words as 'dailygone', 'brose', and 'jorum-jirger' through *St Kilda's Parliament*, the 1981 collection by his friend Douglas Dunn.[67] This is certainly one of Dunn's finest books, though it is sometimes disappointing that his engagement with Scotland's heritage is elegiacally pessimistic or narrow. The poem 'Green Breeks', for instance, is right to censure and locate Scott's upper-class superiority and the way in which that sense of superiority accompanied his presentation of Scottish virtues. Yet, if one were to take the poem as a comprehensive summing-up of Scott, one would have to admit that Dunn has limited the nineteenth-century writer by seeing him largely in terms of class. Dunn's Scottish nationalist statements grew stronger throughout the 1980s, and it is emblematic that the title of his collection of poems should focus on a parliament as far from Westminster as it is possible to get in the British Isles. Yet no more than Dunn is Paulin a simple barbarian; he is at once the poet deploying provincial 'barbarisms' and the sophisticated teacher of English Literature whose essay on 'The British Presence in *Ulysses*' examines Joyce's novel as a critique of

---

[66] Tom Paulin, *Liberty Tree* (London: Faber and Faber, 1983), 70 ('S/He'); Tom Paulin, *Fivemiletown* (London: Faber and Faber, 1987), 16 ('Fivemiletown'); Tom Paulin (ed.), *The Faber Book of Vernacular Verse* (London: Faber and Faber, 1990), pp. xiv, xxii; Douglas Dunn, *The Topical Muse: On Contemporary Poetry*, Kenneth Allott Lecture No. 6 (Liverpool: *Liverpool Classical Monthly*, 1990).

[67] Douglas Dunn, *St. Kilda's Parliament* (London: Faber and Faber, 1981), 87.

British imperialism.[68] At the same time, for all his interest in Irish linguistic barbarism, Paulin has a strong admiration for English civility. His potent Irish loyalties are accompanied, like the local loyalties of Harrison and Dunn, by an international scope (seen clearly in his *Faber Book of Political Verse*). He is a sophisticated barbarian.

Such a description not only links Paulin to Harrison, Dunn, and Heaney, but it also suggests similarities between their work and that of the Australian poet Les A. Murray, whose strategy is best described by the title of his first book of essays, *The Peasant Mandarin* (1978). Though he is suspicious of the 'mandarinizing tendency of Modernism', Murray none the less acknowledges a debt to Modernism's synthesizing impulse ('*the* great thing about Modernism') and 'its admission of a wide range of vernacular speech'.[69] His stance as a figure combining the 'peasant' and the 'mandarin' might again be compared with Eliot's stress on the artist as linking the sophisticated to the primitive, while it also links him with other contemporary poets. Writing of Australia as a 'vernacular republic' whose 'main vernacular tradition is full of Celtic elements', Murray argued with his fellow Australians in 1976 that, if 'the intelligentsia here would adopt the sort of nationalist orientation taken up by their coevals in, for example, Scotland and Wales, the colonial hangovers which subtly cripple much of our life might soon be swept away'.[70] Murray is very much attuned to the work of 'barbarian' writers in what he suggests should be designated 'the Anglo-Celtic archipelago', and, praising Dunn's example, he has written of how 'the formerly feudal and Christian English language' has been turned into an 'estranging culture-language' of new forms of cultural oppression. Writing of Dunn's 'classic achievement' in having redeployed English metropolitan speech against Anglocentric metropolitan values, Murray argues that, 'If one doesn't choose to write in an alternative language, for example Scots, and has no useable tradition of the sort that the received sensibility would scorn as "populist", one has to win the battle for human utterance where he does win it, that is, within the ruling poetic sociolect itself.'[71]

---

[68] Tom Paulin, 'The British Presence in *Ulysses*', in *Ireland and the English Crisis* (Newcastle: Bloodaxe Books, 1984), 92–100.

[69] Les A. Murray, interview with Robert Crawford, *Verse*, 5 (1986), 22.

[70] Les A. Murray, *The Peasant Mandarin: Prose Pieces* (St Lucia: University of Queensland Press, 1978), 155, 144.

[71] Les A. Murray, review of Douglas Dunn, *Selected Poems*, *Verse*, 4/2 (June 1987), 62–3.

Though he has included both Scots and Gaelic vocabulary in his
verse, as well as some distinctively Australian words, Murray writes
his poetry for the most part in an expansive English. It is an English
which continually points to the poet's loyalties and obligations
towards '*remnants defined by a tilt in their speech*' ('The Action'), and to
Murray's non-cosmopolitan background in a rural Australia where
he wishes 'to discern the names of all the humble' ('The Names of the
Humble').[72] Like Harrison (in whose work he has expressed
interest), Murray is aware of the need to escape from the pressures of
a formal education which would 'cut our homes away'. He recalls
that, in his student days, 'Literate Australia was British, or babu at
least', in a climate where 'a major in English made one a minor
Englishman' ('Sidere Mens Eadem Mutato'). Murray's wish for an
Australian 'vernacular republic' is a wish to escape from the perceived mentality of standard Englishness which goes with standard
English language and culture. I have written about this at length
elsewhere, relating his aspirations to his view of his Scottish ancestry
and the way in which he uses this to emphasize that, since Australia is
'not England; we don't need to snub and suppress and deny our
Celtic inheritance—though we mustn't be colonized by it either'.[73]
Like Heaney, Dunn, and many other post-colonial writers such as
Derek Walcott, Murray wishes at once to inhabit and deploy the
English language while maintaining an identity that is, in a sense,
barbarian.

Murray's often controversial opposition to the sophistication of
metropolitan centres is nowhere clearer than in his highly sophisticated verse novel, *The Boys Who Stole the Funeral* (1980), where, as in his
other work, distinctive place-names are deployed to function in lieu of
dialect, anchoring the voice to a particular community. This is a
device which Murray shares with Heaney, Dunn, Harrison, Walcott,
and other poets (such as the Edwin Morgan of 'Canedolia') who
delight in using place-names to articulate the siting of their verse in
territories with a barbarian voice.[74] For Murray, as for these other
poets, place-names are a repository of often suppressed cultures and
languages outside the standard English orbit, those cultures towards

---

[72] Les A. Murray, *Selected Poems* (Manchester: Carcanet, 1986), 44, 20. Unless
otherwise specified, Murray's poems are quoted from this selection.
[73] Robert Crawford, 'Les Murray: Radical Republican and Religious Conservative', *Cencrastus*, Autumn 1987, 36–41; Les A. Murray, *The Peasant Mandarin*, 155.
[74] Edwin Morgan, *Poems of Thirty Years* (Manchester: Carcanet, 1985), 137.

which Modernism constantly gestured. Some of Murray's most striking recent verse celebrates the distinctiveness of his own Australian
territory in a voice that sets itself at a creative angle to standard
English by the strategic use—and even explication of—place-names,
coupled with accents borrowed from the translation and explication of
Aboriginal narratives. Murray achieves a diction that competes with
the expansiveness of the Modernists and includes, but does not
imprison, a wide variety of elements, being at once sophisticated and
philologically, richly barbarous:

> I am driving *waga*, up and west.
> Parting cattle, I climb over the crest
> out of Bunyah, and skirt Bucca Wauka,
> A Man Sitting Up With Knees Against His Chest:
> *baga waga*, knees up, the burial-shape of a warrior.
> Eagles flying below me, I will ascend Wallanbah,
> that whipcrack country of white cedar
> and ruined tennis courts, and speed up on the tar.
> In sight of the high ranges I'll pass the turnoff to Bundook,
> Hindi for musket—which it also took
> to add to the daylight species here, in the prim-
> al 1830s of our numbered Dreamtime
>> and under the purple coast of the Mograni
>> and its trachyte west wall scaling in the sky
>> I will swoop to the valley and Gloucester Rail
>> where boys hand-shunted trains to load their cattle
>> and walk on the platform, glancing west at that country
>> of running creeks, the stormcloud-coloured Barrington,
>> the land, in lost Gaelic and Kattangal, of Barandan.[75]

Murray is a poet well aware of 'the twentieth century's anthropological revolution' which has meant that 'the cultures of mankind are
now on display to the literate and TV-watching Westerner as they
never were in any previous age'.[76] He attempts to move in this world
responsibly, maintaining at the same time a fidelity to the 'homely: |
when did you last hear that word without scorn [?]'.[77] In his place-
naming, his 'vernacular republican' orientation, and his frequent
gestures towards his 'peasant' as well as his 'mandarin' siting, he
deploys an equivalent of dialect.

[75] Les A. Murray, *The Daylight Moon* (North Ryde, NSW: Angus and Robertson,
1987), 86 ('Aspects of Language and War on the Gloucester Road').

[76] Les A. Murray, *Persistence in Folly* (Sydney: Angus and Robertson, 1984), 121.

[77] Murray, *The Daylight Moon*, 61 ('Forty Acre Ethno').

Many of the writers so far mentioned in this chapter read one another's work, or know one another personally. Larkin was Dunn's poetic mentor; Paulin was Dunn's fellow student at Hull and a contributor to Dunn's 1975 survey, *Two Decades of Irish Writing*, where he makes much of Thomas Hardy's advocacy of 'a necessary provincialism'; again, Dunn admires and has a fellow-feeling for the work of Harrison, Heaney, and Murray; Murray admires Dunn's work, and Heaney (like Edwin Morgan) respects that of Murray;[78] Walcott and Heaney are friends, while the *terza rima* of Walcott's splendid *Omeros*, and many of that poem's cadences and attributes, are reminiscent of Heaney's verse. This is not merely a mutual admiration society, but an international grouping of writers whose work may be seen as inheriting something from the demotic side of the essentially provincial movement which was Modernism, as well as sharing preoccupations with other writers of the late twentieth century.

For some, concern with dialect and its equivalents might seem to connect these writers with the dialect work of such English writers as Blake Morrison in *The Ballad of the Yorkshire Ripper* (1987) and the Craig Raine who has written poetry in pidgin and, using MacDiarmid as one of his examples, declared that 'All great poetry is written in dialect.'[79] Yet, Morrison's use of dialect in the title-poem of his 1987 collection appears only an attitude, purposefully struck, though sometimes awkwardly sustained, and the reader is left wondering if there is an implication that users of northern English dialect (found nowhere else in Morrison's work) are somehow more predisposed to violence than speakers of other forms of English. Morrison seems to be trying on dialect as a linguistic clothing which will help motivate a poem; rather than using it as the native voice of a community to which he feels fidelity, he uses it to cast its speakers under suspicion. One worries that there is a too easily perceived equivalence between the physically barbarous acts of the Ripper and the barbarisms of the poem's speech. For most of this chapter an attempt has been made to valorize the word 'barbarian'. Morrison's use of a language which is

[78] Tom Paulin, 'A Necessary Provincialism: Brian Moore, Maurice Leitch, Florence Mary McDowell', in Douglas Dunn (ed.), *Two Decades of Irish Writing: A Critical Survey* (Cheadle: Carcanet, 1975), 242–56; Haffenden, *Viewpoints*, 33; Edwin Morgan, 'A New Tradition', *Cencrastus*, Autumn 1988, 49–50; conversations with Dunn, Heaney, and Murray.

[79] Craig Raine, 'Barbarous Dialects', in *Haydn and the Valve Trumpet: Literary Essays* (London: Faber and Faber, 1990), 89.

barbarous has nothing of this about it. It seems to be on the opposite side to Tony Harrison's 'Rhubarbarians'.

Again, Raine's statement that 'All great poetry is written in dialect' should not be taken as placing him in the barbarian camp. While there are undeniable links between Raine and some of the other poets considered in this chapter (not least of which is the celebration of home), Raine's piece 'Babylonish Dialects' makes it clear, through various examples, that he is using the word 'dialect' not in the sense of 'provincial speech', but in the sense of 'idiolect' — the peculiarities of the speech of an individual rather than that of a community. While one might argue that, at its most extreme, Raine's 'Martian' technique forges a new dialect of English according to which 'Caxtons are mechanical birds' ('A Martian Sends a Postcard Home') and locutions are constantly reminted, the essential point is that no dialect or language of any kind is actually used on Mars.[80] 'Martian' is the peculiar and entertaining invention of Craig Raine. An idiolect rather than a dialect, it expresses no solidarity or fidelity to a particular group; it is an assertion of creative individualism which may reflect a certain uneasiness with a 'standard English' identity. Its purpose is not to enunciate a solidarity but to articulate a sometimes splendid isolation. If there seems to be a particular animus against Tony Harrison in Raine's survey of 'Poetry Today', this may be because there are irreconcilable similarities between these two originally working-class writers, as well as irreconcilable differences of attitude in their various deployments of 'dialect'.[81]

Raine's colleague, Christopher Reid, found a way of moving beyond 'standard English' perceptions by shifting from the outsider-persona of a Martian to the persona of an East European poet in *Katerina Brac* (1985), which plays with 'translatorese', and the persona of a speaker of argot in 'Memres of Alfred Stoker', contained in Reid's collection *In the Echoey Tunnel* (1991). Frank Kuppner had engaged with the 'translatorese' of Chinese poetry, combining it with his own autobiography to produce a voice that is comically barbarous as it both inhabits and negotiates with a seemingly alien medium in *A Bad Day for the Sung Dynasty* (1984), a Scottish poet's book filled with humour on the theme of translation between languages and cultures.

---

[80] Craig Raine, 'A Martian Sends a Postcard Home', in *A Martian Sends a Postcard Home* (Oxford: Oxford University Press, 1979), 1.

[81] Raine, *Haydn and the Valve Trumpet*, 214–37.

At his farewell feast, the Ambassador offers a toast,
But carelessly mis-stresses a monosyllable,
Effectively addressing the Emperor as 'Lustbasket',
And undoing the good work of the previous forty-seven years.[82]

Kuppner's ludic explorations of translation can be related to those of
his fellow Glaswegian Edwin Morgan, who, in addition to producing
a most impressive array of translations from many languages (some
of which are collected in *Rites of Passage* (1976)), frequently uses translation as a strategy and metaphor in his own poetry. Such Morgan
poems as 'The First Men on Mercury', in which speakers of English
and Mercurian gradually exchange languages, are emblematic of an
interest in cultural cross-over or in what might be called '*Crossing the
Border*', to cite the title of Morgan's 1990 collection of essays, a
collection at times richly suggestive of links between Scottish writing
and Modernism.[83] One recent critic has interpreted 'The First Men
on Mercury' in terms of an interchange between speakers of English
and Scots. It is undeniable that Morgan (who has translated both
Mayakovsky and a section of *Macbeth* into Scots) has a strong interest
in Scots-language writing, and likes to complicate the texture of his
English with an admixture of distinctively Scottish forms, while often
signalling a strong fidelity to Scotland in his internationally orientated
work.[84]

Though the positions of all these writers are far from identical,
there are sufficient similarities between them to show that there is
a widespread wish in recent poetry to be seen as in some manner
barbarian, as operating outside the boundaries of standard English
and outside the identity that is seen as going with it. Such a wish
unites post-colonial writers such as Murray and Walcott with writers
working within the 'Anglo-Celtic archipelago'. It joins the post-
colonial and the provincial. Indeed, though the continuing strength of
London metropolitan publishing houses and journals and of Anglocentric literary history may distort the picture, it is surely true that,
for most creative users of the English language today, one of the
fundamental questions is how to inhabit that language without sacrificing one's own distinctive, 'barbarian' identity. The vast majority

[82] Frank Kuppner, *A Bad Day for the Sung Dynasty* (Manchester: Carcanet, 1984),
109.

[83] Morgan, *Collected Poems*, 267; Edwin Morgan, *Crossing the Border: Essays on Scottish
Literature* (Manchester: Carcanet, 1990).

[84] W. N. Herbert, 'Morgan's Words', in Crawford and Whyte, *About Edwin
Morgan*, 67.

of contemporary users of English come from outside the zone of 'court English' advocated as the universal standard by the eighteenth-century teachers of Rhetoric and Belles Lettres and by their successors, just as they come from outside the area where RP or BBC English is the treasured norm. Some will still wish to adopt the tones of a linguistic and literary standard endowed with historical, sociopolitical, and pedagogic authority. Some will wish to rebel against it completely, resorting, in the face of standard English, to their own equivalent of Scots. Others again are in search of a subtle nuancing which will allow them to deploy all the riches of the English language while maintaining a local accent which declares them to be free of its imperial claims. These problems and questions haunt writers from the English cultural centre also. In such works as Andrew Motion's *Independence* (1981) there is a close scrutinizing of the British, particularly the English, imperial legacy and what it has meant. Drawing on Auden, particularly the Auden of *The Orators: An English Study*, Glyn Maxwell, one of the most talented young poets now writing in England, repeatedly interrogates the tones of a nation in which issues of class and imperial domination remain to the fore.[85]

All these writers operate in a climate where the power of the English cultural centre is under increasing challenge—not just from the rival centre of New York, which is coming to dominate English publishing through the take-over of such houses as Chatto, Cape, the Bodley Head, and Hutchinson, but also from the 'provinces' and barbarian regions, home of such innovative houses as Bloodaxe Books. Larkin is too easily seen as merely a part of that English cultural centre. When he writes, in relation to Hull, that 'Poetry like prose happens anywhere', he is making a statement which is as important in its devolutionary impact as Les Murray's contention that, as an Australian writer with a 'Boeotian' ideology, he is

against words like 'provincial' and 'capital' and 'metropolitan' and so on. The capital of the world is anywhere a good writer is writing. Or, really, anywhere an individual is living! I figure that no-one should be made to feel relegated by mandarin centralisms—and so writing from where I write is a model for anyone who wants to do the same thing, from, say, Guyana, or Belize, or Zambia, or wherever they like. They shouldn't have to feel as we used to have to feel that you go to London or New York to make a big splash.

[85] Glyn Maxwell, *Tale of the Mayor's Son* (Newcastle: Bloodaxe Books, 1990); see also Glyn Maxwell, 'Echoes of The Orators', *Verse*, 6/3 (Winter 1989), 25–30, and Glyn Maxwell, interview with David Kinloch, *Verse*, 7/3 (Winter 1990), 79–87.

Nor should they be insulted by anthologies of 'English' poetry which contain only British and American poets' work. 'English' ought to mean either exclusively *English*, or else work from the *whole* English-speaking world.[86]

Murray's unhappiness with the word 'English' here is the unhappiness of the non-English writer too easily kidnapped by the world of the metropolitan centres through the use of a term which may mean simply anglophone or anglographic, but which may also slip into an imperial Anglocentricity. To counter this, what is required is not only the constant assertion on the part of writers of their position in cultural politics, but also, on the part of readers, a constant awareness of the need for devolutionary readings—readings which are alert to all the nuances of the 'provincial', 'barbarian', and 'colonial', to the subtle accents and strategies of the marginalized who have all too often been smoothly absorbed or repressed by being designated 'English Literature'.

Such a necessary sensitivity and alertness to the possibility of de-volutionary readings which release texts and authors from easy, un-considered entrapment in English Literature shows how much the wheel has come full circle since the origins of the university teaching of the subject in eighteenth-century Scotland. Those issues of the provincial and the barbarian versus the proprieties of the ruling centre which were at the heart of the formation of university Rhetoric and Belles Lettres and which have continued to resurface in the writing of the last two centuries are once again the focus of attention. Fittingly, Scottish writers, so active not only in the invention and fostering of Rhetoric and Belles Lettres, but also in the construction of a literature of full Britishness, are now active in the devolutionary impulse which has manifested itself so strongly in recent years. If literature was important in the attempted formation of a British mentality, then it is also playing its part in the search for a post-British identity which has grown in twentieth-century Scotland, particularly in the wake of Hugh MacDiarmid. Even after the 1979 Devolution Referendum had failed to give Scotland a greater share in its own affairs, Christopher Harvie can write of 'Scotland, where a sort of intellectual Unilateral Declaration of Independence occurred during the 1980s'.[87] Part of this phenomenon has certainly been

---

[86] Murray, interview with Robert Crawford, 23.

[87] Christopher Harvie, ' "For Gods are Kittle Cattle": Frazer and John Buchan', in Robert Fraser (ed.), *Sir James Frazer and the Literary Imagination* (Basingstoke: Macmillan, 1990), 254.

literary, with such works as Edwin Morgan's *Sonnets from Scotland* being written as 'a kind of comeback, an attempt to show that Scotland was there, was alive and kicking', after the 1979 Referendum.[88]

Such cultural energies should neither be consigned to a Scottish literary ghetto nor blandly absorbed into English Literature. Rather, they can be seen as part of the wider devolutionary—'barbarian'— cultural movement surveyed in this chapter. Douglas Dunn's poem 'Here and There' (from his 1988 collection *Northlight*) presents the argument in terms which, while particularly Scottish, make sense in many of the other contexts discussed in this chapter.

> '*Provincial*', you describe
> Devotion's minutes as the seasons shift
> On the planet: I suppose your diatribe
> Last week was meant to undercut the uplift
> Boundaries give me, witnessed from the brae
> Recording weather-signs and what birds pass
> Across the year. More like a world, I'd say,
> Infinite, curious, sky, sea and grass
> In natural minutiae that bind
> Body to lifetimes that we all inhabit.
> So spin your globe: Tayport is Trebizond . . .[89]

Where, for Harrison, 'Newcastle is Peru', so, for Dunn, his home village of Tayport in Fife is asserted as having its own integrity and significance in a scale which transcends the divisions between metropolitan and provincial.[90]

Like Murray and Heaney, Dunn is attracted to moving beyond 'that subject people stuff', yet he is also aware of the dangers of becoming lost in his own world as '*An inner émigré*'—a phrase that Dunn's poem borrows, significantly, from Heaney's meditation on culture and aestheticism in 'Exposure'.[91] Dunn's response to the charge that fidelity to a particular culture outside received pronunciation and standard English is parochialism is a response which might stand as a twentieth-century retort to the teachers of Rhetoric and Belles Lettres, though it would be naïve to equate absolutely the

[88] Edwin Morgan, *Nothing Not Giving Messages*, ed. Hamish Whyte (Edinburgh: Polygon, 1990), 141.

[89] Douglas Dunn, *Northlight* (London: Faber and Faber, 1988), 26.

[90] Tony Harrison, *Selected Poems*, 63 ('Newcastle is Peru').

[91] Dunn, *Northlight*, 27 ('Here and There'); Heaney, *Selected Poems*, 136.

position of twentieth-century Scottish writers with that of those
working in the eighteenth century.

> '*Worse than parochial! Literature*
> *Ought to be everywhere* . . .' Friend, I know that;
> It's why I'm here. My accent feels at home
> In the grocer's and in Tentsmuir Forest.
> Without a Scottish voice, its monostome
> Dictionary, I'm a contortionist—
> Tongue, teeth and larynx swallowing an R's
> Frog-croak and spittle, social agility,
> Its range of fraudulence and repertoires
> Disguising place and nationality.[92]

Committed rather than chauvinistic, elegant rather than strident,
Dunn's recent work confirms how much is to be gained both from an
alertness to the devolutionary impulses within 'English Literature'
and from seeing the links between Scottish and other literatures in
English. Elsewhere, attacking notions such as Hugh Kenner's that
'there is "an island called England" and that it has a culture unified
by the English language', Dunn calls attention to the extent to which
the strongest writing of recent times in 'notional Britain has emerged
from social backgrounds, not associated with literature before except
marginally, and from what Londoners call "the provinces"'. He goes
on to link this with the way in which he sees 'the strength of English'
as having passed out of the 'gentrified guardianship' of 'London' and
'cute Oxbridge procedures', to such locations as 'Australia, Ireland,
Scotland, India, the north of England, Canada, the United States,
Africa, the Caribbean countries'.[93] Such a roll-call re-emphasizes the
speaker's point in 'Here and There', that

> Englishness
> Misleads you into Albionic pride,
> Westminstered mockery and prejudice—
> *You're* the provincial, an undignified
> Anachronism.

Readers may disagree about whether Dunn's tone carries a
counter-Albionic pride or a cultural self-respect that might be shared
with those 'multitudes of the people' who are beyond the English

[92] Dunn, *Northlight*, 28 ('Here and There').
[93] Douglas Dunn, 'England Sunk', review of Hugh Kenner, *A Sinking Island: The Modern English Writers*, *Glasgow Herald*, 8 Oct. 1988, 18.

cultural centre, yet 'who speak English or non-English forms of English'. However, his work clearly speaks not only for Scots, but also for that great majority of anglophone speakers, readers, and writers who wish to assert their barbarian freedom from the Anglo-centric pressures present from the origins of the university discipline of English Literature, and frequently still inhabiting it.[94] If the Scots were so instrumental in the construction of that pedagogical edifice, along with its attendant Anglocentric attitudes, they have, in the long run, suffered much from it, seeing their literature constantly absorbed into English Literature or else exiled from it for special attention in a tiny specialist area. Now, when there is a new and critical interest afoot in definitions and constructions of Englishness, and when considerations of 'minority' literatures proliferate, it seems bitterly ironic that Scottish Literature continues to be ignored or bypassed by these studies, which all too often contain a hidden Anglocentricity of the sort exemplified in the introduction to this book. Unless we appreciate the subtle and important part played by Scottish culture in the construction and dissemination of English Literature and in the development of post-Enlightenment writing, and unless we examine the way in which Scottish energies interacted and continue to interact with 'provincial' and 'barbarian' writing in English, we cannot finally devolve our reading of English Literature so as to liberate and listen to the full spectrum of suppressed but persistent local accents. It may be simply because Scottish culture was essential to the first construction of the university discipline of English Literature through which most anglographic writing is now studied that it will be the last strand to emerge from the unravelling, the devolving of that subject. Yet, until that strand has fully emerged, and the connections between it and others have been teased out, the devolving of English Literature will not have been completed; nor can it even be said to have fully begun.

[94] Ibid.

Index